Advanced Functional Membranes
Materials and Applications

Edited by

Inamuddin[1], Tariq Altalhi[2], Mohd Imran Ahamed[3], Mohammad Luqman[4]

[1]Department of Applied Chemistry, Zakir Husain College of Engineering and Technology, Faculty of Engineering and Technology, Aligarh Muslim University, Aligarh-202002, India

[2]Department of Chemistry, College of Science, Taif University, 21944 Taif, Saudi Arabia

[3]Department of Chemistry, Faculty of Science, Aligarh Muslim University, Aligarh-202002, India

[4]Department of Chemical Engineering, College of Engineering, Taibah University, Yanbu, Saudi Arabia

Published by **Materials Research Forum LLC**
Millersville, PA 17551, USA

Published as part of the book series
Materials Research Foundations
Volume 120 (2022)
ISSN 2471-8890 (Print)
ISSN 2471-8904 (Online)

Print ISBN 978-1-64490-180-9
eBook ISBN 978-1-64490-181-6

Distributed worldwide by

Materials Research Forum LLC
105 Springdale Lane
Millersville, PA 17551
USA
https://www.mrforum.com

Manufactured in the United States of America
10 9 8 7 6 5 4 3 2 1

Table of Contents

Preface

A membrane is a selective interphase barrier between two contiguous phases controlling the transport of substances between the two partitions. Most membranes are used in separation processes for liquid and gaseous mixtures such as gas separation, reverse osmosis, pervaporation, filtration, and so on. The quest for constructing high-performance membranes is a rising field with nevertheless many unsolved challenges ahead. Utilization of advanced functional membranes in food processing, sensors, medical devices, water treatment, separations, and industrial applications has provoked material scientist to develop membrane-based technologies. These processes are realistically diminishing energy cost and CO_2 production.

Advanced Functional Membranes: Materials and Applications aims to explore recent developments and applications of advanced functional membranes. The chapters discuss on a variety of functional membranes for energy, environmental, and biomedical applications, etc. Leading researchers and experts contributed book chapters. This book is a reference text that can be used by postgraduates, engineers, scientists, professors and policymakers working in the field of material science, polymer chemistry, engineering, environmental science, and industries. The work reported in the following 10 chapters are as follows:

Chapter 1 describes the brief history of the membrane technology development for various applications including desalination, wastewater treatment and ion-exchange application useful in energy storage and conversion. It also gives a brief site on various membrane processes-based on the material choice, along with discussing the advanced polymer-based membranes for various applications.

Chapter 2 discusses filtration mechanism and different polymeric materials used for development of MF and UF membranes. Membrane characteristics, separation mechanism thereby treating various types of wastewaters is also discussed in detail. The major focus is given to membrane fouling and their cleaning method to improve membrane workability and sustainability.

Chapter 3 highlights the potential of different polymers for functional membranes in various applications such as water and gas separation, medical and emerging technologies including fuel cells, lithium-ions batteries, electroconductive, and optoelectronics. For each application, the polymer membrane technology (types, fabrications, mechanisms), the challenges, and the future direction are also reviewed.

Chapter 4 discusses the membrane separation process, the structural modification of the membranes using biopolymers and green solvents, and the functionalization with organic and inorganic materials. An improvement in the performance and anti-fouling characteristics of the membranes with adequate structural modification and functionalization was evidenced, which allows the process to meet the concepts involved in green technology.

Chapter 5 emphasizes with recent developments and challenges in self-assembled membranes and their applications. It begins by identifying the methods to fabricate the membranes through self-assembly mechanism, followed by a discussion on different types of materials used to form these membranes, their applications, advantages and limitations.

Chapter 6 highlights the various fabrication routes of polymeric/ceramic based porous membranes, having nano-/submicron-holes in their structures, and most importantly highlighting their applications in industrial and biomedical fields.

Chapter 7 explains the recent advancement of functional membranes in the water purification. Modern technologies include membrane as a primary methodology in water treatment. The various sources and configurations of membrane materials, different applications on water treatment are discussed in detail.

Chapter 8 enumerates the importance of membrane-based green energy generation. The application of membranes in the energy production, energy storage batteries are discussed elaborately. The development of membrane originated fuel-cell generation are also detailed.

Chapter 9 overviews the different membranes present to attenuate CO_2 from the atmosphere. Membrane/thin film technology has been conceded as the most appealing technology to mitigate CO_2 due to its cost efficiency, low expenditure of energy as well as comprehensibility in operation. Efforts are inclined to disseminate the advantages, disadvantages, and forthcoming practicability of available literature reported to date.

Chapter 10 describes the advanced functions of polymeric membranes for lab-on-chip technologies. The role of these membranes in each part of the sensors as modern data acquisition devices is also described. The chapter is divided into three sections that discuss membrane fabrication, molecular probes, and platforms for reading sensor devices.

Highlights
- Provides a comprehensive overview of advanced membrane technologies
- Cutting-edge technology in advanced functional membranes
- Challenges and perspectives are covered in each chapter
- Significant contribution to the development and applications in this field

Advanced Functional Membranes

Materials Research Foundations **120** (2022) 1-42

Materials Research Forum LLC

https://doi.org/10.21741/9781644901816-1

Chapter 1

Polymer-Based Membranes

Ashok Shrishail Maraddi and S.K. Nataraj[1,2,*]

[1]Centre for Nano and Material Sciences, Jain University, Jain Global Campus, Bangalore-562112, India

[2]IMDEA Water Institute, Avenida Punto Com, 2. Parque Cientfico Tecnologico de la Universidad de Alcala, Alcala de Henares, 28805 Madrid, Spain

* sknata@gmail.com; sk.nataraj@jainuniversity.ac.in

Abstract

Membrane science and technology is increasingly playing an important role in maintaining environmental health. Polymer based membranes in particular show great potential in removing environmentally hazardous pollutants from the wastewater. Membrane technology has also been widely used for ground, surface and seawater desalination which is serving the humanity by providing high quality drinking water. Polymeric membranes have been categorized based on their pore size and their distribution which make them tailor made and innovative separation media namely, reverse osmosis, nanofiltration, ultrafiltration, microfiltration and forward osmosis. All these membrane process require high performing membranes as separation media. This chapter give detailed account of membrane preparation techniques, membrane modules and their applications in different fields. Further, the ideas covered in this chapter are fundamental to all membrane processes. Transport processes, membrane preparation, and flow separation effects are just few of the topics that have been studied.

Keywords

Membrane Technology, Polymer Membrane, Advanced Membranes, Functional Polymer

Contents

1.   Introduction

Today, the membrane technology is a front runner among several existing separation and purification technologies due to its rapid separation and purification characteristics and ease of adaptation. Membrane science and technology has been constantly evolving over several decades now. As it seem to always be the case, the membrane technology has also its origins in nature [1]. Early membranes were conceptualized by studying biological membranes and their functions as they have efficient mass transport properties. Membrane science and technology gained huge importance as one of the most important, useful and rapidly developing branches in engineering and industrial subject and is an expanding field finds wide applications in the field of research, development, engineering and manufacturing which includes synthesis of active pharmaceutical ingredients (APIs), dairy products, macromolecular separation, petroleum industry, desalination, water/ground water treatment, wastewater treatment applications, emerging pollutants removal, controlled-release of pharmaceutical formulations, gas separation, dialysis, artificial kidney etc. Hence, membrane science and technology connect different disciplines from manufacturing, separation, segregation to product recovery, solvent recovery and at the end waste treatment associated with all afore mentioned process steps via selective transportation of desired ingredients across membrane [2-4].

Membrane separation comprises with a variety of processes that can operate under a variety of physical circumstances and principles that can be adapted to separate miscible and/or immiscible natural components from a wide range of mixtures. Since, separation of selective component from mixture across membrane is purely physico-chemical in nature, the membrane's physical and chemical properties, along with the permeating components play crucial role in its separation efficiency. This influences the particular component transportation across the membrane more readily than other. Abundant available polymers have been adapted to prepare these membranes.

Historically, membranes have been conceptualized following biological mimicking membrane found in living organisms. To suit the different conditions, all initial membranes have been developed using cellulose-based precursors. The first evolution of membranes were prepared and tested by Loeb and Sourirajan in early 1960s using modified Cellulose Acetate [5]. The polymers extensively used for membrane preparation can be sourced from commercial market or prepared from different polymerization fundamentals like addition, condensation or by the copolymerization processes. Design and application of task specific membrane with defined pore structures, morphology and their transportation mechanism depends on the various factors such as precursor used, method followed and the mixtures to be separated. In such case, overall membrane performance can be achieved with careful selection of types of polymers used, classification of membrane to be prepared, parameters to be controlled to tune porosity and the engineering the membrane to workable membrane modules. These all steps are inter-dependent and integral to each other. Defined pore sizes, their shape and distribution across the membrane plays very crucial role in different separation mechanisms and membrane transportation qualities. In general, in porous membranes, separation of one component from the mixture may depend just on differences in diffusivity and rate of diffusivity. A solution-diffusion mechanism is used to move or permeate liquid across dense or non-porous membranes. In specific condition, the transportation or penetration of liquid through membrane is determined by the chemical structure and nature of the polymer, as well as the affinity of a liquid, and the selectivity is determined by the solubility and diffusivity ratios, respectively. In dense membranes, the concentration and activity gradient in the provides the driving factor for liquid transport. However, in case of gas separation rate of gas diffusion of transportation can be influenced by increasing the upstream pressure, however pressure has a little effect on activity on the transportation of liquid phase.

With overview of vast sea of literature and the technological report of offering different advantages with respect to technical, economical superiority, it is evident that the membrane technology has been very successful method over the other existing separation technologies or it is considered to be the main path that has been adapted for separation and purification applications at industrial scale. Membrane technology has gotten a lot of consideration in the last two decades as a result of the global water crisis. Special emphasis has been on using polymer-based membranes as tools in water management and wastewater treatment related environmental engineering. With enormous technological advancements over a period of time, membranes in both pristine and composite forms has shown extremely important adaptability behaviors to the specific application. This gives membrane technology an economical advantage when used at industrial scale. Even though membranes have been in existence for over five decades, the increasing water scarcity and the water-related pollutions is driving the growth of this particular technology at an average compounded annual growth rate (CAGR) of 8% and the US market alone for membrane market growth is forecasted at CAGR of 6%. In all, Polymer-based membranes account for the majority of materials types employed in membrane preparation, as they are abundant in nature with simple preparation techniques to adapt, clean and engineer into different modules, that can be maintained well for its long durability. Herein, the present chapter gives an overview of different polymer-based membranes prepared from different materials to their present applications.

2.   Definition of membrane, structure and morphology

Membrane is as a semi-permeable barrier which selectively transport one or more component from the feed mixture. Semi-permeability nature of the membrane indicates its affinity for specific component of the mixture that can preferably be diffused or transported across the membrane. In general membrane can be prepared with different thickness and flexibility. Considering the end application and volume of mixture to be separated, polymer precursor, porosity required and the module suitable to handle the amount of feed can be optimized. The capacity of a membrane to control the penetration rate of a chemical species over the membrane is one of the key qualities utilized since membranes play such an essential role in chemical technology and are employed in a wide range of applications. The transportation properties of membrane are greatly influenced by feed conditions as well such as pH, temperature, operating pressure and the flow rates.

3.   Types of membranes

Polymer-based membranes are a key component in a wide range of industrial applications, including water purification and, in polyelectrolyte form, fuel cells and batteries, where they control the selective transport of ions and/or water. The precise nature of ion and water transport is frequently confusing, because it is dependent on the complex interplay between polymer structures. Broadly the polymer-based membranes are classified as dense or non-porous membrane and porous membrane.

Dense membranes are generally prepared via (1) solution casting method whereas the porous membrane use (2) non-solvent induced phase separation technique (NIPS) and (3) temperature induced phase separation (TIPS) techniques. Advanced membrane preparation processes such as track-etching, stretching, electrospinning, or sintering have been developed in addition to traditional approaches to make membranes for specialized uses. In NIPS, the phase inversion of a polymer carried out via immersion precipitation of caste polymer solution using non-solvent. Here, the polymer solution is casted on a suitable support material, namely the synthetic polymer base, which is then immersed in a coagulation water bath, containing non-solvent solution. The solvent and non-solvent exchange that occurs during this coagulation exchange process results in the creation of porous polymer membranes.

Permeants or feed components are conveyed selectively via diffusion through dense membranes under the driving force of a pressure, concentration, or electrical potential gradient. The relative transit rate of distinct components of a mixture within the membrane, which is dictated by their diffusivity and solubility in the membrane material, is directly connected to their separation. Nonporous, dense membranes, on the other hand, can separate permeants of similar size if the concentrations of the permeants in the membrane material differ significantly.

3.1 Dense membrane via solution casting method

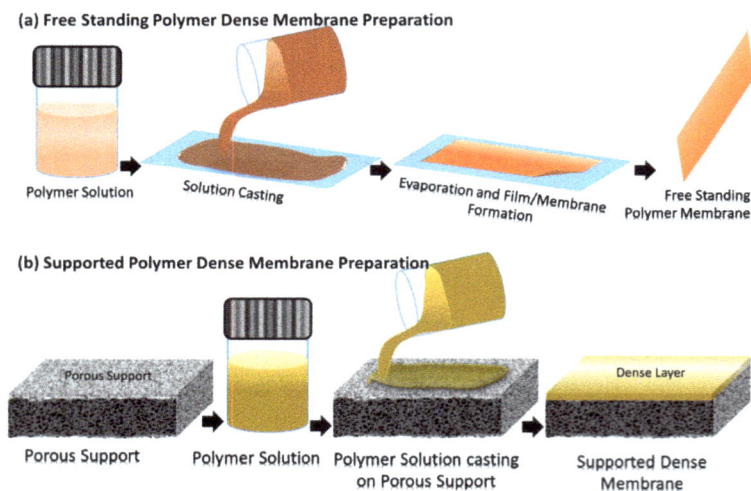

Figure 1 Schematics of dense membrane preparation route both in (a) free standing form and (b) supported dense membrane.

Dense membranes are employed for efficient separation with sufficient permeate flux in practically for all reverse osmosis, gas separation, pervaporation, and membrane distillation processes. Dense membranes further can be classified in to two types based on their physical characteristics namely; (a) free standing dense membrane and (b) supported dense membrane as shown in Figure 1. In solution casting method of self-supported dense membrane preparation follows, solution casting of polymer followed by evaporation, whereas supported membrane is prepared first support is prepared via dry-

Advanced Functional Membranes   Materials Research Forum LLC
Materials Research Foundations **120** (2022) 1-42   https://doi.org/10.21741/9781644901816-1

wet immersion technique followed by dip coating or solution casting leading to complete solvent evaporation. Dense membranes usually with the thickness of 38-110 μm range. The separation here occurs mainly by a solution-diffusion process that includes numerous influencing elements such as temperature, pressure, polymer type, rate of diffusivity, and so on.

3.2   Symmetric porous membranes

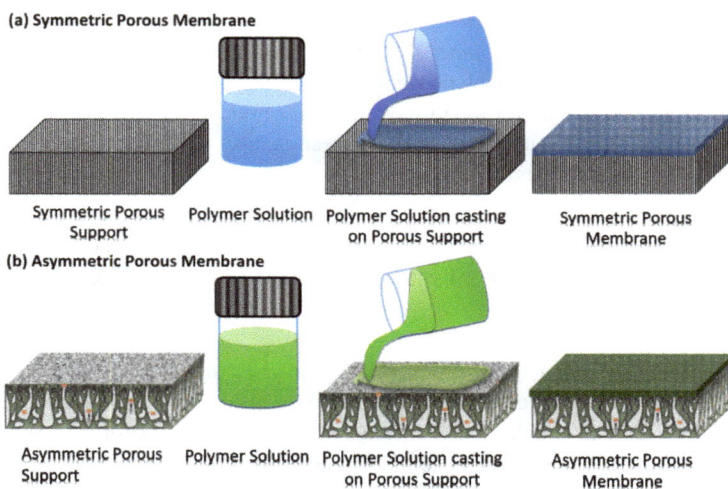

Figure 2 Schematics of porous membrane preparation routes showing morphological features of (a) Symmetric and (b) asymmetric porous membrane.

Porous symmetric polymeric membranes prepared from phase separation generally with track etched support layer having high flow rates and selective layer with excellent retention properties as shown in Figure (a). These membranes are also prepared without separate selective layer from a metastable two-phase liquid dispersion of polymer in solvent and non-solvent systems.

3.3   Asymmetric membranes

In asymmetric membranes, pores are distributed in ascending order from bottom to top of selective layers. The tight pore size regime generally sits on the feed facing selective layer. Among many polymers, polysulfone, polyethersulfone, polyacrylonitrile and cellulose-based systems are co-operative to formation of highly asymmetric membranes. In a classical asymmetric membrane, it is the surface pores that are the tight and the permeation channels are broad to facilitate the transportation of product/permeate as shown in Figure 2(b). In the asymmetric membranes the tightest pores are adhered to the interface of the supporting skin and the selective layer when used in composite polymers. This is also called skin and the asymmetric region of the composite membrane. Asymmetric or anisotropic (from support where the asymmetric region generally originates) regime mainly defined by the degree of asymmetry to be called as tight or loose type of asymmetric membranes. These membranes will have average skin sizes of greater than 1 micron in homopolymer type and varying sizes for composite membranes from NF to RO regime as shown in membrane spectrum in Figure 3. In general, The physical properties of the support layer or permeation layer will be at least twice those of the average skin pore size and thickness [6].

With the advancement of nanocomposite, membrane technology is too benefited by the use of different nanomaterials for membrane applications. Where, the top layer of an asymmetric membrane is functionally induced by different nanoparticles induced to form several selective layer films. In such cases, asymmetric membranes consist of a thin dense layer that can act as a selective skin as well as a porous support, the interstitial spaces generated between the composite layers in the Nano scale range functions as permeation channels. This engineering adaptation of the asymmetrical structure with a dense layer on top that is tight to thin and supported by a porous substrate is strategically crucial to improve separation efficiency. This will also reduction membrane resistance to permeate flow and improve the shelf life.

4.   Preparation of porous membranes and control parameter

4.1   Solvent phase inversion technique

Polymer-membrane science and technology has progressed to the point where it may currently be used in a variety of separation processes, such as water desalination, macromolecular enrichment, and biological component cleansing etc. Synthetic polymer membranes can be made using a variety of procedures that allow for the shaping of a polymer material to achieve the required properties, the most essential of which are morphology, porosity, and distribution. Before casting solution, desired membranes characteristics need to be understood and accordingly precursors need to select. Two elements must be considered while producing polymer-membranes for a certain separation: That consists of 1) the nature of the precursor material (polymer or polymer

Materials Research Forum LLC

https://doi.org/10.21741/9781644901816-1

blend or composite polymers) and 2) the morphology (dense/non-porous or porous) of the membranes. Solution casting of polymer solution to turn in to either porous or non-porous membrane has been successfully adopted.

The most extensively used technology for producing polymeric membranes is non-solvent induced phase inversion. Phase inversion is a demixing method in which a previously homogeneous polymer solution is altered in a controlled manner. Loeb and Sourirajan pioneered the phase inversion procedure in membrane technology in the 1960s, and it is the foundation for most commercially available membrane synthesis. Phase inversion is a demixing method that transforms an initially homogeneous polymer solution from a liquid to a solid state in a controlled manner.

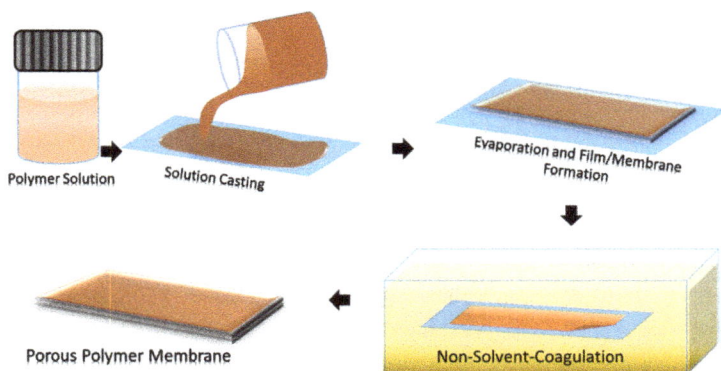

Figure 3 Schematics of non-solvent induced phase inversion method used for preparation of porous membrane.

The polymer solution is created in a suitable solvent or a mixture of a volatile non-solvent, and the solvent is allowed to evaporate, resulting in coagulation bath precipitation as shown in Figure 3. The solution system's core components are polymer and solvent, but various additives can be added to adjust the membrane's porosity, pore dimension, and overall morphology. Nonetheless, when producing the casting solution, the choice of polymer is critical. The solvent, in which the polymer must be soluble in an acceptable concentration, is the next critical parameter to optimize, since it is directly related to the final membrane application. Once the polymer is dissolved, the homogeneous polymer dope solution is cast on the glass plat for free standing membrane and on support porous layer or fabric for composite membrane with the help of doctor's blade or suitable spreading rods. The rods are then stretched apart to achieve consistency throughout the film. Then as cast thin polymer film with/without support will be immersed in gelling bath covering non-solvent (miscible with solvent) to facilitate solvent and non-solvent exchange in polymer matrix. The polymer film starts precipitating out upon solvent/non-solvent exchange, into a porous precipitate membrane within the coagulation bath. The miscibility of solvent and non-solvent interchange within the coagulation bath, results in demixing and precipitation immersion, separating the two phases of polymer solution to polymer-rich and polymer-poor phase. During the phase inversion process, a thermodynamically stable polymer solution is tending to change from liquid to solid state in a controlled manner. Here, the solvent, non-solvent, solvent evaporation and coagulation period determines the structure, size and shape of the membrane pores. Further, as formed membrane can be heat treated to facilitate optimal mechanical and chemical strength to resulted membrane.

In some cases, the transfer of the fragile, ultrathin film onto the microporous support is the method's principal drawback. Thin film is typically transported by sliding the support membrane under the spread film. With extra care taken, the small segments of membrane as thin as 1 to 2 µm can be easily transferred with this approach. On the other hand, thin film of polymer can be directly coated microporous hygienic, defect-free and very excellently microporous support. In all composite thin-film membrane preparation procedures obtaining defect-free films involves substantial attention to set optimization parameters. In all supported membrane preparation procedures, the nature of the microporous support is critical and plays an important role in overall performance and shelf-life of a membrane [7,8].

Furthermore, by modulating various parameters, phase separation of a casting solution into a support polymer can indeed be efficiently controlled to obtain appropriate pore characteristics and morphologies as well as desirable sticky capabilities. These include the use of appropriate co-solvent and non-solvent combinations, controlled evaporation, thermal precipitation by adjusting both the solution and coagulation bath temperatures, and precipitation from the vapour phase by immersing the cast film in a non-solvent vapour phase. As a result, these variables are inextricably related to thermodynamic and kinetic factors. The viscosity of both the casting solution rises significantly as the polymer concentration in the solution increases. As an outcome, the system's period before precipitation gets larger, even though the period changes linearly rather than rapidly.[9] Natural skinned asymmetric crosslinked polyimide (PI) NF membranes suited for organic solvent nanofiltration (OSN) are made at industrial scale using this

Materials Research Forum LLC
https://doi.org/10.21741/9781644901816-1

technology. The preparation of polymer-based hollow fibre membranes involves similar techniques and parameters, apart from module design [10].

4.2   Thermally induced phase inversion technique

Thermally induced phase inversion was first proposed and researched in the 1980s to fabricate microporous membranes, but it received scant attention since NIPS was considered to become a more convenient and adaptable way of creating polymeric membranes. TIPS study is currently gaining popularity recently, leading to the advent of membrane contactors, that offer a number of distinct advantages, including process simplicity, high reproducibility, and a low tendency to form defects, as well as high porosity and the ability to create interesting microstructures with a narrow pore size distribution. Therefore, parameters can be optimized to achieve membrane reproducibility by controlling polymer reassembling during thermally induced evaporation. TIPS emphasizes polymorphism utilizing solvents and process parameters as a distinguishing aspect. Figure 4 illustrates the overall procedure for generating porous membranes using the TIPS approach [11].

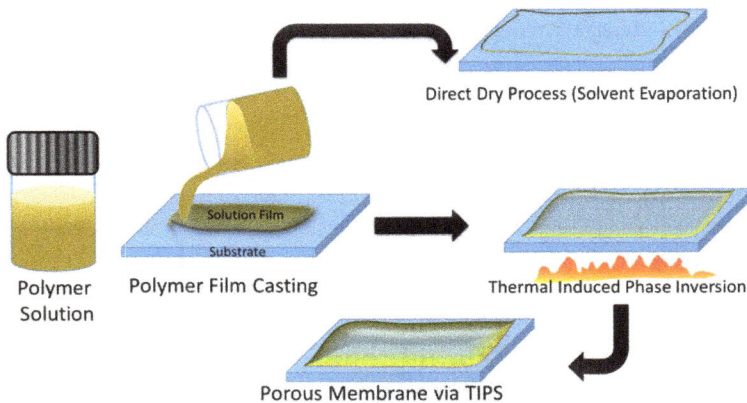

Figure 4 Schematics of porous membrane preparation procedure via Thermally Induced Phase Inversion Technique (TIPS) method.

The TIPS process includes the following steps: 1) Is to dissolve a polymer vigorously in a high-boiling, low-MW solvent at an elevated temperature, typically around or above the polymer's melting point to form a homogeneous melt-blend. 2) Forming the dope polymer solution into the desired shape and size on a suitable substrate, such as with a glass plate for a free-standing flat sheet, a porous support for a composite membrane, or hollow fibres. 3) To cool the cast solution to induce phase separation and polymer precipitation, and 4) To separate the diluent, generally by solvent extraction, to yield a membrane. Hence, Phase inversion is a precise balance of polymer–solvent interaction, cooling rate, chilling media, and temperature gradient.

One of the TIPS method's distinctive qualities is the ability to build membranes from semi-crystalline polymers that may not be ordinarily soluble in solvents at higher temperatures, as previously discussed. TIPS is regaining popularity as a viable platform for fabricating highly porous microporous membranes. There has been intense research in membrane technology for more sustainable techniques to produce polymeric membranes employing green solvents. To achieve long-term goals, green solvents with high volatility have recently become popular, making the TIPS approach more economically friendly. PVDF-type polymers have shown great promise in the development of microporous membranes [12,13].

5.   Membrane separation process spectrum

Polymer-based membranes find several applications because the polymer offers flexibility in membrane design and module adaptation. This makes membrane technology a cost effective, a high pre-concentration factor with a large degree of selectivity and ease of operation [14]. Polymer-based membranes have a wide range of applications due to the ease of the polymer in membrane design and module adaptability. Membrane technology is therefore cost-effective, convenient to use, and also has a high pre-concentration factor with a high degree of selectivity. Membrane separation processes offer a range of membranes, as shown in the Figure 5, that mainly depends on the size and nature of the pollutant to be treated. Microfiltration (MF), ultrafiltration (UF), nanofiltration (NF), reverse or forward osmosis (R-O), and ion-exchange membrane processes such as Electrodialysis are examples of these [15]. Membrane separation is a wide term that refers to the process of deciding the quantity and specific pollutant rejection potentials of a membrane. The pressure required to remove the pollutants from the mixture/feed is used to classify the membrane process spectrum. Microfiltration membranes have the highest pore size and are used to filter out big particles and microorganisms at low pressure. Ultrafiltration membranes contain smaller pores than microfiltration membranes, so they can reject bacteria and soluble macromolecules like proteins at slightly higher pressures in addition to large particles and pathogens [16]. Nanofiltration is a relatively novel membrane category which fits between dense membrane and ultrafiltration pore regimes and operates at a moderately high pressure. Nanofiltration membranes are hence referred to as 'loose' reverse osmosis membranes. Nanofiltration membranes are porous membranes that are often manufactured in an asymmetric manner. As a result, NF are porous membranes with pore sizes of ten angstroms or less that function somewhere between reverse osmosis and ultrafiltration membranes. With pore sizes of less than 1 nm, reverse osmosis membranes are effectively non-porous [17, 18].

5.1   Microfiltration

Microfiltration refers to the membranes category where pore size actively range in the 1 μm to 10 μm. This regime requires minimum operating pressure to generate permeate fluxes. Separation from MF is based largely on molecular separation, and yet process conditions could have a direct effect on the separation efficiency [19]. MF is used for numerous process applications such as diary product separation where is defatted in the manufacturing of whey protein isolates, casein and serum proteins are separated from skim milk, and cheese brines are clarified. The MF membranes are made using polyvinylidene fluoride (PVDF) polymers, poly(ether sulfone) and polyacrylonitrile. The other common applications of MF membranes was found in sugar and sweetener, and industrial bioprocessing industries [20]. MF is used to purify fermentation liquors prior to further refining operations in this scenario. MF is frequently used in the industrial bioprocessing industry to clear bulk fermentation streams and increase the performance of downstream processing stages. The MF process is utilized as a pre-treatment step for other separation processes such as NF and RO in a variety of applications.

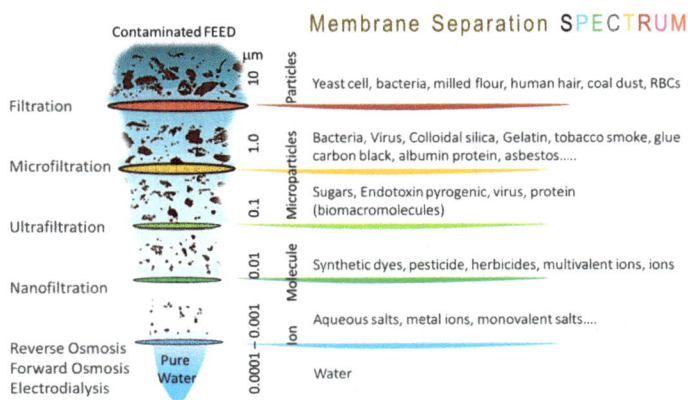

Figure 5 Infographics of membrane separation processes with pore size and separation spectrum.

5.2   Ultrafiltration

The ultrafiltration materials are made from a pore range of from 0.01 to 0.1 in pores and further classified by Daltons (Da) or kilo-Dalton based on their molecular weight-cut-off efficiencies (MWCO) (kD). Typically, MWCOs of UF membranes range between 1,000 Da (1kD) and 100,000 Da (100kD). Like MF, the separation regime of the ultrafiltration process depends mainly on molecular exclusion, but process conditions, such as applied pressure, depend on the type and distribution of the pore [21]. MWCO membrane, on the other hand, can be tuned again for final application. Polyvinyllidene fluoride (PVDF), polyethersulfone (PES) or polysulfone (PS) polymers are the most widespread used polymers for UF applications. In RO and NF thin film composite (TFC) membrane applications, the substantial majority of PS-based UF membranes are used as a supporting membrane [22, 23]. Much farther specific utilization of UF membranes would include UF membranes for the dairy processing, pain industry, resin recovery, bioprocessing industries, domestic wastewater treatment, industrial wastewater treatment, process water recovery and macromolecular augmentation. [24].

5.3   Nanofiltration

Relatively new addition to the membrane separation spectrum is the nanofiltration (NF) process. Tight ultrafiltration regimes have been made reference to as the industrial reference to NF. Magnesium sulphate (MgSO$_4$) and sodium chloride are the main features of NF membranes that are based on their rejection characters (NaCl). The rejected range is generally projected to be ~90-99.5% and ~300-70%, respectively, for MgSO$_4$ and NaCl for NF. The separation mechanism of NF is based in large part on diffusion by the membrane of dissolved species, and pH and chemical loads can be severely affected near or on the surface of the membrane [25]. Recently, the most common use of NF is the demineralization of industrial process intermediaries, as an alternative to both UF and MF processes, combined with concentrating macromolecules. The NF process has evolved over the years as an important separation process for applications such as molecular separation, demineralization, desalination, organic weight loss and waste water treatment. NF is used for removing impurities and enriching primary sweetening components in sugar and sweetener industries [26].

5.4   Reverse osmosis and forward osmosis

The Reverse Osmosis (RO), which categorises the separation range based on rejections features of conventionally sodium chloride, is a commercially most popular membrane in separation ranges. The NaCl range is expected to reach ~96-99.8% for RO membranes. As with NF, the separation with RO is based largely on the spread of dissolved species by the membrane and the osmotic pressure of the process fluid or feed [27]. The RO membranes are generally prepared with the selective thin layer and support for ultrafiltration in thin-film morphology. The thin layer of polyamide is commonly used on the ultrafiltration support of polysulfones. The RO is widely used after UF or NF as an effective tool for various applications for molecular separation, desalination and treatment of wastewater. The use of RO process in-house has revolutionized recent developments in nanotechnology innovation, modified RO composite membranes. Household RO applications also forced its use of RO modules in miniature with greater separation efficiencies in innovative ways [28].

6.   Material choice for polymer based membranes

The selection of an appropriate polymeric materials is a key component of effective, physical and chemical separation membranes. Polymer nature and the membrane process play an important role in obtaining the resulting membrane's enhanced permeability, selectivity and durability. Polymeric membranes, also known as organic membranes, include a family of technology for liquid separation which is industry leader in efficiency and performance. These membranes are generally recognized in four main categories based on efficiency and separation characteristics. An earlier discussion (microfiltration, ultrafiltration, nanofiltration and RO). The polymers utilized in the fabrication of each membrane classes are carefully selected to ensure excellent performance in a specific process environment. Compatibility must be taken into account with the fluid(s) to be processed and cleaning materials required to help maintain hygiene and stability in performance. Polymers include polysulfone/PS/PS, polypiperazine and polyamide, as well as polyvinyl fluoride (pvdf) [25]. Further, list of some of the synthetic polymers used for various type of membranes have been recorded in Table 1.

6.1   Synthetic polymer membranes

Roughly every commercially successful membrane consists of organic synthetic polymers such as polysulfones, sulfone polyether, poly(vinyl diameters), polyacrylonitriles, cellulose acetate, and so forth. Microfiltration, ultrafiltration with large pores and different solution preparation conditions with adopted parameters frequently consist of identical polymer materials. During membrane preparation process coagulation, pore size control will usually be followed to produce various pores and morphologies.

Table 1 List of some of the synthetic polymers and their structures used in membrane preparation.

No	Synthetic Polymer	Structure	Ref.	No	Polymer	Structure	Ref.
1	Polysulfone (PSF)		[29]	7	Polyvinyl alcohol (PVA)		[30]
2	Polyvinylidene fluoride (PVDF)		[31]	8	Polycarbonate (PC)		[32]
3	polyether sulfone (PES)		[33]	9	Polyurethane (PU)		[34]
4	Polyamide (PA)		[35]	10	Polyimide (PI)		[36]
5	Cellulose acetate (CA)		[37]	11	Polyvinylpyrroli done (PVP)		[38]
6	Polyacrylonitrile (PAN)		[39]	12	PIM-1		[40]

Materials from inorganic materials such as ceramics or metal oxides/hydroxides can also be prepared with composite membranes of polymer origin. Further, blending composite membranes seems to be possible to achieve porosity and permeability ranging from microfiltration, ultrafiltration or nanofiltration. In terms of thermally stable, chemical resistant, composite membranes have proven to be an advantage over only polymer membranes. On the other hand, there are also several downsides with ceramic membranes processed for implementation in microfiltration, such as a high cost and mechanical fragility that prevented their widespread use. Likewise, metallic membranes often consist of stainless steel and can be porous very finely. It is mainly used for gas separation, but may also be used as a membrane support or for water filtration at high tempers. However, for large-scale use, metallic membranes pose difficult issues in design and module preparation and pose several logistical problems. The current tendency on membrane development is to use Nano-functionalized membranes. Polymer membranes with active nanoparticles besides bio foliation avoidance are an example of modern membranes that are becoming extremely important because of various proven advantages over polymer membranes alone.

Furthermore, synthetic membranes can be classified as organic and inorganic. Inorganic membranes include various materials, such as zirconia, alumina, titanium, palladium and silver, for example, ceramic, carbon, zeolites and oxides of metals. Organic membranes, like polytetrafuoroethylene, polystyrene and Polyamide and others that form macromolecules, are used for the majority of commodities and industry [41]. In order to achieve increased properties, enhanced film strength and cost effective maintenance of the conventional scale-up protocols, use these synthetic polymers to prepare blends for use as membranes [42, 43]. Inorganic synthetic membranes, even so, seem to be hugely reliant on the mechanism of charge exclusion to promote electrical refusal and mass transport. Charging properties can be easily adjusted in the membrane surface, however, to ensure optimum pollutant refuse and separation performance through different parameters, like feed solution pH and ionic strength conditions. When the optimum load on the surfaces of the pottery membrane is changed, the overlap effect of the double layer will avoid the ionic rejecting patter as distance potential attenuation is seldom assessed to determine the extent of this possible overlap [44,45]. The main advantages of the synthetic polymer membrane are that polymers offer extensive opportunities to change in the form of composites and blends and can also be customized to make them adaptable in any form, such as flat sheet and hollow fibre [46]. Over the years, progress in the science of polymers has permitted a variety of innovative polymers available for membranes with special features that can be used through various morphologies in particular.

6.2   Biopolymer membranes

Table 2 List of some of the important biopolymers used for membrane application.

No	Biopolymers	Structure	Ref.		Biopolymer	Structure	Ref.
1	Chitosan (CS)		[47]	8	Polylactic Acid (PLA)		[48]
2	Sodium alginate (SA)		[49]	9	Polyhydroxyalkanoates (PHAs)		[50]
3	Agar (AG)		[51]	10	*Polyhydroxybutyrate (PHB)*		[52]
4	Agarose (AGR)		[53]	11	Pullulan		[54]
5	Chitin (C)		[55]	12	Nitrocellulose (NC)		[56]
6	Cellulose		[57]	13	Cellulose acetate (CA)		[58]
7	Carrageenan (CAR)		[59]	14	Cellulose triacetate (CTA)		[60]

For numerous material and process applications, biopolymers include a variety of naturally occurring polymers (Table 2) from renewable resources. Biomacromolecules or biopolymers are generally known as substances produced through live organisms. Their unique characteristics and abundance have recently been taken into account in the research and development of sustainable end products. Biopolymers are chain-like molecules composed of repeated chemical blocks, which easily deplete biomass

Materials Research Forum LLC
https://doi.org/10.21741/9781644901816-1

resources in the environment. Many repeat units are active in several product segments today with cellulosic products and their biopolymers derivative. On the other hand, plenty of others continue to stay uncovered and underutilized. Biopolymers have opened up to greater utilization in membrane applications with a rapid development of the understanding of basic biosynthetic pathways and bioprocessing protocols. A range of polysaccharides, proteins, lipids, polyphenols and specialty polymers are derived from the biopolymer backbones. The use of these sections will also help to glug and manage final products. Many other synthetic and biopolymers with an increased interest in green and durable materials are being modified to suit biopolymers. This would add value to environmental compatibility and enhance protection [61]. Also, hydrophilic and biocidal properties of biopolymer membranes with unique wettability have drawn considerable research interests for addressing challenges biocompatibility and biodegradability [62]. Several advancements in thin film and even in the scientific fields of functionality have also led to the remedy of new membranes in conventional membranes, such as fouling, scaling and biofouling. The durability of a membrane was increased, but the maintenance and operating costs were also drastically reduced when used. Building on these reasons, membrane processes are an intensified process that offers a number of advantages, including lower energy and reduced plant volume, as expected by the process intensification strategy.

Biopolymers which have properties along with biocompatibility, biodegradability, portability and ecological responsibility can therefore be a legitimate alternative to the conventional polymers based on petroleum. Biopolymers have shown great potential for water treatment and wastewater treatment, as biopolymers could be used to modify their physical and chemical properties such as pH, temperature, mechanical and thermal stress, etc. [63]. Moreover, biopolymers display low toxicity for human health, high biocompatibility and biodegradability [64]. In large-scale membrane production, various biopolymers such as CA, CS, PLA, PHA, PHB, TPS, PVA and PU are now substituting conventional polymers such as PSF, PES, PI etc. The other important property of biopolymer is that the majority of biopolymers are inherently film-forming and have a high affinity for water and a composite potential for different nanomaterials. However, there remain significant drawbacks in membranes prepared with biopolymers such as low mechanical strength. Nevertheless, biopolymer membranes with enhanced mechanical characteristics have been produced using nanomaterials which can potentially be of immediate prosperity.

6.3   Membrane modules and types of modules

Once Membrane is prepared in large scale, adopting the as prepared either flat sheet or hollow fibre membrane in to engineering module becomes crucial to extract optimal performance from the device. Engineering the membrane into module form is art of extracting maximum from the membrane itself. In the past several innovative pathways have been seen in the way the membrane is converted into engineering devices and components to separate the feed into permeated and resentful streams. Based on their structure and arrangements, commercially successful modules are therefore classified, and the industry uses four different kinds of membrane modules. They are (1) Flat sheet plate and frame modules, (2) Spiral-wound modules, (3) Hollow fibre modules, and (4) Tubular modules.

6.4   Plate and frame module

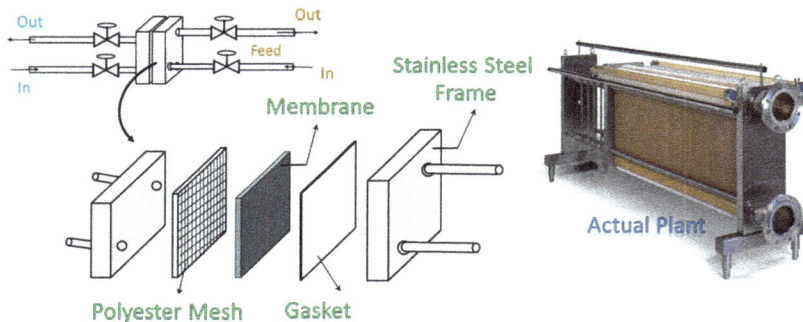

Figure 6 Schematic and actual membrane module and plant prepared using flat sheet membranes.

Yet another platform and frame module are designed in a metal frame with flat sheet membranes and spacers, as shown in Figure 6. Depending on the relation between two solutions, the model also shows three possible modes of plate and frame module. These three modes may work in the same operating conditions differently. Four conditions are employed to simulate the plate and frame module: the volumetric flow of the feed and solution drawing, the concentration of the drawing solution, the flow direction and membrane orientation. The flat and frame module is the simplest device for packaging flat sheet membranes among all membrane modules [65]. Plattened-and-frame modules, like stripes, size and shape of research lab devices holding a single, small membrane, systems with membrane series that can be used to contain plate and frame modules, are designed and developed in various directions. In the plate- and frame module, between two geometric types, the horizontal rectangle and vertical rectangle are designed for feed flow in the cross-current flow direction. A single membrane sheet with a lower pack density is simulated in the plate and frame module [66]. However, this module is easy to operate and handle among all membrane engineering modules for large-scale panel and frame module in industrial applications.

6.5   Spiral wound module

Most commonly used spiral injury modules and successful, commercial membrane modules exist in the membrane market due to their high volume-to-membrane surface. Spiral membrane module consumes an enormous amount of other than membrane engineering material itself. A spiral-wound module has the membrane, the feed and the permeate channels, the membrane spacing, the permeate tube and the membrane housing as the main components. This module strategically places engineering materials such as feed and permeate distances to avoid feeding or permeate fasteners in separation experiments. Although the channel spacer may increase mass transfer near the membrane, the pressure loss along the membrane leaf inevitably increases as shown in Figure 7. During the spiral membrane wound process membrane sheets are glued to form an envelope on three sides with spacers and support layers. More envelopes are attached and rolled up around this permeated tube in order to create feed channels and permeate. The membrane sheets are positioned in a pressure-module containing membrane packaged in a sticked or plastic thread, which stabilises the assembly once it has been finished. Used three or more modules in series are usually connected to a pressure vessel for optimal output. The performance of the modules is influenced by many factors in the spiral wound membrane: 1) propensity and capability to foul; 2) quality control and operating conditions, such as feed pre-treatment, feed concentration, food supply pressures, and permeate recovery, 3) distances that have a large impact on the locally mixed, mass transmission, and the extent that concentration polarization and pressure loss are determined [67,68]. Spiral wound membranes are widely used to remove calcium and other divalent ions from hard drinking water in under-the-sink nanofiltration modules. Many modules in the industrial scale include membrane envelopes wrapped around the central collection pipe, each of which has an area of 1-2 m^2.

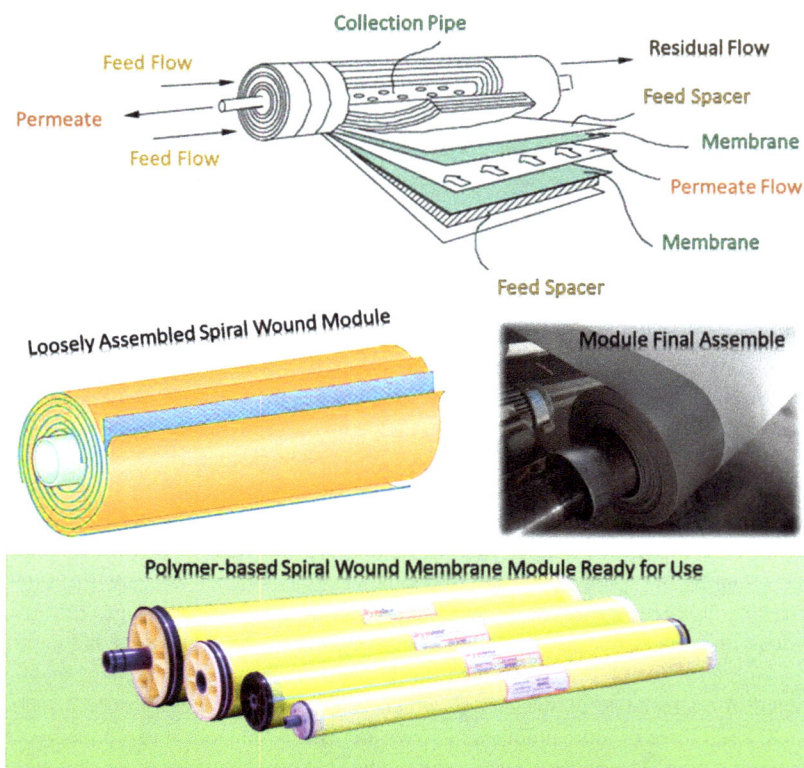

Figure 7 An overview of the spiral wound membrane from schematics to end product.

The design is straightforward when the membrane is wrapped between spacers and supports, consisting of a membrane envelope of spacer and membrane wound around a perforated central collecting tube. Finally, all this spiral wound is placed within a tubular pressure vessel around the collection tube. Feed goes through the membrane envelope axially through the module. One part of the feed penetrates the membrane envelope, spiralling to the centre and exiting through the tube. Nanofiltration and ultrafiltration spiral-wound modules have been applied extensively for applications of reverse osmosis in the desalination, organic molecular separation, and food processing industries

6.3 Hollow fibre module

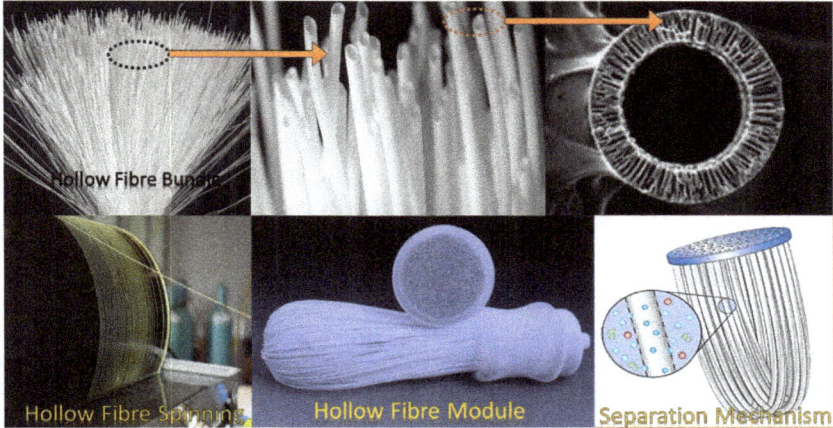

Figure 8 Actual photographs of bundle of hollow fibre membrane, their spinning process, ready to use module and schematics of separation mechanism in hollow fibre modules

The hollow fibre membrane segment (HFM) encompasses more than 50% of the membrane markets used in water treatment as this sector has superior packing density, higher volume surface and self-supporting capacity in the membrane module form. Hollow fibre membranes are an artificial membrane class with HF that acts as a semi-permeable bundling barrier with no other material that enables the use of economically cost-effective engineering. The HFMs have a porous, separating layer coated polymer fibre structure. The polymer density increases in order toward the outer surface of the fibre or vice-versa, depending on the direction to which it flows [69]. This design is easy to make and makes it necessary to contain very large membrane areas within a cost-effective system. Hollow fibre membrane with little diameter can withstand operating conditions in packaged conditions to prevent collapse of the fibre in high applied pressure. Typically, the internal diameters and thick walls of HFM are 50 μm and the outer ones are $100 - 200$ μm. HFMs are becoming more popular because of their high packaging density, mechanical stability and high permeate output compared with all other membrane modules. Contactors with a wall thickness of <1 mm and a pore size <0 have high specific contact surfaces (m^2/m^3), <1 mm internal diameter. and pore size <0.3 microns, a tube in shell configuration is used as shown in Figure 8. HFMs are also known for their physically robustness, prerequisite limited space for use, and simple housing engineering protects the delicate hollow fibre membranes [70].

In their mechanistic separation, hollow fibre membranes are also classified into two types, i.e. 1) inward to outer and 2) outer to inland which indicate feed flow and permeate direction. Types of HFMs are also classified in three types for membrane preparation (doping and polymer solidification) methods. 1) Spinning of the melt: The Polymer is directly heated by melting in an inert atmosphere in a chamber above its melting point and the melting polymer is extruded by a spinneret. As soon as the melt polymer is crossed, the cooling immediately leads to a transition to the phase and solidifies the polymer. This produces a capillary or HFM with a uniform structure. 2) Dry spinning: The Polymer is dissolved in a highly volatile solvent by the dry spinning technique and the solution is extruded by a known spinneret diameter. The polymer solution is heated during extrusion and, due to solvent evaporation, the polymer solidifies which lead to the development of HFM, and 3) Wet spinning: the extruded polymer solution will be immersed in a non-solvent bath during this process. There is an air gap between Spinneret and non-solvent Bath, where membrane formation begins with a particular thickness of the inner wall [71].

6.6   Tubular membranes (ceramic membranes)

Self-supporting membranes are not tubular membranes. They are located within a tube, generally of inorganic materials, made of a special kind of material. In most of the tubular packages, the membrane inside is supported and housed. Ceramics and their composites are the most popular membrane material used in the tube membrane module. These materials are fragile and at times breakable, so that tubular membranes are placed inside the tube, the tubular membrane flow is normally inside a tube. Due to its excellent mechanical stability, chemical and thermal resistivity, ceramic membrane has attracted considerable concern and are cost efficient for some extreme applications where polymeric membranes are not easily maintained. Ceramic membrane has the advantage of its thermal and mechanical properties that make it possible to sinter at a much lower temperature than alumina. Although several studies have demonstrated the benefit of the ceramic membrane, sustainable developments in ceramic membranes have hindered the use of them because of their non-flexible and fragile nature.

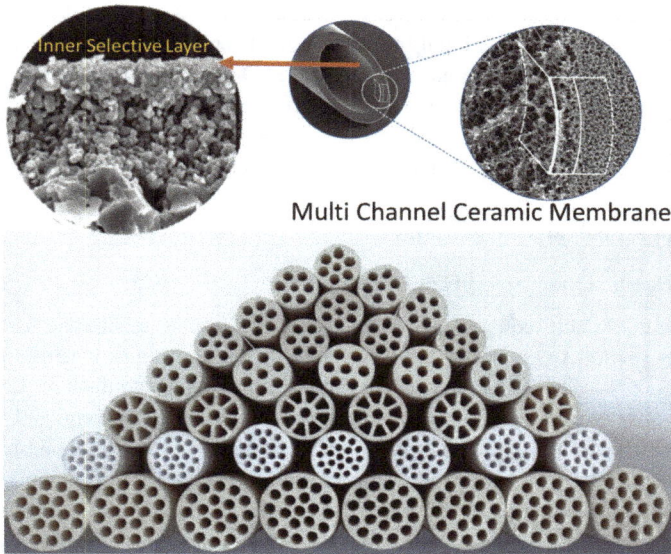

Figure 9 Structure and morphologies of multi-channel alumina-based ceramic membranes.

Ceramic containers, nevertheless, are less expensive, effective and higher performance compared to polymers, applied in extreme conditions such as pH-like, high temperature and harsh solvent conditions [72]. For the removal of suspended matter from the ultrafiltration and microfiltration range ceramic membranes are effectively used. Ceramic membranes are usually made from oxides of aluminium, titanium or silicone that are stable under extreme feeding conditions. The advantages of ceramic membranes are that they are chemically inert and stable at high temperatures in which the polymers are ineffective. These physical and chemical properties make ceramic MF/UF membranes particularly suitable for food, cooling power stations, highly contaminated treatment of waste water, biotechnology and the pharmaceutical industry in which membranes require repeated sterilisation and cleaning of steam by aggressive solutions. These pores are in the pack of different sizes of nanoparticles of a metal oxide at high temperature and pressure as illustrated in Figure 9 and range from 0.01 to 10 mm in ceramic membranes for a microfiltration and ultrafiltration range. In general, these membranes are produced through a slip-coating and sole-gel extrusion process. Other techniques, especially sol gel-powered techniques are used for the fabrication of pores between 0.001 and 0.01, as well as for the manufacture of highly packaged small ceramic pores prepared via soil-gel methods, are suitable for use in gas separation [73]. The traditional method of preparing ceramic membranes is through a multistep process of manufacturing using powder composites of ceramic or metal oxide, which depends on the application and membrane configuration following shaping, heat processing and layer deposition procedures [74].

7. Advanced polymer membranes

7.1 Block copolymer membranes

Controverted structure requirements for advanced antifouling membrane for a range of applications such as CO_2 capture, seawater desalination and other industrial applications require new approaches to membrane science and membrane manufacture. Limitation of pore size and distribution is the main disadvantage with conventional membrane preparation. The molecular manipulation of a porous structure in combination with surface chemistry is now a well- known approach to the creation of emerging physical pores to increase the features of all membranes in certain application applications using a block copolymer self-assembly route [75]. Block copolymers pave the way to overcome the disadvantages of the conventional membranes. These polymers can be used either as porous or pore films, hollow fibres and particles. As the main advantage of the block copolymers for membrane manufacture, this polymer is multi-purpose morphology which can be controlled by adjusting the chemical composition and the length of the block for the desired porous dimensions. With a self-assembled block copolymer route different chemical structures can be obtained in just one polymer and selective dense layers with a particular porosity and composition can be prepared in combination with other nanocomposite materials that find huge applications in gas separation, pervaporation, nanofiltration and fuel cells [76, 77]. Besides its self-assembly property, diblocks and triblocks are equipped with micro-domain isolation, which shows an amazingly complex and periodic geometry which has unlimited nanoscale engineering potential for porous membrane applications [78]. The self-assembly di-block copolymers produce the selective deterioration of one-block spheres, lamellae, cylinders and complex nano-bicontinuous pores. The copolymers of triblock, on the other hand, produce more complex structures with numerous morphological components unknown to science and technology membrane community but that could revolutionize the properties of fluid transport [79]. The first block copolymer membranes was reported half a century ago as charge-mosaic membranes [75]. The whole charge murals are the arrays of loaded domains containing paths for both anions and cations, both anion-exchange and cation-exchange. The first copolymer cast on substrate to achieve its purpose was poly(styrene-b-2-vinylpyridine. In 1975, Odani et.al in addition, in order to create an ordered porous structure, used poly(Styrene–butadiene–styrene) (PSBS). Authors followed two important approaches in this method to pore production: 1) film casting-selective block sacrifice and 2) self-assembly through non-solvent-induced phase separation (NIPS). In applications such as gas separation, ultrafiltration, nanofiltration, pervaporation, fuel cells and batteries, copolymer membranes produced through the method have been used. Recent developments have shown a promising self-assembled copolymers with PEG and fluoro-blocks as innovative insights for proposed anti-fouling applications [80]. On the other hand, oligomerically polymerized polyamide membranes with successful integration of anti-fouling properties have been recently incorporated into oligomeric zwitterionic building blocks [81].

7.2   Functional polymers-based membranes

Functional membranes offer multiple features, both help in the separating pollutants or contaminants from an aqueous solution and also contribute to adsorbing, degrading and/or disabling contaminants. Functional membranes based on polymers are generally designed and adapted to specific applications. These membranes are used to filter out organic and inorganic contaminants like dyes, emerging pollutants, and heavy metals, etc. [82]. The production of membrane using functional polymer involves processing polymers and the subsequent formation of membranes in which methods regularly followed have to be modified in order to achieve stable membranes [83]. This process maintains its original properties while providing new properties such as hydrophilicity, bio-compatibility, anti-static characteristics and performance. Nanomaterials such as clays and other colloidal suspension are also used to optimise the functionality of polymers to optimise different parameters relating to treatment such as adsorption properties, swelling patterns, and rheological/colloidal phenomenon.

7.3   FO membranes

Among other membrane separation processes, the FO process is recovering its momentum. With conventional pressurized techniques at molecular level the use of FO can be achieved with special application. Unlike the pressure-driven membrane processes, FO is an osmotically driven, low-hydraulically operated membrane process. In the absence of the applied pressure, the flow rate of FO was higher than that of the pressurized membrane processes, less membrane fouling and more energy-efficient [84]. As a matter of principle, FO is a spontaneous method where solvent permeates a FO membrane from a solution that has higher chemical water potential (feed solution) (draw solution) [85]. For years now, for FO-operations, RO-membranes have been used in combination with suitable recyclable pulled solutes as effective semimeter membrane from CA, CTA, PA and the Composite polymer [86]. In absence of dedicated FO membrane, In recent years, molecular design of FO membranes with high flux and power density has gotten a lot of interest in research and development [87,88].

7.4 Polyelectrolytes complex membranes

Complex (PEC) polyelectrolyte systems have unique loading and surface properties to be used for separation and purification. PEC-based membranes have already shown promising materials, namely pervaporation and nanofiltration, for diverse membrane separations [89,90]. Polyelectrolyte have a strong tendency to adsorb on the surfaces of the particles in an aqueous suspension [91]. The groups of organic polymers charged are called polyelectrolytes when hydrolyzed and have many of their properties. They can be used to convey neutral particles with surface charge. Polyelectrolytes are cost-effective, quick, performance-efficient, thick and controlled. The method is also easy to use for the modification by composite membranes of commercially available membranes. In terms of molecular structure, weak and strong electrolyte activity, molecular weight and charge density the polyelectrolytes can be differentiated [92]. PECs are able to separate through different mechanisms because of the electrostatic repulsion forces. This kind of repulsion is also maintained in very open structures due to its high charging density. In water treatment applications, PECs can provide extreme thin, but effective barriers. With the incorporation of polyelectrolytes on the existing membrane surface, the development of membranes has been greatly improved. In most cases, polyelectrolyte charge can effectively be applied when the salt ion penetration is rejected via a polyelectrolyte-modified membrane [93]. Polyelectrolytes can also be used with a good control over the distribution of ionised molecules within the thin film in several other existing membrane surface modifications. Polyelectrolyte complex membranes can generally be made from a stock of biopolymers and are highly biocompatible, non-toxic and capable of film formation. PECs also offer promising methods to integrate multi-layer architecture with membranes and surface properties can be modified to adjust membrane load and hydrophilicity [94].

7.5 Polymer-based aerogel membranes

Aerogel are a special kind of functional material with ultra-porous solid gel, with a gas-inducing solid network structure that completely replaces the liquid part [95]. The Aerogel consists of a 3D network like a solid structure in the form of highly cross-linking structures that has a large number of air-filled pores. The 3D structures can be developed using different precursors as illustrated in Figure 10.

Polymer-driven aerogels are recently gaining attention because of their easy handling and their extensive potential for application [96,97]. Aerogels with two characteristics, structural properties and functional properties, can be distinguished from other porous materials. Structural characteristics consist of a gel-like structure of hierarchical, fractal and non-cryptal microstructural structures, crosslinks, granules or powders. The characteristics of functionality resulting from structural properties are unique in terms of bulk characteristics such as hydrophilicity, high-spectrum, high porosity and a high mechanical resistance, low dielectric constant, low sound speed, high-density surface area and high-speed porosity [98].

(a) Polymer Gel   (b) Polymer Gels before Freeze drying   c) Polymer Aerogel

(d)   Polymer-based Nanocomposite Aerogels

Figure 10 Photographs of Aerogels made from different precursors (a-c) polymer aerogel from polymer gel to aerogel, and (d) Polymer-based nanocomposite aerogels.

8.   Membranes for ground water treatment

The remediation of contaminated water gains the highest priority because of an increased water scarcity and water contamination, both at industrial scale and on groundwater resources. Thousands of people around the world are exposed to contaminated water sources such as heavy metal, active pharmaceutical ingredients, organic matter, and emerging pollutants. APIs are contaminated with pollutants [99]. For various types of contaminates with different chemical and anthropogenic sources, innovation is required in membrane technology. Long-term ingestion of heavily metal-contaminated water may be at risk of lung, bladder, kidney and skin damage and cancers according to the World Health Organization (WHO) [100]. Several technologies for heavy metals remediation and organic contaminated groundwater have revealed certain definite results such as: 1) full or substantial pollutant destruction and degradation initiating further investigations in the nature of degraded components, 2) pollutant extraction for further processing or disposal, 3) stabilization in less moving forms of contaminants Therefore, membrane technology can be used in conjugation with other means to rectify unsolved problems in purifying contaminated water sources [101].

9.   Polymer membranes in energy conversion and storage devices

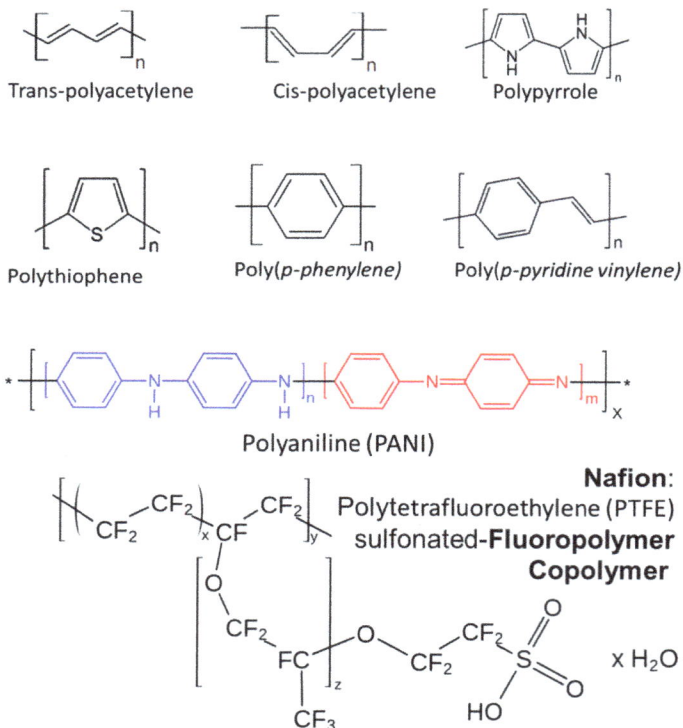

Figure 11 Structures of ion-exchange polymers popularly used in energy storage and conversion devices.

In this era of electronic devices, it is highly desirable to develop lightweight, flexible and folding electrodes and polyelectrolytic membrane with decent mechanical durability and electrochemical activity. In this context, it is still an incredible challenge to find suited polymer-based membrane with suitable ion-exchange characteristics for electrochemical storage systems such as batteries and supercapacitors as well as conversion devices such as FCs. Among several polymers, freestanding robust PANI in the form of a leuchemeraldine base (LEB), emeraldine base (EB), pernigraniline-base (PNB), polypyrrole (PPy), polythiophene (PT), poly(p-phenylene) (PPP), polysulfone/poly (Ether sulfone) (PSf/PES), polyimide/poly(ether imide) (PVPEI) (Figure 11). In addition, most of the polymers can be converted into wearable solid state supercapacitors, high electrochemical performance batteries and fuel cell devices and decent mechanical durability. The lightweight, flexible and folding polymer-based ion exchange membrane represents both an electrode and solid electrolytic materials for electro-chemical energy storage systems of the next generation.

Conclusions and future perspectives

For years, PANI, PPY, and Nafion membranes have been an ion exchange membrane used for energy storage and conversion devices (individually perfluorinated polymer containing sulfone groups). However, in almost every sector of the economy, bio-origin functional materials and polymers emerge as potential materials for various and diverse purposes. Researchers have recently spent considerable time finding the right combination of natural materials for sustainable energy applications. Important applications for energy storage and conversion are now developing as nanoscience and materials processing technologies develop. Volcanic eruptions and rock weathering could produce natural polymers and functional materials, while anthropogenic sources could be obtained by the release of non-refined waste, biomass and algae sources. Recent developments in the field of biomaterial-based polymers have shown efficient utilisation in preparation of an electrodialytic membrane, fluid membrane, nanofibre-like separation membranes, etc.

Acknowledgments

For financial assistance, SKN is grateful to the Ministry of Science and Technology, NANOMISSION GRANT (SR/NM/NT-1073/2016), DST INSPIRE faculty grant (IFA12-CH-84) and DST-Technology Mission Division (DST/TMD/HFC/2K18/124G) of the Government of India. For financial assistance, the Community of Madrid Spain (2017-T1/AMB5610) has sponsored the Talent Attraction Programme.

References

[1] V.T. Stannett, W.J. Koros, D.R. Paul, H.K. Lonsdale, R.W. Baker, Recent advances in membrane science and technology, 32 (1979) 69-121. https://doi.org/10.1007/3-540-09442-3_5

[2] T. Deblonde, C. Cossu-Leguille, P. Hartemann, Emerging pollutants in wastewater: A review of the literature, International Journal of Hygiene and Environmental Health, 214 (2011) 442-448. https://doi.org/10.1016/j.ijheh.2011.08.002

[3] P. Verlicchi, A. Galletti, M. Petrovic, D. Barceló, Hospital effluents as a source of emerging pollutants: An overview of micropollutants and sustainable treatment options, Journal of Hydrology, 389 (2010) 416-428. https://doi.org/10.1016/j.jhydrol.2010.06.005

[4] B. Petrie, R. Barden, B. Kasprzyk-Hordern, A review on emerging contaminants in wastewaters and the environment: Current knowledge, understudied areas and recommendations for future monitoring, Water Research, 72 (2015) 3-27. https://doi.org/10.1016/j.watres.2014.08.053

[5] S. Sourirajan, Separation of Some Inorganic Salts in Aqueous Solution by Flow, under Pressure, through Porous Cellulose Acetate Membranes, Industrial & Engineering Chemistry Fundamentals, 3 (1964) 206-210. https://doi.org/10.1021/i160011a005

[6] J. Ditter, R.A. Morris, R. Zepf, Large pore synthetic polymer membranes, Google Patents, 2002.

[7] A.K. Hołda, B. Aernouts, W. Saeys, I.F. Vankelecom, Study of polymer concentration and evaporation time as phase inversion parameters for polysulfone-based SRNF membranes, Journal of membrane science, 442 (2013) 196-205. https://doi.org/10.1016/j.memsci.2013.04.017

[8] A.K. Hołda, I.F. Vankelecom, Understanding and guiding the phase inversion process for synthesis of solvent resistant nanofiltration membranes, Journal of Applied Polymer Science, 132 (2015). https://doi.org/10.1002/app.42130

[9] H. Mariën, I.F. Vankelecom, Transformation of cross-linked polyimide UF membranes into highly permeable SRNF membranes via solvent annealing, Journal of Membrane Science, 541 (2017) 205-213. https://doi.org/10.1016/j.memsci.2017.06.080

[10] I. Soroko, M.P. Lopes, A. Livingston, The effect of membrane formation parameters on performance of polyimide membranes for organic solvent nanofiltration (OSN):

Part A. Effect of polymer/solvent/non-solvent system choice, Journal of Membrane Science, 381 (2011) 152-162. https://doi.org/10.1016/j.memsci.2011.07.027

[11] D.R. Lloyd, K.E. Kinzer, H.S. Tseng, Microporous membrane formation via thermally induced phase separation. I. Solid-liquid phase separation, Journal of Membrane Science, 52 (1990) 239-261. https://doi.org/10.1016/S0376-7388(00)85130-3

[12] J.T. Jung, J.F. Kim, H.H. Wang, E. di Nicolo, E. Drioli, Y.M. Lee, Understanding the non-solvent induced phase separation (NIPS) effect during the fabrication of microporous PVDF membranes via thermally induced phase separation (TIPS), Journal of Membrane Science, 514 (2016) 250-263. https://doi.org/10.1016/j.memsci.2016.04.069

[13] J.T. Jung, H.H. Wang, J.F. Kim, J. Lee, J.S. Kim, E. Drioli, Y.M. Lee, Tailoring nonsolvent-thermally induced phase separation (N-TIPS) effect using triple spinneret to fabricate high performance PVDF hollow fiber membranes, Journal of Membrane Science, 559 (2018) 117-126. https://doi.org/10.1016/j.memsci.2018.04.054

[14] S. Muthukumaran, S.E. Kentish, G.W. Stevens, M. Ashokkumar, Application of Ultrasound in Membrane Separation Processes: A Review, Reviews in Chemical Engineering, 22 (2006). https://doi.org/10.1515/REVCE.2006.22.3.155

[15] J.H. Jhaveri, Z.V.P. Murthy, A comprehensive review on anti-fouling nanocomposite membranes for pressure driven membrane separation processes, Desalination, 379 (2016) 137-154. https://doi.org/10.1016/j.desal.2015.11.009

[16] E. Salehi, P. Daraei, A. Arabi Shamsabadi, A review on chitosan-based adsorptive membranes, Carbohydrate Polymers, 152 (2016) 419-432. https://doi.org/10.1016/j.carbpol.2016.07.033

[17] M. Zahid, A. Rashid, S. Akram, Z.A. Rehan, W. Razzaq, A Comprehensive Review on Polymeric Nano-Composite Membranes for Water Treatment, Journal of Membrane Science & Technology, 08 (2018). https://doi.org/10.4172/2155-9589.1000179

[18] Q. Liu, G.-R. Xu, Graphene oxide (GO) as functional material in tailoring polyamide thin film composite (PA-TFC) reverse osmosis (RO) membranes, Desalination, 394 (2016) 162-175. https://doi.org/10.1016/j.desal.2016.05.017

[19] S. Kuiper, C.J.M. van Rijn, W. Nijdam, M.C. Elwenspoek, Development and applications of very high flux microfiltration membranes, Journal of Membrane Science, 150 (1998) 1-8. https://doi.org/10.1016/S0376-7388(98)00197-5

[20] L. Fan, J.L. Harris, F.A. Roddick, N.A. Booker, Influence of the characteristics of natural organic matter on the fouling of microfiltration membranes, Water Research, 35 (2001) 4455-4463. https://doi.org/10.1016/S0043-1354(01)00183-X

[21] A. Mehta, A.L. Zydney, Permeability and selectivity analysis for ultrafiltration membranes, Journal of Membrane Science, 249 (2005) 245-249. https://doi.org/10.1016/j.memsci.2004.09.040

[22] A.F. Ismail, M. Padaki, N. Hilal, T. Matsuura, W.J. Lau, Thin film composite membrane — Recent development and future potential, Desalination, 356 (2015) 140-148. https://doi.org/10.1016/j.desal.2014.10.042

[23] Y. Song, F. Liu, B. Sun, Preparation, characterization, and application of thin film composite nanofiltration membranes, Journal of Applied Polymer Science, 95 (2005) 1251-1261. https://doi.org/10.1002/app.21338

[24] X. Shi, G. Tal, N.P. Hankins, V. Gitis, Fouling and cleaning of ultrafiltration membranes: A review, Journal of Water Process Engineering, 1 (2014) 121-138. https://doi.org/10.1016/j.jwpe.2014.04.003

[25] J. Liu, Z. Xu, X. Li, Y. Zhang, Y. Zhou, Z. Wang, X. Wang, An improved process to prepare high separation performance PA/PVDF hollow fiber composite nanofiltration membranes, Separation and Purification Technology, 58 (2007) 53-60. https://doi.org/10.1016/j.seppur.2007.07.009

[26] X.Q. Cheng, Z.X. Wang, Y. Zhang, Y. Zhang, J. Ma, L. Shao, Bio-inspired loose nanofiltration membranes with optimized separation performance for antibiotics removals, Journal of Membrane Science, 554 (2018) 385-394. https://doi.org/10.1016/j.memsci.2018.03.005

[27] S. Lee, C. Boo, M. Elimelech, S. Hong, Comparison of fouling behavior in forward osmosis (FO) and reverse osmosis (RO), Journal of Membrane Science, 365 (2010) 34-39. https://doi.org/10.1016/j.memsci.2010.08.036

[28] M. Xie, L.D. Nghiem, W.E. Price, M. Elimelech, Comparison of the removal of hydrophobic trace organic contaminants by forward osmosis and reverse osmosis, Water Research, 46 (2012) 2683-2692. https://doi.org/10.1016/j.watres.2012.02.023

[29] J.-H. Choi, J. Jegal, W.-N. Kim, Fabrication and characterization of multi-walled carbon nanotubes/polymer blend membranes, Journal of Membrane Science, 284 (2006) 406-415. https://doi.org/10.1016/j.memsci.2006.08.013

[30] L. Deng, T.-J. Kim, M.-B. Hägg, Facilitated transport of CO2 in novel PVAm/PVA blend membrane, Journal of Membrane Science, 340 (2009) 154-163. https://doi.org/10.1016/j.memsci.2009.05.019

[31] X. Cao, J. Ma, X. Shi, Z. Ren, Effect of TiO 2 nanoparticle size on the performance of PVDF membrane, Applied Surface Science, 253 (2006) 2003-2010. https://doi.org/10.1016/j.apsusc.2006.03.090

[32] J.-Y. Lai, M.-J. Liu, K.-R. Lee, Polycarbonate membrane prepared via a wet phase inversion method for oxygen enrichment from air, Journal of Membrane Science, 86 (1994) 103-118. https://doi.org/10.1016/0376-7388(93)E0136-8

[33] J.-F. Li, Z.-L. Xu, H. Yang, L.-Y. Yu, M. Liu, Effect of TiO2 nanoparticles on the surface morphology and performance of microporous PES membrane, Applied Surface Science, 255 (2009) 4725-4732. https://doi.org/10.1016/j.apsusc.2008.07.139

[34] M.B. Karimi, G. Khanbabaei, G.M.M. Sadeghi, Vegetable oil-based polyurethane membrane for gas separation, Journal of Membrane Science, 527 (2017) 198-206. https://doi.org/10.1016/j.memsci.2016.12.008

[35] J. Yin, G. Zhu, B. Deng, Graphene oxide (GO) enhanced polyamide (PA) thin-film nanocomposite (TFN) membrane for water purification, Desalination, 379 (2016) 93-101. https://doi.org/10.1016/j.desal.2015.11.001

[36] H.H. Yong, H.C. Park, Y.S. Kang, J. Won, W.N. Kim, Zeolite-filled polyimide membrane containing 2,4,6-triaminopyrimidine, Journal of Membrane Science, 188 (2001) 151-163. https://doi.org/10.1016/S0376-7388(00)00659-1

[37] M. Elimelech, Z. Xiaohua, A.E. Childress, H. Seungkwan, Role of membrane surface morphology in colloidal fouling of cellulose acetate and composite aromatic polyamide reverse osmosis membranes, Journal of Membrane Science, 127 (1997) 101-109. https://doi.org/10.1016/S0376-7388(96)00351-1

[38] M. Han, Thermodynamic and rheological variation in polysulfone solution by PVP and its effect in the preparation of phase inversion membrane, Journal of Membrane Science, 202 (2002) 55-61. https://doi.org/10.1016/S0376-7388(01)00718-9

[39] Y. Xiuli, C. Hongbin, W. Xiu, Y. Yongxin, Morphology and properties of hollow-fiber membrane made by PAN mixing with small amount of PVDF, Journal of Membrane Science, 146 (1998) 179-184. https://doi.org/10.1016/S0376-7388(98)00107-0

[40] J.M. Luque-Alled, A.W. Ameen, M. Alberto, M. Tamaddondar, A.B. Foster, P.M. Budd, A. Vijayaraghavan, P. Gorgojo, Gas separation performance of MMMs

containing (PIM-1)-functionalized GO derivatives, Journal of Membrane Science, 623 (2021). https://doi.org/10.1016/j.memsci.2020.118902

[41] S. Agarwal, A. Greiner, J.H. Wendorff, Functional materials by electrospinning of polymers, Progress in Polymer Science, 38 (2013) 963-991. https://doi.org/10.1016/j.progpolymsci.2013.02.001

[42] S. Dixit, S. Pal, Recent Advanced Technologies in the Processing of Hybrid Reinforced Polymers for Applications of Membranes, Polymers and Polymer Composites, 24 (2018) 289-305. https://doi.org/10.1177/096739111602400408

[43] T. Malik, H. Razzaq, S. Razzaque, H. Nawaz, A. Siddiqa, M. Siddiq, S. Qaisar, Design and synthesis of polymeric membranes using water-soluble pore formers: an overview, Polymer Bulletin, 76 (2018) 4879-4901. https://doi.org/10.1007/s00289-018-2616-3

[44] J.M. Skluzacek, M.I. Tejedor, M.A. Anderson, NaCl rejection by an inorganic nanofiltration membrane in relation to its central pore potential, Journal of Membrane Science, 289 (2007) 32-39. https://doi.org/10.1016/j.memsci.2006.11.034

[45] A. Lee, J.W. Elam, S.B. Darling, Membrane materials for water purification: design, development, and application, Environmental Science: Water Research & Technology, 2 (2016) 17-42. https://doi.org/10.1039/C5EW00159E

[46] S.N.W. Ikhsan, N. Yusof, A.F. Ismail, W.N.W. Salleh, F. Aziz, J. Jaafar, H. Hasbullah, Synthetic polymer-based membranes for treatment of oily wastewater, (2020) 3-22. https://doi.org/10.1016/B978-0-12-818485-1.00001-0

[47] A. Ghaee, M. Shariaty-Niassar, J. Barzin, T. Matsuura, Effects of chitosan membrane morphology on copper ion adsorption, Chemical Engineering Journal, 165 (2010) 46-55. https://doi.org/10.1016/j.cej.2010.08.051

[48] A. Moriya, T. Maruyama, Y. Ohmukai, T. Sotani, H. Matsuyama, Preparation of poly(lactic acid) hollow fiber membranes via phase separation methods, Journal of Membrane Science, 342 (2009) 307-312. https://doi.org/10.1016/j.memsci.2009.07.005

[49] R.Y.M. Huang, R. Pal, G.Y. Moon, Characteristics of sodium alginate membranes for the pervaporation dehydration of ethanol–water and isopropanol–water mixtures, Journal of Membrane Science, 160 (1999) 101-113. https://doi.org/10.1016/S0376-7388(99)00071-X

[50] P. Tomietto, P. Loulergue, L. Paugam, J.-L. Audic, Biobased polyhydroxyalkanoate (PHA) membranes: Structure/performances relationship, Separation and Purification Technology, 252 (2020). https://doi.org/10.1016/j.seppur.2020.117419

[51] D. Chen, M.A. Hickner, E. Agar, E.C. Kumbur, Optimizing membrane thickness for vanadium redox flow batteries, Journal of Membrane Science, 437 (2013) 108-113. https://doi.org/10.1016/j.memsci.2013.02.007

[52] J. Guo, Q. Zhang, Z. Cai, K. Zhao, Preparation and dye filtration property of electrospun polyhydroxybutyrate–calcium alginate/carbon nanotubes composite nanofibrous filtration membrane, Separation and Purification Technology, 161 (2016) 69-79. https://doi.org/10.1016/j.seppur.2016.01.036

[53] P.-J. Lin, M.-C. Yang, Y.-L. Li, J.-H. Chen, Prevention of surfactant wetting with agarose hydrogel layer for direct contact membrane distillation used in dyeing wastewater treatment, Journal of Membrane Science, 475 (2015) 511-520. https://doi.org/10.1016/j.memsci.2014.11.001

[54] H. Pang, K. Tian, Y. Li, C. Su, F. Duan, Y. Xu, Super-hydrophobic PTFE hollow fiber membrane fabricated by electrospinning of Pullulan/PTFE emulsion for membrane deamination, Separation and Purification Technology, 274 (2021). https://doi.org/10.1016/j.seppur.2020.118186

[55] P. Samoila, A.C. Humelnicu, M. Ignat, C. Cojocaru, V. Harabagiu, Chitin and Chitosan for Water Purification, Chitin and Chitosan2019, pp. 429-460. https://doi.org/10.1002/9781119450467.ch17

[56] Y. Yang, A. Raza, F. Banat, K. Wang, The separation of oil in water (O/W) emulsions using polyether sulfone & nitrocellulose microfiltration membranes, Journal of Water Process Engineering, 25 (2018) 113-117. https://doi.org/10.1016/j.jwpe.2018.07.007

[57] L. Manjarrez Nevárez, L. Ballinas Casarrubias, O.S. Canto, A. Celzard, V. Fierro, R. Ibarra Gómez, G. González Sánchez, Biopolymers-based nanocomposites: Membranes from propionated lignin and cellulose for water purification, Carbohydrate Polymers, 86 (2011) 732-741. https://doi.org/10.1016/j.carbpol.2011.05.014

[58] L.A. Goetz, B. Jalvo, R. Rosal, A.P. Mathew, Superhydrophilic anti-fouling electrospun cellulose acetate membranes coated with chitin nanocrystals for water filtration, Journal of Membrane Science, 510 (2016) 238-248. https://doi.org/10.1016/j.memsci.2016.02.069

[59] A. Prasannan, J. Udomsin, H.-C. Tsai, C.-F. Wang, J.-Y. Lai, Robust underwater superoleophobic membranes with bio-inspired carrageenan/laponite multilayers for the effective removal of emulsions, metal ions, and organic dyes from wastewater, Chemical Engineering Journal, 391 (2020). https://doi.org/10.1016/j.cej.2019.123585

[60] T.P.N. Nguyen, E.-T. Yun, I.-C. Kim, Y.-N. Kwon, Preparation of cellulose triacetate/cellulose acetate (CTA/CA)-based membranes for forward osmosis, Journal of Membrane Science, 433 (2013) 49-59. https://doi.org/10.1016/j.memsci.2013.01.027

[61] D.L. Kaplan, Introduction to Biopolymers from Renewable Resources, (1998) 1-29. https://doi.org/10.1007/978-3-662-03680-8_1

[62] M. Zeng, I. Echols, P. Wang, S. Lei, J. Luo, B. Peng, L. He, L. Zhang, D. Huang, C. Mejia, L. Wang, M.S. Mannan, Z. Cheng, Highly Biocompatible, Underwater Superhydrophilic and Multifunctional Biopolymer Membrane for Efficient Oil–Water Separation and Aqueous Pollutant Removal, ACS Sustainable Chemistry & Engineering, 6 (2018) 3879-3887. https://doi.org/10.1021/acssuschemeng.7b04219

[63] P. Budd, K. Msayib, C. Tattershall, B. Ghanem, K. Reynolds, N. McKeown, D. Fritsch, Gas separation membranes from polymers of intrinsic microporosity, Journal of Membrane Science, 251 (2005) 263-269. https://doi.org/10.1016/j.memsci.2005.01.009

[64] F. Russo, F. Galiano, A. Iulianelli, A. Basile, A. Figoli, Biopolymers for sustainable membranes in CO2 separation: a review, Fuel Processing Technology, (2020) 106643. https://doi.org/10.1016/j.fuproc.2020.106643

[65] B. Lian, G. Blandin, G. Leslie, P. Le-Clech, Impact of module design in forward osmosis and pressure assisted osmosis: An experimental and numerical study, Desalination, 426 (2018) 108-117. https://doi.org/10.1016/j.desal.2017.10.047

[66] B. Gu, D.Y. Kim, J.H. Kim, D.R. Yang, Mathematical model of flat sheet membrane modules for FO process: Plate-and-frame module and spiral-wound module, Journal of Membrane Science, 379 (2011) 403-415. https://doi.org/10.1016/j.memsci.2011.06.012

[67] F. Li, W. Meindersma, A.B. de Haan, T. Reith, Optimization of commercial net spacers in spiral wound membrane modules, Journal of Membrane Science, 208 (2002) 289-302. https://doi.org/10.1016/S0376-7388(02)00307-1

[68] J. Schwinge, P.R. Neal, D.E. Wiley, D.F. Fletcher, A.G. Fane, Spiral wound modules and spacers, Journal of Membrane Science, 242 (2004) 129-153. https://doi.org/10.1016/j.memsci.2003.09.031

[69] M. Pourbozorg, T. Li, A.W.K. Law, Effect of turbulence on fouling control of submerged hollow fibre membrane filtration, Water Research, 99 (2016) 101-111. https://doi.org/10.1016/j.watres.2016.04.045

[70] A. Altaee, A. Braytee, G.J. Millar, O. Naji, Energy efficiency of hollow fibre membrane module in the forward osmosis seawater desalination process, Journal of Membrane Science, 587 (2019) 117165. https://doi.org/10.1016/j.memsci.2019.06.005

[71] K.C. Khulbe, T. Matsuura, Recent progress in polymeric hollow-fibre membrane preparation and applications, Membrane Technology, 2016 (2016) 7-13. https://doi.org/10.1016/S0958-2118(16)30149-5

[72] M. Lee, Z. Wu, K. Li, Advances in ceramic membranes for water treatment, (2015) 43-82. https://doi.org/10.1016/B978-1-78242-121-4.00002-2

[73] S.K. Hubadillah, M.H.D. Othman, T. Matsuura, A.F. Ismail, M.A. Rahman, Z. Harun, J. Jaafar, M. Nomura, Fabrications and applications of low cost ceramic membrane from kaolin: A comprehensive review, Ceramics International, 44 (2018) 4538-4560. https://doi.org/10.1016/j.ceramint.2017.12.215

[74] S.K. Hubadillah, M.H.D. Othman, T. Matsuura, M.A. Rahman, J. Jaafar, A.F. Ismail, S.Z.M. Amin, Green silica-based ceramic hollow fiber membrane for seawater desalination via direct contact membrane distillation, Separation and Purification Technology, 205 (2018) 22-31. https://doi.org/10.1016/j.seppur.2018.04.089

[75] S.P. Nunes, A. Car, From Charge-Mosaic to Micelle Self-Assembly: Block Copolymer Membranes in the Last 40 Years, Industrial & Engineering Chemistry Research, 52 (2012) 993-1003. https://doi.org/10.1021/ie202870y

[76] J.R. Werber, C.O. Osuji, M. Elimelech, Materials for next-generation desalination and water purification membranes, Nature Reviews Materials, 1 (2016) 1-15. https://doi.org/10.1038/natrevmats.2016.18

[77] S.H. Chen, C. Willis, K.R. Shull, Water transport and mechanical response of block copolymer ion-exchange membranes for water purification, Journal of Membrane Science, 544 (2017) 388-396. https://doi.org/10.1016/j.memsci.2017.09.001

[78] A.-V. Ruzette, L. Leibler, Block copolymers in tomorrow's plastics, Nature Materials, 4 (2005) 19-31. https://doi.org/10.1038/nmat1295

[79] H.A. Klok, S. Lecommandoux, Supramolecular materials via block copolymer self-assembly, Advanced Materials, 13 (2001) 1217-1229. https://doi.org/10.1002/1521-4095(200108)13:16<1217::AID-ADMA1217>3.0.CO;2-D

[80] S.P. Nunes, K.V. Peinemann, Ultrafiltration membranes from PVDF/PMMA blends, Journal of Membrane Science, 73 (1992) 25-35. https://doi.org/10.1016/0376-7388(92)80183-K

[81] P.H.H. Duong, K. Daumann, P.-Y. Hong, M. Ulbricht, S.P. Nunes, Interfacial Polymerization of Zwitterionic Building Blocks for High-Flux Nanofiltration Membranes, Langmuir, 35 (2018) 1284-1293. https://doi.org/10.1021/acs.langmuir.8b00960

[82] K. Buruga, H. Song, J. Shang, N. Bolan, T.K. Jagannathan, K.-H. Kim, A review on functional polymer-clay based nanocomposite membranes for treatment of water, Journal of Hazardous Materials, 379 (2019) 120584. https://doi.org/10.1016/j.jhazmat.2019.04.067

[83] J. Yin, B. Deng, Polymer-matrix nanocomposite membranes for water treatment, Journal of Membrane Science, 479 (2015) 256-275. https://doi.org/10.1016/j.memsci.2014.11.019

[84] L. Shen, X. Zhang, L. Tian, Z. Li, C. Ding, M. Yi, C. Han, X. Yu, Y. Wang, Constructing substrate of low structural parameter by salt induction for high-performance TFC-FO membranes, Journal of Membrane Science, 600 (2020) 117866. https://doi.org/10.1016/j.memsci.2020.117866

[85] W. Xu, Q. Chen, Q. Ge, Recent advances in forward osmosis (FO) membrane: Chemical modifications on membranes for FO processes, Desalination, 419 (2017) 101-116. https://doi.org/10.1016/j.desal.2017.06.007

[86] N. Singh, I. Petrinic, C. Hélix-Nielsen, S. Basu, M. Balakrishnan, Concentrating molasses distillery wastewater using biomimetic forward osmosis (FO) membranes, Water Research, 130 (2018) 271-280. https://doi.org/10.1016/j.watres.2017.12.006

[87] T.-S. Chung, X. Li, R.C. Ong, Q. Ge, H. Wang, G. Han, Emerging forward osmosis (FO) technologies and challenges ahead for clean water and clean energy applications, Current Opinion in Chemical Engineering, 1 (2012) 246-257. https://doi.org/10.1016/j.coche.2012.07.004

[88] T. Cath, A. Childress, M. Elimelech, Forward osmosis: Principles, applications, and recent developments, Journal of Membrane Science, 281 (2006) 70-87. https://doi.org/10.1016/j.memsci.2006.05.048

[89] F.-Y. Zhao, Q.-F. An, Y.-L. Ji, C.-J. Gao, A novel type of polyelectrolyte complex/MWCNT hybrid nanofiltration membranes for water softening, Journal of Membrane Science, 492 (2015) 412-421. https://doi.org/10.1016/j.memsci.2015.05.041

[90] S. Nam, Pervaporation and properties of chitosan-poly(acrylic acid) complex membranes, Journal of Membrane Science, 135 (1997) 161-171. https://doi.org/10.1016/S0376-7388(97)00144-0

[91] Q. Zhao, Y.-L. Ji, J.-K. Wu, L.-L. Shao, Q.-F. An, C.-J. Gao, Polyelectrolyte complex nanofiltration membranes: performance modulation via casting solution pH, RSC Adv., 4 (2014) 52808-52814. https://doi.org/10.1039/C4RA09164G

[92] F.M. Lounis, J. Chamieh, P. Gonzalez, H. Cottet, L. Leclercq, Prediction of Polyelectrolyte Complex Stoichiometry for Highly Hydrophilic Polyelectrolytes, Macromolecules, 49 (2016) 3881-3888. https://doi.org/10.1021/acs.macromol.6b00463

[93] Y. Yang, Q. Zhang, S. Li, S. Zhang, Preparation and characterization of porous polyelectrolyte complex membranes for nanofiltration, RSC Advances, 5 (2015) 3567-3573. https://doi.org/10.1039/C4RA13699C

[94] P. Ahmadiannamini, X. Li, W. Goyens, N. Joseph, B. Meesschaert, I.F.J. Vankelecom, Multilayered polyelectrolyte complex based solvent resistant nanofiltration membranes prepared from weak polyacids, Journal of Membrane Science, 394-395 (2012) 98-106. https://doi.org/10.1016/j.memsci.2011.12.032

[95] A.C. Pierre, G.M. Pajonk, Chemistry of Aerogels and Their Applications, Chemical Reviews, 102 (2002) 4243-4266. https://doi.org/10.1021/cr0101306

[96] S. Zhao, W.J. Malfait, N. Guerrero-Alburquerque, M.M. Koebel, G. Nyström, Biopolymer Aerogels and Foams: Chemistry, Properties, and Applications, Angewandte Chemie International Edition, 57 (2018) 7580-7608. https://doi.org/10.1002/anie.201709014

[97] M.H. Mruthunjayappa, V.T. Sharma, K. Dharmalingam, N. Sanna Kotrappanavar, D. Mondal, Engineering a Biopolymer-Based Ultrafast Permeable Aerogel Membrane Decorated with Task-Specific Fe–Al Nanocomposites for Robust Water Purification, ACS Applied Bio Materials, 3 (2020) 5233-5243.

[98] Y. Si, J. Yu, X. Tang, J. Ge, B. Ding, Ultralight nanofibre-assembled cellular aerogels with superelasticity and multifunctionality, Nature Communications, 5 (2014). https://doi.org/10.1038/ncomms6802

[99] B. Van Der Bruggen, C. Vandecasteele, T. Van Gestel, W. Doyen, R. Leysen, A review of pressure-driven membrane processes in wastewater treatment and drinking water production, Environmental Progress, 22 (2003) 46-56. https://doi.org/10.1002/ep.670220116

[100] A.M. Nasir, P.S. Goh, M.S. Abdullah, B.C. Ng, A.F. Ismail, Adsorptive nanocomposite membranes for heavy metal remediation: Recent progresses and challenges, Chemosphere, 232 (2019) 96-112. https://doi.org/10.1016/j.chemosphere.2019.05.174

[101] M.A. Hashim, S. Mukhopadhyay, J.N. Sahu, B. Sengupta, Remediation technologies for heavy metal contaminated groundwater, Journal of Environmental Management, 92 (2011) 2355-2388. https://doi.org/10.1016/j.jenvman.2011.06.009

Advanced Functional Membranes   Materials Research Forum LLC
Materials Research Foundations **120** (2022) 43-71   https://doi.org/10.21741/9781644901816-2

Chapter 2

Advanced Functional Membranes for Microfiltration and Ultrafiltration

Ria Majumdar[1]*, Umesh Mishra[1], Biswanath Bhunia[2]

[1]Department of Civil Engineering, National Institute of Technology Agartala, Jirania, India-799046

[2]Department of Bio Engineering, National Institute of Technology Agartala, Jirania, India-799046

* civil.ria2017@gmail.com

Abstract

In the era of freshwater scarcity, the world seeks to limit overexploitation of all accessible resource of freshwater and also to recover potable water by treating wastewater. Various domestic, agricultural and industrial activities lead in generating high amount of wastewater from which water can be recovered for human needs. Over many years, conventional filtration process has gained success to some extent in effluents treatment for disposal. Yet, improvement in filtration process is necessary for advanced wastewater treatment in order to make the whole process handy and cost-effective. The functionality of membrane can be enhanced by incorporating polymers with novel functions in advanced membrane separation methods like microfiltration (MF) and ultrafiltration (UF) for reclaiming potable water from wastewaters, catalysis and biomedical applications. This book chapter mostly focuses on the use of different polymeric substances for MF and UF membrane preparation and their application in various sectors. Basic principle, novel manufacturing techniques for polymeric membrane, process design i.e., synthesis of polymers as designed membrane feedstock, surface modifications, synergistic fabrication of different polymers for advanced functionality, antifouling and antibacterial properties of membrane are also discussed.

Keywords

Filtration, Membrane Technology, Polymer Membrane, Wastewater, Purification, Potable Water

Abbreviations

BOD   Biological oxygen demand

Materials Research Forum LLC

https://doi.org/10.21741/9781644901816-2

CA	Cellulose acetate
CEB	Chemically enhanced backwashing
CNC	Cellulose nanocrystals
CNF	Cellulose nanofibers
COD	Chemical oxygen demand
CPCB	Central Pollution Control Board
HCl	Hydrochloric acid
HNO_3	Nitric acid
H_3PO_4	Phosphoric acid
H_2SO_4	Sulphuric acid
MF	Microfiltration
MLD	Million liters per day
MW	Molecular weight
MWCO	Molecular weight cut off
NaOH	Sodium hydroxide
NC	Nanocellulose
NF	Nanofiltration
PAN	polyacrylonitrile
PE	Polyethylene
PES	polyethersulfone
PP	polypropylene
PTEE	Poly tetrafluoroethylene
PVC	Polyvinyl chloride
PVDF	polyvinylidene fluoride
TDS	Total dissolved solid
TMP	Transmembrane pressure
TOC	Total organic carbon
TSS	Total suspended solid
RO	Reverse osmosis
UF	Ultrafiltration

Contents

1. Introduction

Water is a crucial need for mankind and thus all human activities are associated with it. It plays a noteworthy role in the socio-economic growth of any population. Most countries around the world are still facing shortage to meet the need of potable water. Even though, water covers almost 71% of the Earth's surface, hardly 2.5% of this amount is fresh water out of which 1% is accessible for consumption and the rest is found in snowfields and glaciers [1]. It is obvious that the world is facing scarcity of freshwater distribution among various industries, urban and rural areas [2]. Consequently, developing countries and agricultural activities are heavily affected by the lack of sufficient water resources for livestock production and irrigation purposes. The evidence of such circumstances is observed across the world, especially in Asia, Africa, Middle East, and Latin America. On the other hand, freshwater resources takes several years to replenish again which is not sufficient for growing needs of human activities [3]. Therefore, it is necessary to search for an alternative water resource to mitigate the insufficiency of freshwater and that could be sea water, wastewater, etc. Huge amounts of wastewaters are generated everyday domestically, industrially and from agricultural areas with the increasing activities of populations. As per Ministry of Water Resources, River Development and Ganga Rejuvenation, wastewater generation from different industries (Food, dairy and beverages; Textile, bleaching and dyeing; Sugar; Distillery; Chemical, Tannery; Pulp and paper; and others) of India was 501 MLD (million litres per day) [4]. However, based on Central Pollution Control Board (CPCB) generation of sewage was 61948 MLD out of which only 23277 MLD were treated. As a result, around 38000 MLD of untreated sewage was directly discharged into nearby land or water bodies [2].

Generation of wastewater is inevitable as it generates during essential activities of the value chain in all parts of life. Wastewater contains higher concentration of minerals, pathogens and toxic compounds which make it undrinkable and worthless for reuse without treatment. The favorable way of resolving these problems is separation process. UF and MF are the typical pressure driven membrane-based well-established separation system in industrial scale [5]. Such separation technologies are mainly based on the application of polymeric membrane and represent a sustainable approach towards resolving the particulate and macromolecule-based separation process. MF and UF can be a suitable alternative as no chemicals or phase changes are required like other separation methods i.e. adsorption, distillation, fractionation, extraction, etc., [6]. Addition to that, the relative simplicity of operation, less energy consumption, low weight, minimum space requirements, modularity, easy scale-up, and possibilities of carrying out the continuous separation process are the advantages of MF and UF [7, 8]. The utilization of these filtration methods is emerging in industrial scale since the last 50 years. The markets for membrane science are highly diverse i.e., from chemical to medicine or pharmaceutical industry. However, this chapter mainly focuses on wastewater treatment and gas separation by filtration technology.

The objective of this chapter is to discuss the preparation of different polymer-based functional membrane with novel properties and their application in treatment of water and wastewater. It also gives general ideas on synthesis, fabrication and synergistic combination of various polymeric materials in membrane preparation. Basic principle of membrane filtration process, their manufacturing technology, process design, antifouling and antibacterial properties of membrane are also discussed.

Advanced Functional Membranes   Materials Research Forum LLC
Materials Research Foundations **120** (2022) 43-71   https://doi.org/10.21741/9781644901816-2

2. Basic principle of micro- and ultrafiltration

Filtration is a separation process by which insoluble materials are removed from the gaseous or liquid solution based on their physical as well as chemical properties by use of a membrane or filter [9]. This membrane or filter medium allows the liquid or gas to pass through and retains the solid particles. Therefore, membrane is described as a barrier that segregates two phases from each other by limiting the movements of compounds through it in a control manner [3,10-12]. Such compounds include ions, molecules, and other small particles. The desired products will be either the clarified liquid (also known as 'filtrate') or the solid retained/collected in the filter medium ('retentate' or 'residue') [13]. Filtration is used in many laboratories to minimize the complexity of sample, improve clarity of viscous solutions, and to prepare the sample for analysis. Depending on the filtration method applied, particles or molecules are removed based on their shape, charge or size.

Conceptualization of filtration process by membrane technology was known since 18th century. Unlike other unit separation process in bio-chemical engineering (adsorption, distillation, etc.), the superiority of membrane technology is based on distinctive separation principle, i.e., specific transport phenomena of filter medium. Its degree of selectivity mainly relies on pore size distribution and thus classified as– microfiltration (MF), ultrafiltration (UF), nanofiltration (NF), and reverse osmosis (RO) [14-16]. Table 1 illustrates comparison between different pressure-driven filtration processes as per membrane pore size. Fig 1 shows retention and permeation of various sizes of particles via membrane. Membrane technology has recently gained success due to the non-requirement of additives, efficient isothermal performance at low heat and thus less consumption of energy as compared to other thermal separation process. In addition to that, downscaling and up scaling of membrane technology, integration and fabrication into other reaction process are simple. The main objective of membrane technology is to remove bacteria, viruses, organic material, microorganisms, and particulate matter from water, wastewater and polluted gas. However, this excludes the removal of taste, odor and colour of water and gas. The first application of membrane on a large scale was with MF and UF.

2.1   Microfiltration

It is one of most desired separation processes for wastewater treatment and its background was started in 1920s. Commercially available first MF membrane was prepared from nitrocellulose in 1926. Though, in 1940s the usage of MF membranes increases, however, it was little limited in the small-scale industries and laboratories till 1960s. In-line or dead-end mode of filtration is the most commonly used process for MF. Cross-flow mode was introduced in the year 1970s to MF and thus its application increases to large industrial scale. After few years, third type of MF was emerged as semi-dead-end filtration. In 1990-

93, MF/UF was installed for first time in order to treat surface water [17-18]. MF is generally used for the segregation of suspended particles with diameter in the range 0.1 - 10μm from a fluid mixture. This range of particles includes a wide variety of natural and industrial materials. The solute separated by MF is typically larger than those separated by UF and NF and thus negligible for RO. The pore diameter and permeate flux is also greater for MF than UF, NF and RO.

Table 1 Membrane separation process based on pore size distribution.

	Particle size (μm)	Molecular weight (Da)	Pressure bar	Separation mechanism	Particle characteristics	Removable particles	Material retained	Ref
MF	0.1 – 1	10^5 - 10^6	0.5 – 1	Molecular sieve	Macromolecular to cellular	Water, Bacteria, cells, polymers, colloids, humic acids, oil emulsions	Bacteria, suspended particles	[3, 15, 16, 19, 20, 40, 75]
UF	0.01 – 0.1	10^3 - 10^5	1 – 10	Solution diffusion	Molecular to macromolecular	Small colloids, humic acids, protein, viruses, pesticides, polysaccharides	Glucose, lactose, salt, micropollutants	[3, 15, 16, 19, 20, 35, 76]
NF	0.001 – 0.01	100 - 1000	5 – 70	Molecular sieve	Ionic to molecular	Pesticides, endocrine disruptors, salts	Colloids, macromolecules	[3, 15, 16, 20, 77-80]
RO	0.0001 – 0.001	100	10 – 100	Molecular sieve	Ionic	All contaminants including monovalent ions	Dissolved salts	[15, 16, 20, 81-83]

2.2 Ultrafiltration

UF is a powerful and novel separation process that is used for the fractionation of micro-solutes and water from macromolecules, colloids and protein. It is carried out by a finely porous filter medium (membrane) of size ranging in 50 to 1 nm. The MWCO for UF membranes ranges 1,00,000 to 2,00,000Da [15]. Membranes used in UF are characterized by pore size in the range 0.001-0.1μm able to retain species. The retentate for UF includes sugars, polymers, proteins, biomolecules and colloidal particles whereas salts whilst solvent will pass through the membrane. Both MF and UF process mechanism works on the principle of separating the molecules using porous membrane as per their pore size distribution [19]. UF often operated in tangential direction of upstream surface of membrane. However, membrane fouling by protein accumulation is a major limitation of UF membranes.

Advanced Functional Membranes   Materials Research Forum LLC
Materials Research Foundations **120** (2022) 43-71   https://doi.org/10.21741/9781644901816-2

The required driving force for transport facility through MF and UF filter medium is a pressure gradient which forces the suspended solid and solutes to pass through the membrane where they are accumulated as permeate and retentate. As MF and UF do not typically reject salts, the pressure gradient is very less (1 -10 bar) as compared to RO.

The rejection efficiency R of a membrane is evaluated by:

$$R = 1 - \frac{C_p}{C_r} \tag{1}$$

Where C_p denotes concentration of permeate and C_r represents retentate concentration in mol.m^{-3}. R is the function of pore size, particle size and porosity.

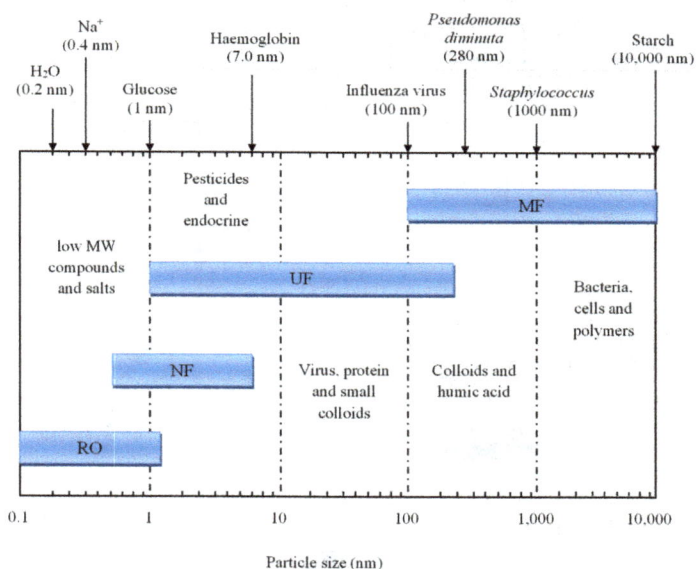

Figure 1 Retention and permeation of particle size via membrane [16, 20]. Redrawn. Permission is taken from Elsevier.

The rejection phenomena of MF and UF membranes are generally demonstrated by minimal molecular weight cut off (MWCO) that is expressed as MW of smallest species retained by the membrane. Alternatively, the permeate flux across porous filter medium is affected by pressure gradient applied on the membrane, membrane resistance and the viscosity of fluid to be passed. The volumetric flux is determined by-

Materials Research Forum LLC
https://doi.org/10.21741/9781644901816-2

$$Q_v = \frac{P(\Delta p - \Delta \pi)}{l} \tag{2}$$

Where Q_v is volumetric flux, P is co-efficient of permeability, l is membrane thickness in m, Δp and $\Delta \pi$ is hydrostatic and osmotic pressure gradient (Pa) respectively.

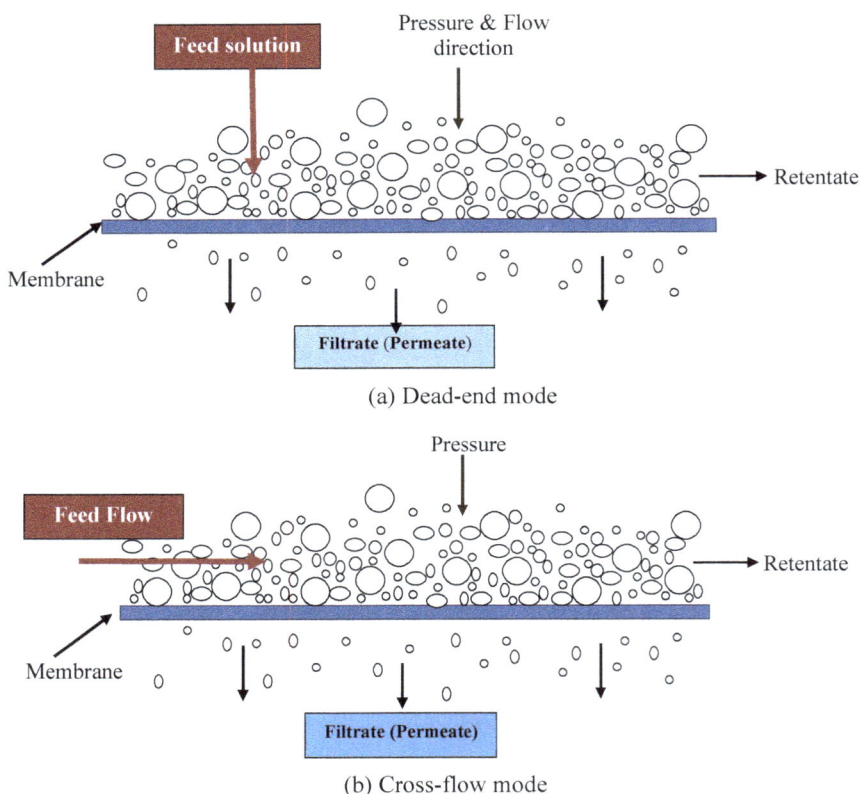

(a) Dead-end mode

(b) Cross-flow mode

Figure 2 Filtration mode for MF and UF membrane [20].

MF and UF membranes can be employed for both cross-flow and dead-end filtration [21] as shown in Fig 2. In dead-end mode, solution is fed perpendicularly across the filter medium under sufficient pressure and rejected particles (retentate) are accumulated on the surface. To maintain the flow, pressure increases to some extent that membrane should be replaced. In contrast, feed solutions are supplied tangentially through the membrane in cross-flow filtration in order to get clean, particle free permeate [22, 23]. Even though, the infrastructure require for cross-flow mode is more difficult, but membrane lifespan and flux rate are relatively higher than dead-end filtration [24].

Dead-end mode is most favorable choice for batch studies, but for continuous process cake formation occurs due to the deposition of retentate which eventually reduces the membrane flux [25, 26]. This cake layer can be removed by backwashing (flow from opposite direction of feed stream) and thus the operation must be stopped [27]. In contrast, cross-flow mode allows continuous feed stream and the tangential flow direction helps in keeping membrane clean by scouring velocity.

3. Polymeric materials for membrane preparation

Significant advantages of polymeric substances than inorganic materials (metals and ceramic) for membrane preparation in order to execute separation process include:

- high flexibility
- segments and functional groups for specific applications
- high selectivity for target separation
- less expensive
- low energy consumption

MF and UF membranes are usually fabricated by natural polymers such as cellulose derivatives and synthetic polymers i.e. synthetic polymer like cellulose acetate (CA), polypropylene (PP), polyacrylonitrile (PAN), perfluoropolymers, polyamides, polysulfones, polyvinylidene fluoride (PVDF), polyethersulfone (PES), etc. [6]. Lots of literature reviews are reported about the structure and novel functions of advanced polymeric membrane. MF and UF membranes are generally prepared using these polymers and by phase inversion methods [28].

Aliphatic polyamides like nylon-6, nylon 4-6, and nylon 6-6 are extensively used as MF membranes and characterized by good thermal, chemical, and mechanical stability. Amorphous polymers are generally used to prepare UF membranes owing to their convenient resolution and control of small pore size. Polymeric materials which have a glass transition temperature are generally employed. In particular, polyacrylonitrile (PAN) widely used for general aqueous systems because of its resistance to solvents and chemicals, also exhibits relatively low protein binding owing to its hydrophilic properties. Polysulphone (PS), also widely used in the production of UF membranes, has good mechanical strength and resistance to heat and pH; in contrast it exhibits poor resistance to solvents.

3.1   Biopolymers

Utilization of organic polymers for membrane preparation has extensively increasing due to their unique properties. Their shape, size and structure can be altered as required for a specific application. In spite of not being a novel material for membrane preparation, the renewable characteristics of cellulose are of great interest. In addition, cellulose has stability in wide range of solvents [29]. Nanocellulose (NC) i.e. cellulose nanofibers (CNF) and cellulose nanocrystals (CNC) for MF and UF membranes can be a sustainable alternative towards molecular separation process [30].

3.2   Amphiphilic polymers

Amphiphilic polymers includes both hydrophilic (water loving) and hydrophobic (water-hating) polymers. Though hydrophilic substances are incorporated for the manufacture of MF and UF membrane, but it is not suitable as water molecules act as plasticizers affecting thermal stability and mechanical strength. Crystalline polymer possesses high thermal stability and chemical resistance since the crystalline domain contributes to effect of cross-linking between amorphous domains and hinders the free rotation of polymer segments. MF membranes generally consist of crystalline polymers. Polytetrafluoroethylene (PTEE), poly (vinylidene fluoride) (PVDF), polypropylene (PP), are typical hydrophobic polymers widely used as MF materials. Hydrophobic materials are also widely used to minimize adsorption phenomena which reduce permeate fluxes and create difficulties in membrane cleaning.

3.3   Thermoplastic polymers

Thermoplastic polymers, also known as thermo-softening plastics, are defined as a polymer material which becomes moldable or pliable at a specific elevated temperature and solidifies at low heat. Example of thermoplastic polymers includes polypropylene (PP), polystyrene, polyvinyl chloride (PVC), polyethylene (PE), etc. Superiority of such polymer over other polymeric substance includes:

- These are light weight
- it has high strength
- it is recyclable
- it is high molecular weight polymer
- these can be reshaped by heating and cooling

- intermolecular force acting between polymer chains becomes weak upon heating and thus allows a liquid with high viscosity

4.   Membrane filtration technology

Membrane filtration technology is an effectual separation process as compared to other conventional methods and has gained huge interest in industrial scale over last few decades. Besides solid, liquid and gas separation, membrane can also separates negative or positive charge, neutral or bipolar [1]. In the separation process, membrane or filter medium firstly act as a semi-permeable barrier and then it transports the particles through it. As an effective process, membrane technology is most favorable technique for oily wastewater treatment (by separating oil from the wastewater), in fuel cell industries, food industries, pharmaceutical, desalination, etc.

4.1   Membrane characteristics

Membranes are generally thin films. On the basis of its structure, membranes are classified into various categories namely isotropic (or symmetrical), non-isotropic (or asymmetrical), homogeneous and non-homogeneous, porous and non-porous [31]. Table 2 shows various characteristics of membrane applicable for separation process.

Table 2 Characteristics of membrane and membrane technology.

Membrane process	Characteristics	Ref
Separation	Particles, liquids, ions, molecules, gases, etc.	[20, 84, 85]
Flow pattern	Dead-end filtration, cross-flow filtration	[20, 85]
Recycle process	Backwashing	[20, 85]
Shape	Hollow fiber, flat sheet, tubular, capillary, capsule	[29, 31, 84-86]
Preparation method	Phase inversion, sol-gel process, track-etching, extrusion, stretching, micro-fabrication, interface reaction,	[29, 31, 86]
Driving force	Pressure, temperature, concentration, voltage	[29, 84]
Membrane material	Inorganic materials (Ceramics, metal, oxides), Organic polymers, composite materials	[31, 86]
Membrane cross-section	Isotropic, Anisotropic, mixed matrix composite, bi-layer, multi-layer,	[31, 86]
Structure	Porous, non-porous, charged	[84]

Symmetrical membranes are characterized by identical structures and its transport phenomena occur over entire membrane cross-section. Conversely, structure and transport characteristics of asymmetric membrane vary over the membrane cross-section; typically, a relatively dense thin selective layer (0.1-1µm) is supported by much thicker (100-200µm) porous substrate. The mass flux is mainly evaluated by the thickness of selective layer, whereas porous sub layer provides mechanical strength and has minimal effect on the separation process. Fig 3 shows a schematic diagram of various types of membranes according to their structure.

An ideal porous membrane is defined as dense polymer made up of cylindrical pores and can be prepared by track-etching. This procedure involves irradiation of a thin polymer layer (about 10µm thick) with splitting particles from a nuclear reactor or other radiation source. The highly energetic ions penetrate the polymeric film and disintegrate the polymer chains leaving 'tracks' in the membrane material (tracking). In the etching step, the tracked film is immersed in a solution with extremely high or low pH in which tracks are converted into cylindrical pores with a uniform diameter and a narrow pore size distribution. The exposure time of film to radiation controls the number of membrane pore (pore density), while the etching time determines the pore diameter which can range between 0.2 µm to 10 µm. Membrane prepared by track-etching procedure are symmetrical and permeate flux is proportionate to the membrane thickness; consequently, they have to be thinner than asymmetric microporous membranes in order for having comparatively high flux. Polycarbonate and polyester films are the usual materials used for track-etched membranes. The porosity of these membranes is of the order of 10%.

4.2   Membrane process and separation mechanism

Separation process occurs with membrane as a result of gradient through the membrane and thus membrane separation technology may be subdivided into two categories based on gradient types – pressure driven and non-pressure driven. The first category occurs due to the pressure gradient across the membrane which is called transmembrane pressure (TMP). On the other hand, concentration gradient is the base of non-pressure driven membrane filtration; for example, forward osmosis and dialysis. Cross-flow microfiltration, a pressure driven process, has a wide range of application in concentrating, separating, and purifying colloids, macromolecules, suspended particles from mixture.

Passive transport is defined as a type of membrane transport with any requirement of energy for passing the particles across the membrane. Such phenomena occurs as a result of driving force which is created by a difference of concentration, pressure or by an electric field [32]. Table 3 shows the classification of membranes according to passive transport characteristics. On the other hand, active transport phenomena of a membrane require

Advanced Functional Membranes

Materials Research Foundations **120** (2022) 43-71

Materials Research Forum LLC

https://doi.org/10.21741/9781644901816-2

external energy to pass particles through membrane. It helps to transport molecules which are unable to move against gradient or any other resistance of the process.

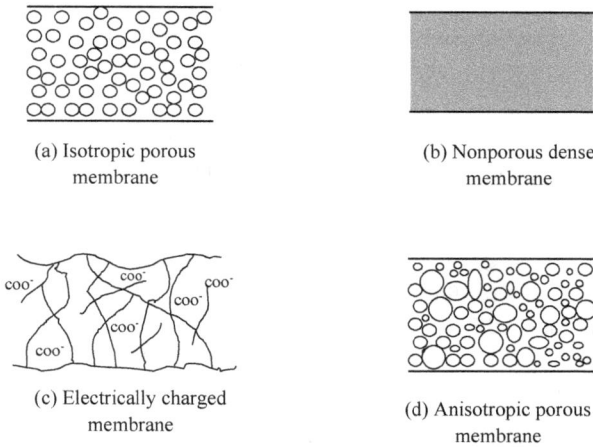

(a) Isotropic porous
membrane

(b) Nonporous dense
membrane

(c) Electrically charged
membrane

(d) Anisotropic porous
membrane

Figure 3 Schematic diagram of various polymeric membranes [10]

Table 3 Characteristics of membrane technology via passive transport.

Characteristics		Criteria				References
Membrane type		Non-porous	Micro-porous	Meso-porous	Macroporous	[3, 15]
Pressure required (bar)		15-75	5-15	2-5	1-3	[3, 16]
Pore diameter (d_p)		0.3-0.6	$d_p \leq 2$ nm	$d_p = 2$-50 nm	$d_p = 50$-500 nm	[10, 15]
Transport phenomena	Concentration	Pervaporation	Dialysis	Dialysis	-	[10, 29]
	Pressure	Gas separation Reverse osmosis (RO)	NF	UF	MF	[10, 29]
	Electrical fields	Electro-dialysis	-	Electro-dialysis	-	[10, 29]

Non-porous membranes are defined as dense films in which permeate pass through by pressure, concentration or electrical field gradient. It is mostly used for molecular separation, RO and NF of gas phase. The selectivity and transport rate of a non-porous membrane can be controlled by the interconnection within membrane material and permeate [33]. The efficiency of membrane separation is described by diffusion or solution selectivity. As very limited numbers of molecules or mixtures can transport via non-porous membranes, therefore, high-selectivity of transport characteristics can be achieved. A substitute of molecule-selective technique by non-porous membrane is the application of special transport process i.e., facilitated transport by affine carriers.

For porous membranes, the selectivity and transport rate are affected by size exclusion and the viscous flow. However, the interaction between solute and membrane pore surface gradually minimizes membrane performance. In addition, selective adsorption can be applied with meso- and macroporous membranes as a sustainable alternative approach towards separation process. Porous membranes can be used for very specific separation-based operations as per size and shape differences. It is obvious that membrane selectivity and permeability can be fully influenced by concentration polarization. The trans-membrane gradient significantly decreases with the increase in concentration of rejected molecules or particles on the surface of membrane (membrane fouling). Membrane fouling is the unwanted deposition or adsorption of materials in/on separation layer. High yield, throughput and purity of any product after filtration i.e., ideal membrane performance can be obtained by adaption of process conditions and membrane materials. That is why, optimization of membrane module configuration and process design is necessary before applications.

5. Application of MF and UF polymeric membrane

MF and UF membranes have broad range of applications in wastewater treatment, food industries, pharmaceutical industries, pulp and paper industries, beverage production, etc. [34]. Besides solid, liquid and gas separation, membrane can also separates negative or positive charge, neutral or bipolar [1]. In the separation process, membrane or filter medium firstly act as a semi-permeable barrier and then it transports the particles through it. As an effective process, membrane technology is most successful process for treatment of oily wastewater (by separating oil from the wastewater), in fuel cell industries, food industries, pharmaceutical, desalination, etc. Table 4 shows various applications of MF and UF membranes in the separation process.

5.1   Oily waste water treatment

Wastewater purification is an emerging issue for environmental, social and economic growth of an area and thus for a country. Huge amount of emulsified solution as a part of oily wastewater are produced on daily basis from kitchen, food processing factories, oil manufacturing industries, pharmaceuticals, petrochemical industries, etc. [15]. Separation of oil from solution is necessary and quite challenging. For oily-wastewater treatment, polymeric membrane fabricated with hydrophilic particles can be a great choice for facilitating the separation of oil that can easily pass the water across membrane [1]. MF and UF polymeric membranes are the substantial approach for oil-water separation, however, UF shows better rejection efficiency than MF membranes [35]. Hydrophilic particles embedded in polymeric membrane are able to repulse the oil droplets and allow water to pass through the membrane [36-39]. Sludge of oily wastewater includes phenols, benzofluorene, phenanthrene, aromatic hydrocarbons, toluene, alkanes, benzene, anthracene, pyrene, etc. [40-42]. MF membranes have high flux and thus cause a greater risk oil passage. On the other hand, UF membrane possesses higher recovery of water, lower tendency of fouling formation and easy removal of fouling.

5.2   Food processing industry

Utilization of MF and UF polymeric membranes in food processing industries is an essential tool for separation process and was gained popularity over 30 years. Literature shows that almost 20-30% of total cost for membrane preparation was from food industries. Membrane filtration technology is extensively used for clarification of alcoholic beverages, water purifications, production of protein content rates, clarification of fruit juice, concentration of beverages and fruit juice, recovery of oil-seed protein, sugar recovery from candy, etc. [43].

5.3   Poultry slaughterhouse wastewater

Huge amounts of water are consumed in the poultry slaughterhouse in order to process live birds to meat thereby generating large quantity of wastewater. The estimation of water consumption is around 26.5 litres per bird [44]. The wastewater from this slaughterhouse contains carbohydrates, protein, fats, blood, feather, skins, etc. It is contaminated with grit, inorganic materials, pathogenic bacteria and fecal matters. Therefore, such wastewater is highly rich in BOD and COD.

Table 4 Application of MF and UF polymeric membrane.

Treated items	Type of membrane	Polymer used	Filtration mode	Remarks/Results	Refs.
Oil emulsions	MF	PTEE	Cross-flow	High oil emulsion removal	[39]
Poultry slaughterhouse wastewater	UF	UF-25-PAN	-	99% fats, 98% suspended matters, 94% BOD and COD removal	[44]
Synthetic emulsified Oily-wastewater	MF	PVDF	Cross-flow	95% removal of organic matters	[87]
Vegetable oil factory	UF	UFPHT20-6338 (Polysulphone)	Cross-flow	91% COD, 87% TOC, 100% TSS, 40% Cl⁻, 85% PO_4^{3-} removal	[88]
Municipal wastewater	MF	PP, Polysulfonether, Fluoro-polymer	Dead-end	Coliforms and phosphorus removal – removal of contaminants less than detection limit	[89]
Textile wastewater	UF	PVDF	-	20% colour and 60% COD removal	[90]
Oily wastewater	UF-RO	Polyacrylonitrile, polyamide	Cross-flow	Reduction of 100% oil and grease, 98% TOC, 95% TDS, 100% turbidity and 98% COD	[91]
Urban wastewater	MF-RO	Polysulfone	-	Removal of pharmaceutical and pesticides to discharge limit	[92]
Textile wastewater	MF-NF	Polyetherimide	Cross-flow	Indigo blue-dye removal, water recovery 40%	[93]
CO_2 separation	MF	Poly(ethylene glycol) and poly(dimethyl siloxane)	-	50% removal efficiency	[94]
Refinery wastewater	UF-RO	Polyacrylamide and Polyaluminium chloride	Cross-flow	Rejection of 97% salt	[95]
Metal finishing industry	UF-RO	PVDF	Cross-flow	90-99% removal of different contaminants	[96]

Advanced Functional Membranes   Materials Research Forum LLC
Materials Research Foundations **120** (2022) 43-71   https://doi.org/10.21741/9781644901816-2

6.   Membrane fouling and cleaning

6.1   Membrane fouling

Membrane fouling is the main limitation encountered at the time of practical application of separation process. It occurs when macromolecules, particles, biomolecules, salts, colloids, organic matters, microbes, protein, etc. are retained or accumulated on/in the membrane pores and surface thereby minimizing the performance and permeate flux [45-47]. When foulants (material causing fouling) are settled down in membrane pores then it cannot be removed, thus fouling is considered as an irreversible process. Large amount of foulants deposition causes a formation of cake layer which eventually decreases membrane performance by blocking the pores [48]; however, this type of fouling is reversible [45]. As results, higher pressure is required to free up transportation of particles through membrane. The more the fouling, the higher the pressure desired [45]. Therefore, fouling has adverse effect on overall membrane performance such as huge consumption of energy, decrease in membrane filtration area, more down time, etc.

The correlation between applied pressure and permeate flux is demonstrated by Darcy's law. The flux for an unfouled and fouled membrane is given by equation 3 and 4 respectively.

$$J_w = \frac{\Delta P - \Delta \pi}{\mu R_m} \tag{3}$$

$$J_w = \frac{\Delta P - \Delta \pi}{\mu (R_m + R_f)} \tag{4}$$

Where J_w is permeate flux, μ denotes viscosity of liquid, R_m and R_f represents membrane resistance without or with fouling respectively. It can be seen from equation 4 that, when fouling increases the denominator also increases and therefore the permeate flux decreases.

According to the type of foulants, the membrane fouling can be classified into various categories such as organic fouling, biofouling, colloidal and inorganic fouling [49]. Colloid fouling can be either inorganic, organic or composites. Colloids may include polysaccharides, microorganisms, lipoproteins, biological debris, silts, oils, clay, manganese oxides, iron, etc.

Fouling of membrane is also differentiated by the mechanism of its appearance and thus divided into four types (as shown in Fig 4) -

- Complete pore blocking
- Partial pore blocking
- Internal pore blocking
- Cake layer formation

(a) Complete blocking

(b) Intermediate blocking

(c) Standard blocking

(d) Cake pore blocking

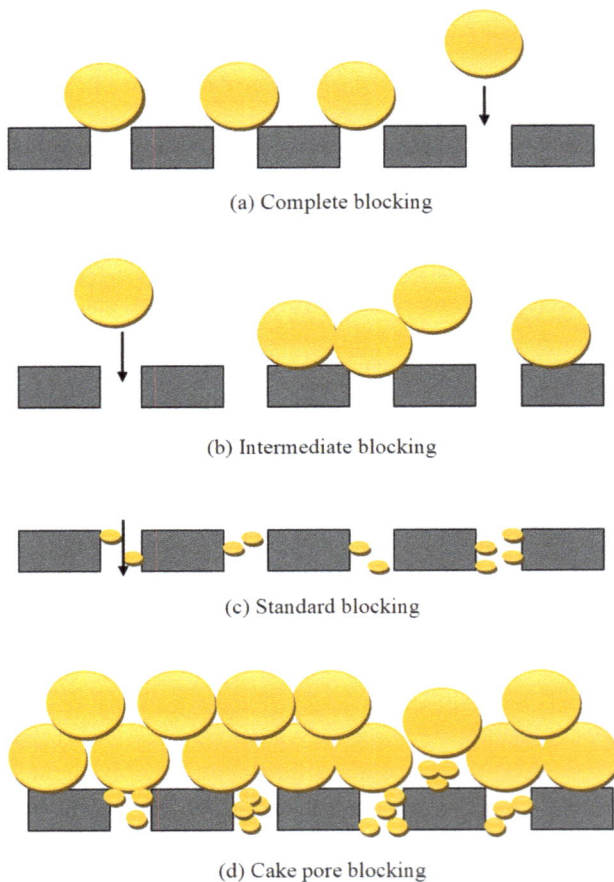

Figure 4 Comparison between various types of fouling [20]. Redrawn. Permission is taken from Elsevier.

Several factors affecting the mechanism of fouling of membrane are size of solute or particle; structure of membrane (MF and UF membrane); interaction between solvent, solute, and membrane; and membrane porosity. Recently many research work has focused on proteins [50], algae [51], colloidal materials [52], oil [53,54], and organic matter foulants [55]. The functionality of MF and UF membrane solely depends on their fouling

Advanced Functional Membranes   Materials Research Forum LLC
Materials Research Foundations **120** (2022) 43-71   https://doi.org/10.21741/9781644901816-2

properties and various researchers have improved many strategies for its minimization or removal. One of such strategy includes cleaning of membrane.

6.2   Membrane cleaning

As time passes, foulant materials stick to the membrane surface. Therefore, cleaning is necessary in order to increase the membrane performance i.e., permeability and selectivity by minimization or removal of fouling material. A cost-effective and efficient cleaning process plays a significant role in wastewater treatment. Various physical, chemical and synergistic applications were studied for cleaning of membrane fouling and this depends on the membrane as well as feed solution [3]. Membrane cleaning for MF is quite limited whether many works were reported on UF membrane cleaning. The incorporation of hydrophilic materials into polymeric membrane increases the hydrophilicity which helps in resisting the formation of fouling [31]. This is carried out by immobilization of hydrophilic polymers by plasma or photo parization [56-58], or surface modification [59], or inclusion of hydrophilic polymer to the casting solution [60]. Cleaning of fouled membrane can be broadly classified as chemical, physical, physico-chemical and biological (or biochemical) cleaning [3].

6.2.1   Chemical cleaning

When the membrane fouling is in irreversible condition, chemical cleaning is employed to it. The principle of this process relies on the interaction between membrane material and foulants, cleaning material and membrane, foulants and cleaning material. This basic knowledge helps in selecting the best suited chemical for foulant cleaning [61]. It is considered as the most suitable and efficient method for enhancement of membrane lifespan, product safety and to maintain continuous filtration operation. The effect of such cleaning process includes loosening and dissolution of foulant, avoidance of new foulant creation, and no adverse impact to the membrane materials. During this process, cleaning agents added to the residue at place which helps in weaken the bonds of foulant to material and then cross-flow filtration is applied to remove the fouling. Cleaning agents for chemical cleaning includes chelating agents, acids, disinfectants, bases (or alkalis), surfactants, etc. Each agent has its unique role for application. For example, acids are used for removal of inorganic foulants such metal oxides and salt precipitates. Hydrochloric acid (HCl), nitric acid (HNO_3), sulphuric acid (H_2SO_4) and phosphoric acid (H_3PO_4) are the commonly used acids for chemical cleaning [62, 63]. On the other hand, Alkalis or bases are used for eliminating organic foulants from the membrane. Most commonly used alkali or base is sodium hydroxide (NaOH). However, phosphates and carbonates are also employed for this operation [3,64].

6.2.2 Physical cleaning

Physical cleaning of membrane fouling is a mechanical process and includes pneumatic cleaning, periodic back flushing, ultrasonic cleaning and sponge-ball cleaning. Periodic back flushing applies a pressure in the permeate side to cause a backward movement of fouling material and thus helps in loosening the bond of materials. It is most commonly used for gel and cake layer removal [65]. In the ultrasonication cleaning process, ultrasounds are employed to create agitation in the aqueous medium and thus transmitting energy as of turbulence to the membrane material. Such energy helps in unloading the foulant from the membrane surface. Ultrasonic cleaning is an effective physical cleaning process and mostly depends on the velocity of cross-flow, operational temperature, power of ultrasound wave and duration of pulse [66-68].

As the name suggests, sponge-ball cleaning is carried out by sponge balls (made up of polyurethane) for wiping out the membrane surface. The balls are inserted into the permeator, scrubbed the pores by moving and wipe off the foulants. Pneumatic cleaning includes air scouring and air lifting. In the process, air is supplied under a certain pressure which generates a shear force between the foulants and membrane materials thereby executing the cleaning process.

6.2.3 Physico-chemical cleaning

This cleaning process is a combination of physical as well as chemical process for foulant removal. Therefore, chemical agents are added during physical cleaning process for enhancement of the removal efficiency i.e., better cleaning performance. Chemically enhanced backwashing (CEB) is the most significant process for physico-chemical cleaning process. Another example includes the application of ultrasonication during chemical cleaning which reports 95% removal efficiency [69-70].

6.2.4 Biological or biochemical cleaning

Biochemical or biological cleaning is carried out by utilizing different bioactive agents such as enzymes, a single molecule or enzyme mixture [71-72]. There is a possibility of damage of membrane material during chemical, physical or physico-chemical cleaning, but biological cleaning is less toxic to membrane material thereby considering it as a sustainable cleaning approach [72-74].

7. Recommendation for future research

Membrane technology is gradually achieving huge success in the field of water and wastewater treatment and gas separation. Over past years, numerous works has been done in this area, however, there is still some limitation and further improvement is necessary. As fouling and huge energy demand is a key limitation in non-equilibrium pressure driven process, constant research is required for achieving a solution of this problem either by incorporating anti-fouling agents to the membrane or introducing a cost-effective pretreatment method. On the other hand, recovery of solute can make the whole process cheaper. Future scope of research must focus on alternative process for recovery of salt-based draw solutes.

Conclusion

There are numerous applications of MF and UF membrane technology for wastewater purification and potable water generation to minimize the scarcity. This chapter summarizes basic principle of filtration methodologies (MF and UF membrane filtration), several polymeric substances for membrane preparations, membrane design and separation mechanisms, their application in various waste water treatment and lastly surface modifications to improve flux, performance and antifouling properties by incorporation of other advanced functional polymeric materials to make the filtration process handy and cost-effective.

References

[1] F. Yalcinkaya, E. Boyraz, J. Maryska, K. Kucerova, A review on membrane technology and chemical surface modification for the oily wastewater treatment, Materials 13 (2020) 493. https://doi.org/10.3390/ma13020493

[2] Central Pollution Control Board, Untreated sewage flow into river Yamuna, 2015.

[3] E. Obotey Ezugbe, S. Rathilal, Membrane technologies in wastewater treatment: a review, Membranes 10 (2020) 89. https://doi.org/10.3390/membranes10050089

[4] R.D.a.G.R. Ministry of Water Resources, DISCHARGE OF SEWAGE INTO GANGA RIVER, 2017.

[5] S. Muthukumaran, K. Baskaran, Comparison of the performance of ceramic microfiltration and ultrafiltration membranes in the reclamation and reuse of secondary wastewater, Desalin. Water Treat. 52 (2014) 670-677. https://doi.org/10.1080/19443994.2013.826333

[6] A. Cassano, A. Basile, Membranes for industrial microfiltration and ultrafiltration, Advanced Membrane Science and Technology for Sustainable Energy and Environmental Applications, Elsevier, 2011, pp. 647-679. https://doi.org/10.1533/9780857093790.5.647

Materials Research Forum LLC
https://doi.org/10.21741/9781644901816-2

[7] W. Ma, Q. Zhang, D. Hua, R. Xiong, J. Zhao, W. Rao, S. Huang, X. Zhan, F. Chen, C. Huang, Electrospun fibers for oil–water separation, Rsc Adv. 6 (2016) 12868-12884. https://doi.org/10.1039/C5RA27309A

[8] S.S. Madaeni, M.K. Yeganeh, Microfiltration of emulsified oil wastewater, J. Porous Mater. 10 (2003) 131-138. https://doi.org/10.1023/A:1026035830187

[9] C. Tien, Principles of filtration, Elsevier, 2012.

[10] M.T. Ravanchi, T. Kaghazchi, A. Kargari, Application of membrane separation processes in petrochemical industry: a review, Desalination 235 (2009) 199-244. https://doi.org/10.1016/j.desal.2007.10.042

[11] R.W. Baker, Membrane technology and applications, John Wiley & Sons, 2012. https://doi.org/10.1002/9781118359686

[12] M. El-Ghaffar, H.A. Tieama, A review of membranes classifications, configurations, surface modifications, characteristics and Its applications in water purification, Chem. Biomol. Eng. 2 (2017) 57-82.

[13] G. Srikanth, Membrane separation processes: Technology and business opportunities, Chem. Eng. World 34 (1999) 55-66.

[14] A. Sagle, B. Freeman, Fundamentals of membranes for water treatment, The future of desalination in Texas 2 (2004) 137.

[15] M. Padaki, R.S. Murali, M.S. Abdullah, N. Misdan, A. Moslehyani, M. Kassim, N. Hilal, A. Ismail, Membrane technology enhancement in oil–water separation. A review, Desalination 357 (2015) 197-207. https://doi.org/10.1016/j.desal.2014.11.023

[16] F. Yalcinkaya, A review on advanced nanofiber technology for membrane distillation, J. Eng. Fibers Fabr. 14 (2019) 1558925018824901. https://doi.org/10.1177/1558925018824901

[17] A. Zirehpour, A. Rahimpour, Membranes for wastewater treatment, Nanostructured Polymer Membranes; John Wiley & Sons Ltd.: London, UK 2 (2016) 159-207. https://doi.org/10.1002/9781118831823.ch4

[18] R. Grant, Membrane separations, Mater. Manuf. Process 4 (1989) 483-503. https://doi.org/10.1080/10426918908956311

[19] J.Z. Leos, A.L. Zydney, Microfiltration and ultrafiltration: principles and applications, Routledge, 2017. https://doi.org/10.1201/9780203747223

[20] H.T. Madsen, Membrane filtration in water treatment–removal of micropollutants, Chemistry of Advanced Environmental Purification Processes of Water, Elsevier, 2014, pp. 199-248. https://doi.org/10.1016/B978-0-444-53178-0.00006-7

Advanced Functional Membranes
Materials Research Foundations **120** (2022) 43-71

Materials Research Forum LLC
https://doi.org/10.21741/9781644901816-2

[21] N.F. Gray, Filtration methods, Microbiology of Waterborne Diseases, Elsevier, 2014, pp. 631-650. https://doi.org/10.1016/B978-0-12-415846-7.00035-4

[22] T. Furukawa, K. Kokubo, K. Nakamura, K. Matsumoto, Modeling of the permeate flux decline during MF and UF cross-flow filtration of soy sauce lees, J. Membr. Sci. 322 (2008) 491-502. https://doi.org/10.1016/j.memsci.2008.05.068

[23] S. Ripperger, J. Altmann, Crossflow microfiltration–state of the art, Sep. Purif. Technol. 26 (2002) 19-31. https://doi.org/10.1016/S1383-5866(01)00113-7

[24] A. Massé, H.N. Thi, P. Legentilhomme, P. Jaouen, Dead-end and tangential ultrafiltration of natural salted water: Influence of operating parameters on specific energy consumption, J. Membr. Sci. 380 (2011) 192-198. https://doi.org/10.1016/j.memsci.2011.07.002

[25] S. Lee, A. Fane, R. Amal, T. Waite, The effect of floc size and structure on specific cake resistance and compressibility in dead-end microfiltration, Sep. Sci. Technol. 38 (2003) 869-887. https://doi.org/10.1081/SS-120017631

[26] K.J. Howe, A. Marwah, K.-P. Chiu, S.S. Adham, Effect of coagulation on the size of MF and UF membrane foulants, Environ. Sci. Technol. 40 (2006) 7908-7913. https://doi.org/10.1021/es0616480

[27] S.A.A. Tabatabai, M.D. Kennedy, G.L. Amy, J.C. Schippers, Optimizing inline coagulation to reduce chemical consumption in MF/UF systems, Desalin. Water Treat. 6 (2009) 94-101. https://doi.org/10.5004/dwt.2009.653

[28] B. Bolto, J. Zhang, X. Wu, Z. Xie, A review on current development of membranes for oil removal from wastewaters, Membranes 10 (2020) 65. https://doi.org/10.3390/membranes10040065

[29] M. Ulbricht, Design and synthesis of organic polymers for molecular separation membranes, Curr. Opin. Chem. Eng. 28 (2020) 60-65. https://doi.org/10.1016/j.coche.2020.02.002

[30] A. Mautner, K.-Y. Lee, P. Lahtinen, M. Hakalahti, T. Tammelin, K. Li, A. Bismarck, Nanopapers for organic solvent nanofiltration, Chem. Commun. 50 (2014) 5778-5781. https://doi.org/10.1039/C4CC00467A

[31] S.H. Mohamad, M. Idris, H.Z. Abdullah, A.F. Ismail, Short review of ultrafiltration of polymer membrane as a self-cleaning and antifouling in the wastewater system, Adv. Mat. Res., Trans Tech Publ, 2013, pp. 318-323. https://doi.org/10.4028/www.scientific.net/AMR.795.318

[32] A.T. Fane, R. Wang, Y. Jia, Membrane technology: past, present and future, Membrane and Desalination Technologies, Springer, 2011, pp. 1-45. https://doi.org/10.1007/978-1-59745-278-6_1

[33] S.C. George, S. Thomas, Transport phenomena through polymeric systems, Prog. Polym. Sci. 26 (2001) 985-1017. https://doi.org/10.1016/S0079-6700(00)00036-8

[34] R. Singh, J. Tembrock, Effectively control reverse osmosis systems, Chem. Eng. Prog. 95 (1999) 57-64.

[35] T. Bilstad, E. Espedal, Membrane separation of produced water, Water Sci. Technol. 34 (1996) 239-246. https://doi.org/10.2166/wst.1996.0221

[36] J. Kong, K. Li, Oil removal from oil-in-water emulsions using PVDF membranes, Sep. Purif. Technol. 16 (1999) 83-93. https://doi.org/10.1016/S1383-5866(98)00114-2

[37] S.R.H. Abadi, M.R. Sebzari, M. Hemati, F. Rekabdar, T. Mohammadi, Ceramic membrane performance in microfiltration of oily wastewater, Desalination 265 (2011) 222-228. https://doi.org/10.1016/j.desal.2010.07.055

[38] V. Singh, M. Purkait, C. Das, Cross-flow microfiltration of industrial oily wastewater: experimental and theoretical consideration, Sep. Sci. Technol. 46 (2011) 1213-1223. https://doi.org/10.1080/01496395.2011.560917

[39] A. Ezzati, E. Gorouhi, T. Mohammadi, Separation of water in oil emulsions using microfiltration, Desalination 185 (2005) 371-382. https://doi.org/10.1016/j.desal.2005.03.086

[40] B. Mrayyan, M.N. Battikhi, Biodegradation of total organic carbons (TOC) in Jordanian petroleum sludge, J. Hazard. Mater. 120 (2005) 127-134. https://doi.org/10.1016/j.jhazmat.2004.12.033

[41] M.V. Reddy, M.P. Devi, K. Chandrasekhar, R.K. Goud, S.V. Mohan, Aerobic remediation of petroleum sludge through soil supplementation: microbial community analysis, J. Hazard. Mater. 197 (2011) 80-87. https://doi.org/10.1016/j.jhazmat.2011.09.061

[42] M. Kriipsalu, M. Marques, A. Maastik, Characterization of oily sludge from a wastewater treatment plant flocculation-flotation unit in a petroleum refinery and its treatment implications, J. Mater. Cycles Waste Manag. 10 (2008) 79-86. https://doi.org/10.1007/s10163-007-0188-7

[43] G. Hong, W. Xiong, Application of membrane separation technology in food industry of China [J], Membr. Sci. Technol. 4 (2003).

[44] D. Yordanov, Preliminary study of the efficiency of ultrafiltration treatment of poultry slaughterhouse wastewater, Bulg. J. Agric. Sci. 16 (2010) 700-704.

[45] T.F. Speth, R.S. Summers, A.M. Gusses, Nanofiltration foulants from a treated surface water, Environ. Sci. Technol. 32 (1998) 3612-3617. https://doi.org/10.1021/es9800434

[46] Q. She, C.Y. Tang, Y.-N. Wang, Z. Zhang, The role of hydrodynamic conditions and solution chemistry on protein fouling during ultrafiltration, Desalination 24 (2009) 1079-1087. https://doi.org/10.1016/j.desal.2009.05.015

[47] H. Shon, S. Vigneswaran, I.S. Kim, J. Cho, H. Ngo, Fouling of ultrafiltration membrane by effluent organic matter: A detailed characterization using different organic fractions in wastewater, J. Membr. Sci. 278 (2006) 232-238. https://doi.org/10.1016/j.memsci.2005.11.006

[48] A.W. Mohammad, C.Y. Ng, Y.P. Lim, G.H. Ng, Ultrafiltration in food processing industry: review on application, membrane fouling, and fouling control, Food Bioproc. Tech. 5 (2012) 1143-1156. https://doi.org/10.1007/s11947-012-0806-9

[49] G. Amy, Fundamental understanding of organic matter fouling of membranes, Desalination 231 (2008) 44-51. https://doi.org/10.1016/j.desal.2007.11.037

[50] D. Sioutopoulos, A. Karabelas, V. Mappas, Membrane fouling due to protein—Polysaccharide mixtures in dead-end ultrafiltration; the effect of permeation flux on fouling resistance, Membranes 9 (2019) 21. https://doi.org/10.3390/membranes9020021

[51] M. Devanadera, M. Dalida, Fouling of Ceramic Microfiltration Membrane by Soluble Algal Organic Matter (Saom) from Chlorella Sp. and Aeruginosa M. and its Mitigation Using Feed-Pretreatment, Proceedings of the 14 th International Conference on Environmental Science and Technology, Rhodes, Greece, 2015, pp. 3-5. https://doi.org/10.1016/j.jcis.2016.10.013

[52] J.M. Dickhout, J. Moreno, P. Biesheuvel, L. Boels, R. Lammertink, W. De Vos, Produced water treatment by membranes: a review from a colloidal perspective, J. Colloid Interface Sci. 487 (2017) 523-534.

[53] P. Li, S. Zhang, Y. Lv, G. Ma, X. Zuo, Fouling mechanism and control strategy of inorganic membrane, E3S Web of Conferences, EDP Sciences, 2020, p. 04047. https://doi.org/10.1051/e3sconf/202019404047

[54] S. Huang, R.H. Ras, X. Tian, Antifouling membranes for oily wastewater treatment: Interplay between wetting and membrane fouling, Curr. Opin. Colloid Interface Sci. 36 (2018) 90-109. https://doi.org/10.1016/j.cocis.2018.02.002

Materials Research Forum LLC
https://doi.org/10.21741/9781644901816-2

[55] M.M. Bazin, Y. Nakamura, N. Ahmad, Chemical Cleaning of Microfiltration Ceramic Membrane Fouled by Nom, J. Teknol 80 (2018). https://doi.org/10.11113/jt.v80.12156

[56] S. Béquet, J.-C. Remigy, J.-C. Rouch, J.-M. Espenan, M. Clifton, P. Aptel, From ultrafiltration to nanofiltration hollow fiber membranes: a continuous UV-photografting process, Desalination 144 (2002) 9-14. https://doi.org/10.1016/S0011-9164(02)00281-3

[57] Z.-P. Zhao, J. Li, D. Wang, C.-X. Chen, Nanofiltration membrane prepared from polyacrylonitrile ultrafiltration membrane by low-temperature plasma: 4. Grafting of N-vinylpyrrolidone in aqueous solution, Desalination 184 (2005) 37-44. https://doi.org/10.1016/j.desal.2005.04.036

[58] C. Qiu, F. Xu, Q.T. Nguyen, Z. Ping, Nanofiltration membrane prepared from cardo polyetherketone ultrafiltration membrane by UV-induced grafting method, J. Membr. Sci. 255 (2005) 107-115. https://doi.org/10.1016/j.memsci.2005.01.027

[59] A. Asatekin, A. Menniti, S. Kang, M. Elimelech, E. Morgenroth, A.M. Mayes, Antifouling nanofiltration membranes for membrane bioreactors from self-assembling graft copolymers, J. Membr. Sci. 85 (2006) 81-89. https://doi.org/10.1016/j.memsci.2006.07.042

[60] A. Rahimpour, S.S. Madaeni, Polyethersulfone (PES)/cellulose acetate phthalate (CAP) blend ultrafiltration membranes: preparation, morphology, performance and antifouling properties, J. Membr. Sci. 305 (2007) 299-312. https://doi.org/10.1016/j.memsci.2007.08.030

[61] C. Liu, S. Caothien, J. Hayes, T. Caothuy, T. Otoyo, T. Ogawa, Membrane chemical cleaning: from art to science, Pall Corporation, Port Washington, NY 11050 (2001).

[62] J.C.-T. Lin, D.-J. Lee, C. Huang, Membrane fouling mitigation: Membrane cleaning, Sep. Sci. Technol. 45 (2010) 858-872. https://doi.org/10.1080/01496391003666940

[63] N. Porcelli, S. Judd, Chemical cleaning of potable water membranes: A review, Sep. Purif. Technol. 71 (2010) 137-143. https://doi.org/10.1016/j.seppur.2009.12.007

[64] Y.-j. Zhao, K.-f. Wu, Z.-j. Wang, L. Zhao, S.-s. Li, Fouling and cleaning of membrane-a literature review, J. Environ. Sci. 2 (2000) 241-251.

[65] N. Yigit, G. Civelekoglu, I. Harman, H. Koseoglu, M. Kitis, Effects of various backwash scenarios on membrane fouling in a membrane bioreactor, Survival and Sustainability, Springer, 2010, pp. 917-929. https://doi.org/10.1007/978-3-540-95991-5_87

[66] J. Li, R. Sanderson, E. Jacobs, Ultrasonic cleaning of nylon microfiltration membranes fouled by Kraft paper mill effluent, J. Membr. Sci. 205 (2002) 247-257. https://doi.org/10.1016/S0376-7388(02)00121-7

[67] M.-W. Wan, F. Reguyal, C. Futalan, H.-L. Yang, C.-C. Kan, Ultrasound irradiation combined with hydraulic cleaning on fouled polyethersulfone and polyvinylidene fluoride membranes, Environ. Technol. 34 (2013) 2929-2937. https://doi.org/10.1080/09593330.2012.701235

[68] H. Kyllönen, P. Pirkonen, M. Nyström, Membrane filtration enhanced by ultrasound: a review, Desalination 181 (2005) 319-335. https://doi.org/10.1016/j.desal.2005.06.003

[69] S. Popović, M. Djurić, S. Milanović, M.N. Tekić, N. Lukić, Application of an ultrasound field in chemical cleaning of ceramic tubular membrane fouled with whey proteins, J. Food Eng. 101 (2010) 296-302. https://doi.org/10.1016/j.jfoodeng.2010.07.012

[70] A. Maskooki, S.A. Mortazavi, A. Maskooki, Cleaning of spiralwound ultrafiltration membranes using ultrasound and alkaline solution of EDTA, Desalination 264 (2010) 63-69. https://doi.org/10.1016/j.desal.2010.07.005

[71] A. Maartens, P. Swart, E. Jacobs, An enzymatic approach to the cleaning of ultrafiltration membranes fouled in abattoir effluent, J. Membr. Sci. 119 (1996) 9-16. https://doi.org/10.1016/0376-7388(96)00015-4

[72] A. Maartens, P. Swart, E. Jacobs, Enzymatic cleaning of ultrafiltration membranes fouled in wool-scouring effluent, pancreas 7 (1998) 35.5.

[73] Z. Wang, J. Ma, C.Y. Tang, K. Kimura, Q. Wang, X. Han, Membrane cleaning in membrane bioreactors: a review, J. Membr. Sci. 468 (2014) 276-307. https://doi.org/10.1016/j.memsci.2014.05.060

[74] Z. Allie, E. Jacobs, A. Maartens, P. Swart, Enzymatic cleaning of ultrafiltration membranes fouled by abattoir effluent, J. Membr. Sci. 218 (2003) 107-116. https://doi.org/10.1016/S0376-7388(03)00145-5

[75] R. Wakeman, C. Williams, Additional techniques to improve microfiltration, Sep. Purif. Technol. 26 (2002) 3-18. https://doi.org/10.1016/S1383-5866(01)00112-5

[76] X. Shi, G. Tal, N.P. Hankins, V. Gitis, Fouling and cleaning of ultrafiltration membranes: A review, J. Water Process Eng. 1 (2014) 121-138. https://doi.org/10.1016/j.jwpe.2014.04.003

Materials Research Forum LLC
https://doi.org/10.21741/9781644901816-2

[77] A.W. Mohammad, Y. Teow, W. Ang, Y. Chung, D. Oatley-Radcliffe, N. Hilal, Nanofiltration membranes review: Recent advances and future prospects, Desalination 356 (2015) 226-254. https://doi.org/10.1016/j.desal.2014.10.043

[78] A. Schäfer, A.G. Fane, T.D. Waite, Nanofiltration: principles and applications, Elsevier, 2005.

[79] B. Van der Bruggen, C. Vandecasteele, Modelling of the retention of uncharged molecules with nanofiltration, Water Res. 36 (2002) 1360-1368. https://doi.org/10.1016/S0043-1354(01)00318-9

[80] D.L. Oatley-Radcliffe, M. Walters, T.J. Ainscough, P.M. Williams, A.W. Mohammad, N. Hilal, Nanofiltration membranes and processes: A review of research trends over the past decade, J. Water Process Eng. 19 (2017) 164-171. https://doi.org/10.1016/j.jwpe.2017.07.026

[81] J. Schaep, B. Van der Bruggen, C. Vandecasteele, D. Wilms, Influence of ion size and charge in nanofiltration, Sep. Purif. Technol. 14 (1998) 155-162. https://doi.org/10.1016/S1383-5866(98)00070-7

[82] C. Fritzmann, J. Löwenberg, T. Wintgens, T. Melin, State-of-the-art of reverse osmosis desalination, Desalination 216 (2007) 1-76. https://doi.org/10.1016/j.desal.2006.12.009

[83] L. Malaeb, G.M. Ayoub, Reverse osmosis technology for water treatment: state of the art review, Desalination 267 (2011) 1-8. https://doi.org/10.1016/j.desal.2010.09.001

[84] A. Gul, J. Hruza, F. Yalcinkaya, Fouling and Chemical Cleaning of Microfiltration Membranes: A Mini-Review, Polymers 13 (2021) 846. https://doi.org/10.3390/polym13060846

[85] J.C. Crittenden, R.R. Trussell, D.W. Hand, K. Howe, G. Tchobanoglous, MWH's water treatment: principles and design, John Wiley & Sons, 2012. https://doi.org/10.1002/9781118131473

[86] M. Ulbricht, Advanced functional polymer membranes, Polymer 47 (2006) 2217-2262. https://doi.org/10.1016/j.polymer.2006.01.084

[87] Y. Wang, X. Chen, J. Zhang, J. Yin, H. Wang, Investigation of microfiltration for treatment of emulsified oily wastewater from the processing of petroleum products, Desalination 249 (2009) 1223-1227. https://doi.org/10.1016/j.desal.2009.06.033

[88] T. Mohammadi, A. Esmaeelifar, Wastewater treatment using ultrafiltration at a vegetable oil factory, Desalination 166 (2004) 329-337. https://doi.org/10.1016/j.desal.2004.06.087

Advanced Functional Membranes   Materials Research Forum LLC
Materials Research Foundations **120** (2022) 43-71   https://doi.org/10.21741/9781644901816-2

[89] R. Gnirss, J. Dittrich, Microfiltration of Municipal Wastewater for Disinfection and Advanced Phosphorus Removal: Results from Trials with Different Small-Scale Pilot Plants, Water Environ. Res. 72 (2000) 602-609. https://doi.org/10.2175/106143000X138184

[90] V. Buscio, M.J. Marín, M. Crespi, C. Gutiérrez-Bouzán, Reuse of textile wastewater after homogenization–decantation treatment coupled to PVDF ultrafiltration membranes, Chem. Eng. J. 265 (2015) 122-128. https://doi.org/10.1016/j.cej.2014.12.057

[91] A. Salahi, R. Badrnezhad, M. Abbasi, T. Mohammadi, F. Rekabdar, Oily wastewater treatment using a hybrid UF/RO system, Desalin. Water Treat. 28 (2011) 75-82. https://doi.org/10.5004/dwt.2011.2204

[92] S. Rodriguez-Mozaz, M. Ricart, M. Köck-Schulmeyer, H. Guasch, C. Bonnineau, L. Proia, M.L. de Alda, S. Sabater, D. Barceló, Pharmaceuticals and pesticides in reclaimed water: efficiency assessment of a microfiltration–reverse osmosis (MF–RO) pilot plant, J. Hazard. Mater. 282 (2015) 165-173. https://doi.org/10.1016/j.jhazmat.2014.09.015

[93] C.F. Couto, W.G. Moravia, M.C.S. Amaral, Integration of microfiltration and nanofiltration to promote textile effluent reuse, Clean Technol. Environ. Policy 19 (2017) 2057-2073. https://doi.org/10.1007/s10098-017-1388-z

[94] S.R. Reijerkerk, M.H. Knoef, K. Nijmeijer, M. Wessling, Poly (ethylene glycol) and poly (dimethyl siloxane): Combining their advantages into efficient CO2 gas separation membranes, J. Membr. Sci. 352 (2010) 126-135. https://doi.org/10.1016/j.memsci.2010.02.008

[95] D. Wang, F. Tong, P. Aerts, Application of the combined ultrafiltration and reverse osmosis for refinery wastewater reuse in Sinopec Yanshan Plant, Desalin Water Treat. 25 (2011) 133-142. https://doi.org/10.5004/dwt.2011.1137

[96] I. Petrinic, J. Korenak, D. Povodnik, C. Hélix-Nielsen, A feasibility study of ultrafiltration/reverse osmosis (UF/RO)-based wastewater treatment and reuse in the metal finishing industry, J. Clean. Prod. 101 (2015) 292-300. https://doi.org/10.1016/j.jclepro.2015.04.022

Advanced Functional Membranes
Materials Research Foundations **120** (2022) 72-110

Materials Research Forum LLC
https://doi.org/10.21741/9781644901816-3

Chapter 3

Materials and Applications for Functional Polymer Membranes

M.M.S. Mohd Sabee[1,b], A. Rusli[1,c], N. Ahmad[1,d], Z.A. Abdul Hamid[1,a*]

[1]Biomaterials Research Niche Group, School of Materials and Mineral Resources Engineering, Universiti Sains Malaysia, Engineering Campus, 14300 Nibong Tebal, Pulau Pinang, Malaysia

[a*]srzuratulain@usm.my, [b]meersaddiq@gmail.com, [c]arjulizan@usm.my, [d]nurazreena@usm.my

Abstract

This chapter covers an inclusive overview of the polymeric membranes with advanced functions in numerous applications such as water and gas separation, medical and emerging technologies include fuel cells, lithium-ions batteries, electroconductive, and optoelectronics. The membrane's performance and behavior in terms of selectivity, permeability, and separation process for these applications are determined by the materials used in the membrane's construction. Thus, in this chapter, the potential of different polymers for functional membranes are discussed including their applications based on their suitability in terms of types, fabrications, and mechanisms. For each application, the polymer membrane technology, the challenges, and the future direction are also discussed. The membrane technology is also always evolving, especially the development of functional polymer membranes for a variety of applications, so that quality of life is improved.

Keywords

Membrane, Water/Gas Separation, Medical, Fuel Cell, Lithium-ions Battery, Electric Conductive, Optoelectronic

Contents

Materials Research Foundations **120** (2022) 72-110
https://doi.org/10.21741/9781644901816-3

Advanced Functional Membranes
Materials Research Foundations **120** (2022) 72-110

Materials Research Forum LLC
https://doi.org/10.21741/9781644901816-3

1. Introduction

Membranes are selective barriers that qualify components of varying sizes or physical/chemical qualities to be separated. The selectivity and permeability of the membranes utilized determine the efficiency of a membrane separation process [1]. Generally, membranes can be categorized based on different nature of materials (natural or synthetic), type of materials (polymer or inorganic) and structure (porous or non-porous) [2]. Natural membrane is a membrane in nature such as biological membrane, consists of phospholipid bilayer which involving biochemical reaction for the ions and chemicals transportation, whereas, synthetic membrane is basically artificially created membrane typically made up of cellulose, filter and synthetic polymer for laboratory or industry applications. On the other hand, the organic membrane is a membrane constructed from the organic polymer materials which include polyethylene (PE), polypropylene, polytetrafluorethylene (PTFE), and cellulose acetate. Meanwhile, inorganic membrane refers to membrane made by ceramics, metals, zeolites, or silica [3]. Apart from that, porous and non-porous membranes are dissimilar in terms of selectivity, permeability and separation process, where porous membrane is normally used for micro- and ultrafiltration via particle size differentiation, whereas, non-porous membrane is usually applied for nanofiltration and reverse osmosis using pressure or concentration gradient [2].

In this chapter, the focus is on the polymeric membranes in various applications. Polymeric membranes, also known as organic membranes, are a class of liquid separation technologies that are among the best in the industries in terms of performance and efficiency. Because of their great manufacturability, low cost, and abundance, polymeric materials have played a significant role in membrane development, and they will preserve to do so in the future, as demonstrated by the chemistry-processing-structure-performance paradigm [4]. Polymeric materials versatility to be tailor made with various functionalities have elevated their usage in the membranes field. Furthermore, polymer membranes are more appealing because of their simple production and processability. This is owing to the broad range of barrier structures and features that polymer materials can create [5].

In fact, polymer-based membranes are also utilized to control the selective transport of ions and/or water in applications such as water purification, batteries, and fuel cells. The precise nature of ion and water transport is frequently confusing, and it is dependent on the intricate interplay between polymer structure and dynamics that allow movement [6]. Therefore, the relevance of polymeric materials in membrane technology is highlighted in this special chapter focusing on water and gas separation, medical and emerging technologies (fuel cells, lithium-ions batteries, electroconductive, optoelectronic) applications. Whereas, the later part of this chapter is discussed on the challenges and future direction of polymer membrane technology highlighted on the aforementioned applications.

Advanced Functional Membranes
Materials Research Foundations **120** (2022) 72-110

Materials Research Forum LLC
https://doi.org/10.21741/9781644901816-3

2.   Functional polymer membrane for water and gas separation applications

Membranes are used in water purification and treatment, detoxification of polluted water and desalination of seawater and brines. Water treatment and detoxification involve the removal of contaminants such as natural organic matter (dissolved and non-dissolved), nutrient ions (phosphate and nitrate) and emerging pollutants including organics, chemical dyes and heavy metals [7]. Meanwhile, seawater desalination involves the removal of dissolved salts to produce saltwater appropriate for domestic, human consumption, or industrial/agricultural applications.

In water treatment sectors, membrane technology is dominating and is considered easier operation, more economical, versatile, low-energy intensive, and efficient compared to other techniques such as chemical precipitation, adsorption, electrolysis and conventional coagulation [7]. The use of a membrane in water treatment also allows large quantities of material to be separated at a molecular scale without phase changes producing high-quality water with low chemical sludge effluent.

Membranes are also used to clean exhaust gases and are commercially used in purification and gas separation mainly hydrogen recovery from gas streams, vapor/gas separation, air dehydration, and treatment of natural gas and biogas [8].

2.1   Types of water and gas membranes

In water treatment and gas separation processes, a porous membrane or dense membrane can be used. However, dense membranes are less frequently chosen due to the low flow of the permeate across the membrane [7]. The cross-section of these membranes can be isotropic (i.e: symmetric) or anisotropic (i.e: asymmetric). Isotropic membranes such as porous, nonporous dense and electrically charged membranes are chemically homogeneous in composition. The porous membranes can be found in different classes according to their pore sizes.

Meanwhile, two main types of anisotropic membranes are composite and phase-separation membranes. The phase separation membranes have a homogeneous chemical composition as in isotropic membranes however, the porosity and pore sizes are different across the membrane thickness. Composite membranes, for instance, coated films and thin-film are structurally and chemically homogeneous in which a thin surface layer is supported by a thicker porous structure. Both layers are typically made from different classes of polymeric materials [9]. Nowadays, most of the dense membranes are fabricated with a dense top layer on a porous structure [7].

For industrial gas and liquid separation, thin-film composite membranes are usually applied. This type of membrane commonly consists of high permeability protective or sealing layers (to prevent damage during handling), selective polymer layer (to provide the desired separation properties), gutter layer made up of a high permeable polymer (to improve interlayer compatibility and support layers) and support layers that highly permeable (to provide mechanical strength to the membrane) [10].

2.2   Water and gas membrane materials

For water treatment, membranes can be produced from inorganics materials such as titanium dioxide, zeolites or ceramics that possess advantages compared to polymeric membranes type in terms of mechanical, chemical, and thermal stabilities, fouling-resistance, reusability and photocatalytic ability [9]. However, their large-scale applications are limited due to brittleness (fragile) and high operational cost.

Industrially established membranes for water filtration using microfiltration (MF), nanofiltration (NF), and ultrafiltration (UF) membranes are mainly made from organic polymers such as polysulfone (PSU), polyetherimide (PEI), polyacrylonitrile (PAN), polyethersulfone (PES), polyvinyl chloride (PVC), polycarbonates (PC), polyamides (PA), polypropylene (PP), polyvinylidene fluoride (PVDF), and cellulose acetate (CA) [9]. Table 1 lists the types of membranes used for different water treatment processes based on their pore sizes, membrane materials and applications. Preparation methods of these membranes may be different to produce different pore sizes to suit different applications. Membrane distillation is different from water filtration in which the driving force for separation is thermal and the requirement for more hydrophobic, more porous and higher chemical/thermal resistance membrane materials in the former [11].

Polymeric membranes are extensively used in water treatment and desalination because of the high modification flexibility, simple pore fabrication procedure, high efficiency in separating particles and dispersed oil, easy operation and fabrication and relatively lower cost compared with inorganic membranes [7]. However, the use of polymeric membranes in water filtration processes leads to problems with poor wettability, pH-dependency and fouling tendency mainly due to their inherent hydrophobicity. Table 1 shows the type of membranes in different water treatment processes.

To overcome the problems, surface modifications, functionalization or synthesis/fabrication of well-defined structures can be done. Other functionalized polymers (e.g: polyacrylic acid or polymethyl methacrylate) have been incorporated as a second component in the membranes material to enhance the membranes' hydrophilicity. Organic nanofillers also are added to the polymeric membranes to improve their performance. Clay is one of the preferred fillers due to its water purification capability and other advantages including economical, easy to synthesize with enhanced thermo-mechanical properties [7].

In addition, functionalization of the clay and/or polymer can be done to enhance performance as well as mechanical, thermal, and chemical stability to suit specific needs in target membrane applications. Functionalization of clay can be conducted using surfactant, coupling agent or salt (intercalation modification). Surface modification of clay using surfactants such as alkyl ammonium salts makes the clay surface more hydrophobic for better dispersion stabilization.

Table 1 Types of membranes used in different water treatment processes [9,10].

Process	Pore sizes (µm)	Driving force	Filtrate	Membrane materials	Applications
Microfiltration (MF),	0.1-5	Induced Pressure	Particles, asbestos, cellular materials (red blood cell and bacteria)	PP, PVDF, PVC, PTFE	Desalination
Ultrafiltration (UF),	0.01-0.1	Induced Pressure	Large particles, microorganisms, dissolved bio-macromolecules (pyrogen, protein and viruses)	PSU, PES, PVC, PTFE, CA	Wastewater treatment, protein separation, food processing, water remediation, recovery of surfactants in industrial cleaning
Nanofiltration (NF),	0.001-0.01	Induced Pressure	Organic molecules, viruses, a range of salts, divalent ions	PSU and PES (supporting substrates), CA	Soften hard water, remove organic matter, wastewater treatment and recovery
Reverse osmosis (RO)	0.0001-0.001	Induced Pressure	Aqueous organic solids (minerals, salt ions, and metals ions) and organic substances	PSU and PES (supporting substrates), CA PA	Brackish seawater and groundwater desalination, potable water production
Membrane distillation	0.1-0.6	Thermal	Salt	PP, PTFE, PVDF, zeolites, ceramics (support)	Desalination

Surface functionalization with the polymer can be carried out by either physical adsorption polymers or chemical grafting. Physical adsorption using anionic or non-ionic polymers causes adsorption of clay physically on the surface of the polymer leads to alteration of physical and chemical properties. Chemical grafting leads to covalent bonding of the polymer and the clays. Further explanation on each of the methods is reported by Buruga et al. (2019).

Multiple functionalities can help not only to sieve the contaminants from the aqueous solution but also able to absorb, degrade or deactivate the contaminants. In fact, functional polymer-clay membranes can be used for heavy metal removal, dye removal, microbial/fungal removal, desalination of seawater and as superabsorbent for pollutants removal [7]. These membranes are also attractive for their antifouling properties. However, the modifications lead to higher costs and difficulty in fabrication and scale-up.

In gas separation membrane, the main commercial application and the types of polymer used as selective layers are listed in Table 2.

Table 2 Main commercial membrane gas separation applications [10,12,13].

Application	Gas separation	Selective polymeric layer
Nitrogen production	N_2/O_2	PSU, polyphenylene oxide (PPO) polyimides (PI), substituted polycarbonates
Hydrogen recovery	H_2/N_2, H_2/CO, H_2/CH_4	PI, PSU, aramids
Vapor recovery	C_2H_4/N_2, C_3H_6/N_2, C_2H_4/Ar, gasoline/air, CH_4/N_2	Silicone rubber
Natural gas treatment	H_2S/CH_4, CO_2/CH_4, He/CH_4	PI, CA
Air dehydration	H_2O/air, H_2O/N_2	PSU

Similar to membranes used in water treatment, membranes for gas separation are also made from two main categories, synthetic polymeric materials and inorganic materials such as ceramic, carbon and metal (especially palladium). The separation of gases will occur based on different transport mechanisms, which depends on the physical properties of the gases and intrinsic membrane material properties [8].

Even though many different polymeric membranes are available which are less brittle, ease on mechanical handling, and cheap in production as compared to ceramic membranes, the polymeric membranes may not be useful in harsh oxidizing agents and high temperatures conditions [14]. For these conditions, the inorganic membrane may be more suitable. For example, the use of palladium for selective transport of hydrogen or perovskites for selective transport of oxygen [2]. Due to this reason, selective polymeric layers that are used for gas separation are commonly added with inorganic materials such as carbon, zeolite or metal-organic frameworks to harness processability and increase solubility selectivity [10].

2.3   Fabrication and functionalization of water and gas membranes

Membranes can be constructed to obtain the desired selectivity and permeability suit the requirements for the component to be separated. Generally, the performance is dependent on membrane's fundamental properties such as selectivity, flux rate, chemical, thermal or mechanical stability under operating conditions, adaptability to operating conditions and resistance toward fouling. In addition, the membranes should have suitable surface morphology in terms of surface roughness, pore size and structure as well as physicochemical properties [7].

For water treatment applications, isotropic membranes can be produced using various methods such as polymer-based track-etched (i.e: irradiated with heavy ions to create tracks through the film), phase inversion (homogeneous polymeric is transformed to solid-state by immersion precipitation as commonly done on PVDF since PVDF is soluble in common solvents) or stretched polymer film (polymers are heated beyond their melting point, followed by extruded into a thin film and then stretched which are commonly applied for PP and PTFE) [9].

The nanofiller can be incorporated into polymers by various methods including melt intercalation, phase inversion (producing well-mixed nanomaterial in a polymer matrix), interfacial polymerization (producing a thin layer with nanocomposites membrane substrates or a thin layer of nanocomposite on the membranes' surface), surface grafting or surface coating [9]. Surface modification of either polymer or nanofiller (e.g: clay) to improve the performance of the membranes can be carried using various fabrication methods. However, fabrication of these functional polymer-nanofiller composite membranes requires advanced technologies and organized design/architecture as well as a good membrane formation [7] for a successful intended application.

For gas separation, membranes are commonly produced by phase inversion since this method can produce thin membranes (approximately 100 nm or less) and defect-free membranes with sufficiently high surface areas for practical applications. Furthermore, asymmetric membranes with dense, thin film on a porous substrate can be produced using this method [12].

The membranes produced from various fabrication methods as mentioned earlier are commonly flat sheet or cylindrical configurations either hollow fibre membranes (diameter below 300µm), capillary membranes (diameter of 300-1000 µm) or tubular membranes (diameter of few cm) [12]. These membranes are typically assembled into various geometries/modules to maximize the surface area that can be assimilated in a certain volume.

The flat sheet geometry is frequently mounted intro spiral wound modules or less frequently into envelope-type modules while hollow-fibre configuration would be assembled into hollow fiber modules. For gas separation, the hollow fibre module is more commonly chosen than spiral wound modules due to higher specific surface area leading to higher productivity, easier fabrication method and lower production cost [12,15]. Spiral wound module is preferred for water purification and RO applications since it induces less pressure drop and reduces the feed pre-treatment needed to prevent fouling compared with hollow fiber modules [12]. The spiral wound module is also used in the separation of CO_2 industrial waste or flue gases with high permeances [2].

2.4   Mechanisms for water/gas separation

For MF, NF, UF and RO, transport through membranes is driven by pressure difference while in gas separation, the driving force is based on chemical potential (or concentration) difference [7]. Based on the phase separation's mechanism, membranes can be categorized into three: (i) the sieve mechanism (different size of solute particles), (ii) the solution-diffusion mechanism (different in solubility and diffusivity of the membrane materials) and, (iii) the electrochemical mechanism (different in charges of the solute components) [7].

In the separation of gas by a porous membrane, the kinetics effect is important and depends on average free path length value, which is the distance of a gas molecule can move between collisions with other gas molecules. If large pores are used (larger than average free path length), separation may not occur but only viscous hydrodynamic flow since the collision between molecules happens more often compared with particles and pore walls. Meanwhile, if the pores are below the average path length, the molecules collide more frequent with the pore wall compared to each other, allowing the movement of light gas molecules faster than heavier gas molecules. Permeability, which is determined by the mass of the gas molecules, defines selectivity [2].

2.5   Challenges and future direction of functional polymer membrane for water and gas separation applications

Gas and water separation can be utilized to increase the quality of life and environment. For example, carbon dioxide removal from coal-fired power plant flue gas is helpful to lessen its effect on climate change [2]. In addition, the development of bio-inspired and environmentally responsive membranes with other special features such as antifouling/self-cleaning, antibacterial/antifungal and ability to withstand extreme conditions are of interest.

However, continuous effort needs to be conducted in altering the structure and/or compositions of the membrane to enhance selectivity and permeability, membrane lifetime, chemical/thermal/mechanical stability and antifouling capability. Tremendous research has been conducted in developing membranes for especially functionalized polymer composites for water treatment. However, most of the research has been conducted at a laboratory scale, thus requires to be extended to large-scale production and industrial applications.

Optimization of separation can be done by combining membrane technology with other technologies, for instance adsorption and thermal separation. The use of membrane modelling and computer simulation, in addition to experimental approaches, will become crucial in membranes development, modules and overall processes [2].

In gas membrane applications especially, there is an inter-relationship between selectivity and permeability as well as between mechanical properties and gas transport [10]. Plasticization and physical aging are commonly encountered issues in mix-gas separation performance over time [12]. Through innovative materials selection, processing methods, fabrication and systematic studies, the performance of the membranes can be improved to enhance the overall quality of life and the environment.

3.   Functional polymer membrane for medical applications

Chronic kidney disease has been known as a worldwide public health problem. Patients with chronic kidney disease survived by haemodialysis treatment apart from peritoneal dialysis and renal transplantation. The use of membrane as separation in blood purification in haemodialysis treatment has kicked start the recognition usage of functional polymer membrane in the medical field. Functional polymer membrane has found its usage in various medical applications such as for supporting scaffold in tissue engineering as guided bone regeneration (GBR), ophthalmology applications, plasma collection oxygenation blood during cardiac surgery and as therapeutic drug delivery system since then. The following section discussed on the medical application of various type of functional polymer membrane, the fabrication techniques, the mechanism of dense and porous membrane and challenges and future direction of functional polymer membrane for medical applications.

3.1   Types of functional polymer membrane for medical applications

Kidneys are the organs in renal system that regulate the blood chemistry via removing harmful substances, preserving essential substances in the blood and also regulate blood's ionic concentration. However, patients diagnosed with chronic kidney disease (CKD) or end-stage renal disease (ESRD) will encounter accumulation of metabolic waste products in the body. This failure eventually resulted in deficient performance of other organs and complications related to renal failure [16]. With the increase in number of patients with CKD or ESRD, the need for haemodialysis membrane is expected to constantly increase. During haemodialysis, the membrane in the dialyzer acts as a medium for separation between the blood and fluid compartment. Basic requirement for the membrane is to be able to remove solutes as well as fluid accumulation in the body, while maintaining electrolyte balance of the patient, good biocompatibility with minimal effect on blood components and ample chemical and mechanical stability to allow sterilization by different approaches [17].

To achieve the basic requirements materials for membrane, the material should possess particular characteristics such as naturally hydrophilic, biocompatible, semipermeable, asymmetric, improved pore size as well as surface chemistry [18]. Although, the current treatment for CKD and ESRD is through haemodialysis treatment, but haemodialysis also has been incorporated with acute/chronic side effects that threatens the patients. This is because during haemodialysis, the interactions between blood and the membrane leads to various body system activation such as coagulation of blood components, leukocytes, cytokine production and existent of free oxygen radicals [16]. Therefore, materials biocompatibility is crucial when selecting materials for membrane.

Cellulose was the first generation for dialysis membrane. However, its major drawbacks were its poor biocompatibility which triggered by the free hydroxyl (OH) groups in the molecular units of the cellulosic polymer [19]. Thus, to overcome this issue, a number of modifications have been introduced to increase biocompatibility such as esterification on the OH groups with benzyl or diethylaminoethyl groups, or via replacing the OH groups with acetyl groups such as diacetate, triacetate, or cellulose acetate [17,19]. In a study by Sirolli et al. (2000), cellulosic membrane chain was grafted with polyethylene glycol (PEG) and the performance was compared to cellulose diacetate and polymethylmethacrylate. The findings from this study showed that PEGylated cellulose is a useful mean to increase the biocompatibility of the cellulosic polymer [19]. However, it was also concluded that

Advanced Functional Membranes   Materials Research Forum LLC
Materials Research Foundations **120** (2022) 72-110   https://doi.org/10.21741/9781644901816-3

biocompatibility of the derivatized cellulosic membranes has wide range variability which may be due to the structural modification compared to the degree of OH replacement.

Moving forward to a more recent study, there are a few methods that has been proposed to improve the biocompatibility of polymeric membrane by using polymer blending, biopolymers, surface chemical modification or biomimetic membranes. Kaleekkal et al. (2015) prepared polysulfone (PSf)/ sulfonated polyethersulfone (SPES) blend to evaluate the performance and its biocompatibility. In this study polyethersulfone (PES) was sulfonated resulting in a hydrophilic SPES which are then blended with PSf. PSf/ SPES blend showed higher ultrafiltration rate (UFR), extended blood clotting time, eliminated platelet adhesion, decreased protein adsorption and reduced complement activity. These findings indicate an increased in biocompatibility when compared to pristine PES membrane [20]. Although PES membrane offers outstanding oxidative, thermal, chemical and mechanical stability, its hydrophobicity remains a challenge in haemodialysis treatments. Hydrophobic surface results in a severe fouling of the membrane materials. As discussed earlier, sulfonated PES gives a hydrophilicity to PES, another method to improve hydrophilicity is to develop mixed matrix membranes (MMM). PES filled with nanofiller of poly(citric acid)-grafted-multi wall carbon nano tube (MWCNT) was fabricated as mixed matrix membrane for haemodialysis. This study showed the improvement of surface hydrophilicity and porosity of PES/MWCNTs MMMs. PES/ MWCNTs MMMs also showed better protein resistance and antifouling properties when compared to PES membrane [21]. For surface chemical modification, polyvinylpyrrolidone (PVP) and polyetherimide (PEI) with heparin immobilization was synthesized to improve blood compatibility. Immobilization of heparin resulted in more hydrophilic membranes properties with significant platelet and protein resistant. These findings suggest an anti-thrombogenic characteristic was observed on the PEI/PVP immobilize heparin membrane [22].

Diseases such as periodontal disease, congenital defects, tumours extraction, pathological bone tissues defects or alveolar bone resorption which may lead to total or partially lost bone [23-24] requires guided bone regeneration (GBR) treatment to prevent soft tissue invasion and provide capacity for guiding new bone formation/growth into the bone defect. For this application, the membrane is required to provide physiological barrier towards fibroblast infiltration, restrain the proliferation of progenitor cell and act as scaffold for bone regeneration [25]. GBR membranes are divided into two groups: resorbable membranes and non-resorbable membranes. High density polytetrafluoroethylene (d-PTFE), expanded polytetrafluoroethylene (e-PTFE, Gore-Tex®), and titanium reinforced expanded polytetrafluoroethylene (Ti-e-PTFE) are among the widely used non-resorbable membranes [25-27]. It was reported that e-PTFE was among the first non-resorbable membranes for dental use due to its stability, chemical and biologically inert, resistant to enzymatic and microbiological attack. Also, e-PTFE features with macro porosity encourages bone regeneration and improved wound stability [27]. However, it has been reported that the main drawbacks of PTFE were the need for secondary surgical [27-29].

Unlike non-resorbable membranes, resorbable membranes made of either natural or synthetic based polymers. Synthetic polymers used include poly(lactid acid) (PLA), poly(glycolic acid) (PGA), poly(Ɛ-caprolactone) (PCL), poly(hydroxyl butyric acid) and poly(hydroxyl valeric acid) [30], while natural polymers are collagen (from human or animal tissues) [25], calcium alginate, chitosan [31-32] and nanocellulose [33-34]. Synthetic resorbable membranes has the advantages of processability, drug-encapsulating ability, tuned biodegradation, and low rigidity [30] while natural polymers are known for their non-antigenicity, biocompatibility, and tuneable resorption via cross-linking treatment [35]. However, resorbable membranes has the disadvantages of low mechanical strength and wide variable degree of resorption which significantly interrupt the amount of bone formation [29]. To overcome these shortcomings of resorbable membranes, bioactive composite, the combination of bioactive ceramic phase and biodegradable polymeric phase are proposed to give mechanical stability for maintenance and induce bone formation bioactivity [24]. Among the most sought bioactive ceramics that has been investigated was hydroxyapatite (HA) [36], bioactive glass [37-38], tricalcium phosphate [24], precious metals [33], clay silicates, zirconia and carbon nanotubes [39]. For instance, study by Fu et al. (2017) was concluded that the synthesized bilayer PLGA/nano-HA with different surfaces and structures showed excellent fibroblastic barrier and favourable osteogenesis effect. The porous layer was claimed to increase osteoblast homing, proliferation and differentiation [29]. In an *in vivo* study, a comparison between polycaprolactone/tricalcium phosphate (PCL/TCP) membrane and collagen membrane, it was reported that PCL/TCP membrane showed a statistically increased in percentage bone area fraction (BAF%) compared to collagen membrane after 2 months in rabbit model. Whereas, in the first month of evaluation, there was no statistical difference observed. The researcher postulated that the significant observation in second month time frame was due to the higher stiffness and slower degradation of PCL/TCP [24]. Dascalu et al. (2019) synthesis a cellulose acetate/ hydroxyapatite (CA/HA) membrane and concluded that CA/HA composite membrane showed that the hydrophilicity, water permeation, protein resorption is suitable for bone regeneration applications. While *in vitro* using osteoblast, cell showed good cell adhesion and viability [40].

In cardiovascular applications, polymer blend and composites has been utilised due to its similarity with the human myocardium. These polymer blend and composites showed improved hydrophilicity and better cell interaction. In this field, these polymers are used as bioresorbable stent, as coatings in cardiac stent, cardiac patches to repair the myocardium, heart valves and etc. In cardiovascular applications, natural polymers, for instance alginate, silk, gelatine, collagen and chitosan has been utilised. While synthetic polymers that offers versatility in their physical and chemical properties such as polyesters (ex: PLA, PLLA, PDLA, PCL), polymethylmethacrylate (PMMA), polytetrafluoroethylene (PTFE) and polyethylene (PE) has been used in cardiovascular treatment [41]. For example, Castilho et al. (2017) investigated on fabricating ultrafine fibre scaffolds of hydroxyl-functionalized (poly(hydroxymethylglycolide-co-ℰ-caprolactone) (pHMGCL) to improve the biocompatibility and architecture to promote cell retention and help cardiac cell growth. It was reported that pHMGCL has improved hydrophilicity when compared to PCL with tuneable degradation rate [42]. PLLA and PLLA/PLGA blend has been explored as coatings in cardiac stents to offer biodegradable stents coatings [43].

In drug delivery system, functional polymer is of interest as a promising transdermal drug delivery (TDD). The advancement in transdermal drug delivery is due to the simpler and non-invasive administration. TD also increases drug bioavailability and reduces the degradation effects of metabolism encountered in the oral route, reduces toxic nature with oral delivery due to low bioavailability and liver damage. Examples natural polymers used as TDD are chitosan and cellulose [44]. While examples of synthetic polymers used in TDD are PCL, PMMA, polycarbonate (PC), PLGA, vinyl polymer and polyvinylidene fluoride (PVDF) [44-45]. In a study by Grilloa et al. (2019), a hybrid drug delivery system has been synthesized. In this finding, benzocaine-loaded poly- ℰ-caprolactone nano particles were incorporated in Poloxamer 407-based hydrogel and to investigate the drug permeation in transdermal diffusion membrane, Strat-M. Results show that the nano particles exhibited increased drug encapsulation efficiency, efficient physicochemical stability as well as rheological properties. These nano particles also showed permeation across Strat-M membrane. These findings suggest that the incorporation of hydrogels nano particles in transdermal synthetic membranes are capable to modulate local anesthetic permeation rate [46].

The polymers discussed are of great interest in biomedical applications due to its numerous advantages regardless natural or synthetic polymers. To improve the functionality of these polymer membranes, many researchers has loaded the membranes with antibiotics, non-antibiotic antibacterial agents, growth factors to enhance osteogenic progenitor cell differentiation such as bone morphogenetic proteins, salicylic acid base poly(anhydride-esters) or platelet-rich fibrin [47].

3.2 Fabrication techniques to fabricate functional polymer membrane for medical applications

In this subsection, the three most common and versatile fabrication techniques utilized to prepare functional polymer membrane for medical applications will be described. The fabrication techniques are electrospinning, solvent casting, and phase inversion.

3.2.1 Electrospinning

Electrospinning method has the advantage of feasibility to obtain micro- or nano-scale fibres with high surface area, controlled porosity and ease of functionalization for numerous applications, superior physicochemical as well as ease of processing. Briefly, this process involves the extrusion of a polymer through an opening prior to external driving force. To date, different modification has been done on the conventional electrospinning techniques to fabricate functional polymer membrane that suits specific medical applications. For example, Fu et al. (2017) combines the electrospinning and phase inversion techniques to fabricate porous and dense layer of poly(lactic-co-glycolic acid)/nano HA membrane with barrier function. This bilayer membrane showed significant tissue ingrowth and optimum therapeutic result for bone tissue regeneration [29]. This technique has also been utilized to incorporate growth factors in silk chitosan/fibroin/nano HA core-shell nanofibrous membranes. In this study, bone morphogenic protein-2 (BMP-2) was incorporated in the core of the membranes. The results obtained from histological and immunohistochemical staining confirmed the formation of bone and osteogenesis [48]. It has also been reported that polymeric material and living cell were directly electrospun to form membranes. This modified technique is called 'cell electrospinning' [49-52].

3.2.2 Solvent Casting

Solvent casting is a conventional fabrication method that has been used to fabricate membranes, composite films, or blend. In short, this technique involves polymers and/or fillers dissolved in a solvent where both materials are soluble. In fact, water is the commonly used solvent for hydrophilic polymers [53-55] while for hydrophobic polymers, organic solvents are used. Mota et al. (2012) fabricated novel guided tissue and bone generation of chitosan/ bioactive glass nanoparticle (CHT/BG-NPS) membrane using solvent casting. In this study acetic acid was used as the solvent for the preparation of CHT/BG-NPS membrane [56]. In solvent casting, the significant parameter that affect the membrane properties are the type of solvent used. In a comparison study of the different solvent used effect on tensile strength of poly(vinyl acetate) membrane by Hansen et al. (2013), it was concluded that when ethanol was used as the solvent showed lower elongation and higher strength rather than membrane cast from chloroform or toluene as solvent that showed lower strength and higher elongation [57]. Solvent casting has the advantage of simplicity whereby it does not imply further thermal or mechanical stress to dry the membrane and it is a less time-consuming technique. A modified approach of solvent casting with freeze casting was proposed by Gorgieva et al. (2018) fabricate a bilayer membrane with micro structural diversity of chitosan/gelatine bilayer membranes with mineralized interface. The study concluded that the obtained membrane showed non- and highly micro-porous and pore interconnecting region [58]. Due to the ease of processing, many functionalized polymer membranes utilize this method for medical applications.

3.2.3 Phase Inversion

Phase inversion techniques was first introduced by Loeb and Sourirajan to fabricate asymmetric polymer membrane. Since then, this technique has caused vigorous growth in membrane technology. In phase inversion technique, a homogenous polymer solution becomes thermo-dynamically unstable and separates between polymer-rich and polymer-lean phases [59]. Recently, cellulose acetate (CA) hybrid membrane was synthesized with enhanced permeability and excellent anti-fouling property was fabricated by combination of phase inversion and chemical reaction. In this study, CA hybrid membrane was fabricated using Loeb-Sourirajan techniques followed by introducing various solvents and organic acids in chemical reaction method. The objective of the study was to improve the morphological structure and anti-fouling property of the membrane [60]. Phase inversion techniques were also reported to enable the production of membrane with surfaces of different structures, a dense and porous surface. A study by Liu et al. (2021) succeeded to fabricate membrane with dense and porous surface of PLA membranes via phase inversion stimulated by water droplets at various temperatures. Cytocompatibility on 3T3 cells showed that the pore size obtained at water droplet temperature of 25 °C are sufficient for cell proliferation and promotes 3D culture of 3T3 cells on the membrane [61].

3.3   Challenges and future direction of functional polymer membrane in medical applications

The major challenges for functional polymer membrane in medical applications is to achieve biocompatibility properties without sacrificing mechanical and structural requirements. For haemodialysis membrane for example, the major concern is the proinflammatory molecules activation once the blood is exposed to the membrane and that current available dialysis membrane are not applicable to eliminate some larger toxic molecules. Thus, to overcome the wide spread of disadvantages of polymer membranes in various medical applications and to cater specific needs of each application, improvements in term of materials selections and improved/modified techniques has been explored. In terms of materials selections, new polymer blends and composites has been evaluated to optimize the significant properties of each material. Incorporation of various inorganic materials such as HA and carbonate apatite as fillers are also actively research to improve properties of the polymeric membranes. In terms of fabrication techniques, new hybrid techniques have been utilised to improve yield, architecture, porosity, hydrophilicity/ hydrophobicity, flexibility and etc. The combination of techniques used enables the feasibility to control numerous properties and improved functionality of the polymer membrane.

4. Functional polymer membranes for emerging technologies

Functional polymer membrane for emerging technologies can be divided into four major applications, fuel cell membrane, membrane for lithium battery, electrically conductive membrane, and optoelectronic thin film. The following sub sections discussed the functional polymeric membranes for the emerging technologies.

4.1 Fuel cell membrane

Fuel cell is defined as an electrochemical cell that uses a pair of redox processes to transform the chemical energy of a fuel (typically hydrogen) and an oxidizing agent (commonly oxygen) into electricity. It is unique as it can use a wide range of fuels and feedstocks, as well as can power large systems (i.e: power plant) and small systems (i.e: computer). Fuel cell can be utilized in a variety of applications such as transportation, industrial/commercial/residential structures, and long-term grid energy storage in reversible systems [62].

In fact, fuel cell has several advantages compared to conventional combustion-based technologies. Fuel cell can run with excellent efficiencies (> 60%) and lower emissions than combustion engines, as well as can convert chemical energy in the fuel directly to electrical energy [63]. Furthermore, the operation of hydrogen fuel cell emits only water, addressing critical climate challenges as no carbon dioxide (CO_2) emissions which resulted significant decrease in air pollutants that produce smog which directly cause health problems. Other than that, fuel cell is also silent during its operation as it only has few moving parts compared to conventional combustion [64].

4.1.1 Fuel cell part

Polymer electrolyte membrane (PEM) fuel cell basically consists of membrane electrode assembly (MEA), which consists of the membrane itself, layers of catalyst, and gas diffusion layers (GDL). Whereas, the parts used to assemble an MEA into a fuel cell are gaskets (seal the MEA to avoid gas leakage), and bipolar plates (assemble PEM into a stack and passages for the fuel and air) [65].

Firstly, the membrane is a specifically treated material which conducts only positively charged ions while blocking electrons. It is the critical component to fuel cell because it must allow the required ions only to travel between the cathode and anode. The chemical process would be disrupted if other chemicals passed through the electrolyte. Basically, the membrane used in transportation applications is exceedingly thin (under 20 microns) [66].

Secondly, catalyst layer is applied on both membrane's sides (one side on anode layer and the other on cathode). Nanometer-sized platinum particles are spread on a large surface area carbon substrate for the catalytic layers. This supported platinum catalyst is sandwiched between the membrane and the GDL, together with an ionomer (ion-conducting polymer). The platinum catalyst on anode allows hydrogen to be separated into electrons and protons. Whereas, the platinum catalyst on cathode reduces oxygen via interaction with protons produced by anode, resulting in the formation of water. On the other hand, the ionomer that is incorporated into catalyst layers permits protons to pass through [67].

Thirdly, the GDL is located outside the catalytic layers and aid in the transport of reactants as well as the elimination of water. In fact, each GDL is usually made up of a carbon sheet with polytetrafluoroethylene (PTFE) partly coated carbon fibers. Gases diffuse quickly via the GDL's pores. The hydrophobic PTFE keeps these pores open, preventing excessive water buildup. While, the microporous layer is a thin coating of large surface area carbon mixed with PTFE that is applied to inner surface of the GDL in many circumstances. This microporous layer can assist water retention (to preserve membrane conductivity) other than water release (to keep the pores open for oxygen and hydrogen diffuse into electrodes) [68].

In order to create useable output voltage, MEAs are commonly incorporated by connecting them on top of each other (series stacked). In order to keep nearby cells apart, each cell is placed between the two bipolar plates. These plates can be constructed of metal, carbon, or composite materials, give both electrical conduction and physical strength to the stack. A flow field or set of channels is machined into the plate to permit gases to flow, commonly found on the plate's surface. Other than that, a liquid coolant can be circulated through additional tubes inside each plate. In addition, each MEA is sandwiched between two bipolar plates, however to achieve a gas-tight seal, the gaskets have to be applied at the edges of the MEA. Typically, these gaskets are constructed of a rubbery polymer [69]. Figure 1 shows the schematic of general fuel cells mechanism.

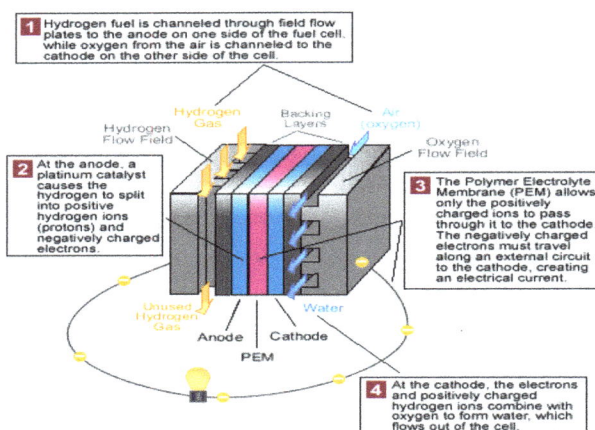

1. Hydrogen fuel is channeled through field flow plates to the anode on one side of the fuel cell, while oxygen from the air is channeled to the cathode on the other side of the cell.

2. At the anode, a platinum catalyst causes the hydrogen to split into positive hydrogen ions (protons) and negatively charged electrons.

3. The Polymer Electrolyte Membrane (PEM) allows only the positively charged ions to pass through it to the cathode. The negatively charged electrons must travel along an external circuit to the cathode, creating an electrical current.

4. At the cathode, the electrons and positively charged hydrogen ions combine with oxygen to form water, which flows out of the cell.

Figure 1 The general mechanism of fuel cells [70]. Permission is granted by ACS Publications.

4.1.2 Fuel cell membrane

The membrane is a thin barrier which serves as a dam because of its selective permeability where the chemical composition determines its property. In fact, membrane in fuel cell technology can be categorized into several types based on its separation processes include microfiltration (MF), ultrafiltration (UF), nanofiltration (NF), and reverse osmosis (RO). MF is a large pores membrane which used to separate solid particles (around 0.2 to 10 μm) and suspensions with high outflow (approximately 0.1 to 5 bar). Typically, MF is used in biological and food industries to separate bacteria and suspended particles. While, UF is usually used in water-oil emulsion isolation, dye bath purification via electrophoresis and, water purification and disinfection. Normally, low molecular weight of solute can easily pass through UF as it is used to condense emulsion or macromolecules. Whereas, NF can be used to separate particles ranging 2 to 10 μm which usually applied in water treatment, and food and milk industries (isolation of minerals). On the other hand, RO has a retention rate higher than 95% because the diameter of the pores is very dense and less than nanometer in size. Due to that, the flow rate is low while the pressure used is high around 60 to 80 bar. RO is basically used in salt/sea water desalination, purification of used water, and ulta-puring water for pharmaceutical and electronic intended [71]. Figure 2 shows the classification of the membrane based on its different separation processes.

Advanced Functional Membranes

Materials Research Forum LLC

Materials Research Foundations **120** (2022) 72-110

https://doi.org/10.21741/9781644901816-3

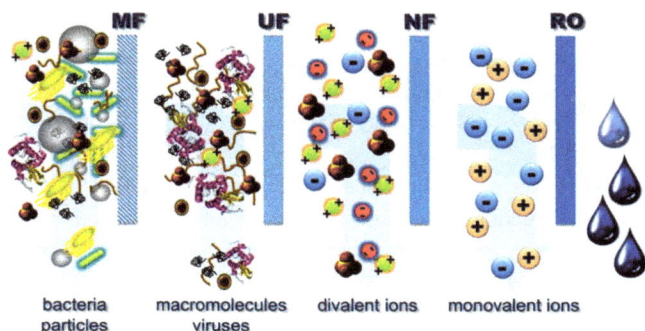

Figure. 2 Classification of membrane by different separation processes [72]. Permission is granted from Elsevier.

On the other hand, several polymers are used in manufacturing membrane for fuel cell technology include cellulose acetate, polysulfone, polyethersulfone, and polyacrylonitrile. Cellulose is a natural, linear and inflexible polymer which gives good mechanical resistance to the membrane. Cellulose acetate is produced via the acetylation process using acetic anhydride, in which increasing the degree of acetylation, basically, increasing its amount of retention while decreasing the permeability. Other than that, the production of cellulose acetate via acetylation is also promising in increasing the mechanical resistance and resistance to hydrolysis of the membrane. Normally, this membrane is asymmetrical and has a retention rate around 99.5%, and range of pressure between 103 to 140 bar. The advantages of cellulose acetate membrane include ease of preparation, relatively high permeability and excellent mechanical property, while the disadvantages are sensitive to pH changes and reduce the salt retention when increasing the temperature [73].

Other polymers used in fuel cell membrane are polysulfone and polyethersulfone. These polymers are basically used for MF or UF type of membrane because of their homogenous property, which gives chemical, thermal and mechanical resistance. The benefits of this membrane include resistance to heat (up to 125°C), chlorine, and acid or alkali washings, as well as ease of production in many different geometries [74]. Whereas, polyacrylonitrile has a high crystallinity, glass transition of 87°C, and melting point at 319°C, which basically suitable for many applications such as shipping, construction, gas filtration system, and UF membrane. In fact, the nitrile groups in acrylonitrile units are extremely polar, as a result, they are strongly attracted and causing the polymer chains to approach one another which in turn, increases the resulting polymer strength. The advantages of polyacrylonitrile membrane are high thermal, chlorine and solvent resistance, high hydrophobicity, as well as ease of manufacture and modification [75].

4.1.3 Classifications

Currently, there are few types of functional polymer membrane fuel cells in market such as polymer electrolyte, alkaline, molten carbonate, direct methanol, phosphoric acid, and solid oxide.

Polymer electrolyte membrane (PEM) fuel cells are having excellent power density, lighter and smaller compared to other fuel cells. In PEM, solid polymer is used as an electrolyte, while carbon electrodes with a platinum catalyst are assembled in PEM fuel cells. PEM fuel cells are most commonly employed and well suited in transportation and some stationary applications [76]. However, the usage of a platinum catalyst to separate the electrons and protons of hydrogen increases the expense of the system. Furthermore, if the hydrogen is obtained from a hydrocarbon fuel, the catalyst is also particularly sensitive to carbon monoxide which lead to poisoning, necessitating the use of an extra reactor in order to remove the carbon monoxide [77].

The other type is direct methanol (DM) fuel cells. Methanol is commonly combined with water and delivered directly to the anode in DM fuel cells. In fact, DM fuel cells avoid a lot of fuel storage issues compared to other fuel cell systems because methanol has a higher energy density compared to hydrogen. The liquid property of methanol made it easy to transport and deliver to the consumers which utilizing recent infrastructure. DM fuel cells are frequently installed to power portable devices such as computers and cellular phones [78].

Other than that, alkaline fuel cells are one of the earliest technologies invented to produce electrical energy and water in spacecraft industry. The electrolyte in alkaline fuel cells is potassium hydroxide in solvent of water, and the anode and cathode are be made of various non-precious metals. In fact, these fuel cells are similar as PEM fuel cells, however, instead of an acid, they use an alkaline as a membrane. Furthermore, the great efficiency of alkaline fuel cells is attributed to the fast rate of electrochemical reactions in the cell where they can achieve more than 60% in space applications [79]. However, higher temperature operation, carbon dioxide (CO_2) poisoning, membrane durability and conductivity, power density, as well as anode electrocatalysis are challenges for alkaline fuel cells [80].

Another fuel cell type is phosphoric acid (PA) fuel cells, where the electrolyte is liquid phosphoric acid, which is located in Teflon-bonded carbide matrix. Whereas, the electrodes are carbon electrodes with a platinum catalyst. Although PA fuel cells are primarily utilized for stationary power generation, they have also been applied to power big vehicles [81]. Given the same weight and volume, PA fuel cells are less potent as compared to other type of fuel cells, therefore, they are often bulky and heavy. Furthermore, PA fuel cells are also expensive which require costly platinum catalyst loadings compared to other types of fuel cells [82].

On the other hand, for electrical utility, industrial, and military uses, molten carbonate (MC) fuel cells are recently produced for coal power plant and natural gas. MC fuel cells operate at high temperature which use a molten carbonate salt combination which sustained in a porous and inert ceramic lithium aluminium oxide matrix (electrolyte). The use of non-precious metals as catalysts can be incorporated at cathode and anode since MC fuel cells work at high temperature about 650°C, which reduce the expenses [83]. However, the main drawback of MC fuel cells technology is its lack of endurance. The high temperature employed in these cells, along with corrosive electrolyte, hasten the component corrosion which leads to breakdown and resulting in reducing cell life. Therefore, scientists are actively searching for corrosion-resistant components as well as designs that can prolong cell life without sacrificing its performance [84].

Other than that, there are solid oxide (SO) fuel cells type, where the electrolyte is hard and non-porous ceramic composition. Basically, SO fuel cells operate at extreme temperature up to 1000°C. The usage of a high-temperature operation eliminates the requirement of a precious-metal catalyst, which lowering costs. It also allows SO fuel cells to reform fuels internally, permitting them to use a wider range of fuels while reducing the expense of adding a reformer into the system [85]. However, there are drawbacks to operate at high temperature where it causes a slow starter and necessitates thermal shielding to keep the heat in and protect the people, which is fine for utility applications but not for transportation. Furthermore, material durability is also a concern due to high working temperature. Therefore, difficulty of this technology is the fabrication of low-cost materials with excellent durability at cell operating temperature. Currently, scientists are actively looking into the possibility of building lower-temperature SO fuel cells that operate under 700°C, less expensive, and fewer durability issues [86].

Lastly, reversible fuel cells, like conventional fuel cells, generate electricity from oxygen and hydrogen while also releasing water and heat byproducts. Reversible fuel cell systems can divide water into oxygen and hydrogen fuel using electricity from solar, wind, or other sources which known as electrolysis. Apart from producing electricity when required, reversible fuel cells can also store extra energy in the form of hydrogen during periods of high-power generation from other technologies, for instance, vigorous winds result in excess available wind power. This energy storage potential is crucial for renewable energy systems that are intermittent [87].

4.2   Membranes for lithium-ion battery

Lithium-ion (Li-ion) battery is a rechargeable battery type. Li-ion batteries are commonly utilized in portable electric and electronic vehicles, medical devices, and their use in military and aerospace applications is on the rise. Basically, Li-ions travel from negative to positive electrode via an electrolyte during discharge, and vice versa while charging. The positive electrode of Li-ion battery is constructed from an intercalated lithium compound which allows for a higher energy density and specific capacity, while the negative electrode is commonly made of graphite. Other advantages of Li-ion battery include prolong life cycle, high efficiency and comparatively low self-discharge rate [88]. Figure 3 shows the schematic of a Li-ion battery which consists of graphitic carbon negative electrode and Li-intercalated compound positive electrode.

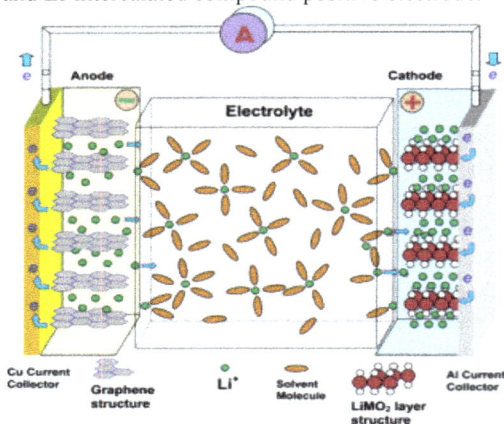

Figure. 3 Schematic of Li-ion battery mechanism [89]. Permission is granted from ACS Publications.

However, there are problems encountered with Li-ion battery include operational temperature, materials availability and cost. Li-ion battery also arises a safety issue because it contains flammable electrolytes that, if damage or wrongly charged, can cause explosion and fire. Therefore, the separator membrane is an important part of Li-ion battery because it acts as physical barrier between negative and positive electrodes, preventing electrical short connections as well as serves as an electrolyte reservoir for ions transport. The membrane, basically, is placed between the two electrodes and must have a strong ionic conductivity and excellent thermal and mechanical stability. Other than that, this membrane has to be porous in order to permit the electrolyte to flow through [90].

4.2.1 Lithium-ion battery membrane

There are several types of membrane separators used in Li-ion battery include composite, microporous, electrospun, nonwoven, polymer blends, and membranes with external surface modification. Whereas, types of membrane used in Li-ion battery based on different types of electrolytes are polymer gel, microporous filter, and polymer solid ion conductor.

Solid polymer such as polyethylene oxide (PEO) is dry, solid and contains no organic liquid. The transportation of Li-ions between the sites is basically aided by segmental movement of the PEO matrix for the mechanism of Li-ions transport. Based on this concept, the good conductivity of PEO can attribute to Li-ions transport at the amorphous region. Thus, there are two approaches to enhance Li-ions conductivity, PEO crystallization is reduced, and the interaction between Li-ions and PEO chains is weakened [91].

On the other hand, semi-crystalline polyolefin polymers such as polyethylene and polypropylene are used for microporous membrane. These polymers are having fuse function, where they can be flowable if the temperature achieves their melting point, which decreases the ionic conductivity and shut the electrochemical reactions of the battery to protect from thermal runaway. It is very crucial for cell over-charge and external short-circuit protection. Moreover, some functional groups can be attached via surface modification on both polymers to increase their compatibility of membrane-electrodes interfaces [92].

Other than that, liquid electrolyte is incorporated in the polymer matrix to generate the polymer gel. In fact, polymer gel has more shape flexibility compared to liquid and polymer chains usually hold liquid electrolyte in the polymer gel. The copolymer of polyvinylidene fluoride-hexafluoropropylene (PVDF-HFP) introduces amorphous regions in the polymer gel that improve ionic conductivity, while the crystalline domains increase the mechanical strength. Other than that, poly(methyl methacrylate) (PMMA) polymer gel has good compatibility and normally used as a layer to reduce the evaporation and leakage of the liquid electrolyte. This strong affinity, as a result, increases electrolyte retention and ions transport which improves cell performance. On the other hand, poly(acrylonitrile) (PAN) polymer gel provides high conductivity and forms stable dimension, as well as increases cycling performance because of the low diffusion resistance [93].

4.2.2 Classification

In fact, various types of Li-ion battery in terms of shape, size and composition are utilized based on their specific applications.

Firstly, lithium cobalt oxide (LCO) battery which made of cobalt and lithium carbonate. LCO battery also known as Li-ion cobalt or lithium cobaltate usually utilized in laptops, electronic cameras and cell phones because of its high specific energy. LCO consists of cobalt oxide cathode and graphite carbon anode, where the lithium ions migrate from anode to cathode during discharge, while the flow is reversed during charging. However, LCO has a number of disadvantages, including a short battery life and low specific power. Furthermore, according to Battery University, LCO is not as safe as other types of Li-ion battery. Despite this, LCO is still a popular choice for portable electronic devices because of its high specific energy property [94].

Secondly, lithium manganese oxide (LMO) battery which is also known as lithium manganate or Li-ion manganese, as well as Li-manganese or manganese spinel. Historically, LMO technology battery was found in early 1980s with the first publication reported in the Materials Research Bulletin. In 1996, Moli Energy manufactured the first commercial Li-ion battery utilizing LMO as the cathode material. LMO battery, in fact, has a higher temperature stability and safer compared to other Li-ion battery types. Therefore, LMO is frequently found in medical equipment and devices, and can also be used in electric vehicles, power tools, and other applications [95].

Other than that, phosphate is basically used as a cathode in lithium iron phosphate (LPO) battery, which often known as Li-phosphate battery. LPO has low resistance qualities, which significantly improve its thermal stability and safety. Furthermore, durability and a long lifecycle are other advantages of LPO where fully charged LPO can be kept with no effect on the charge's overall lifespan. In addition, considering the lengthy battery life of LPO, it is frequently the most cost-effective option. As a result, LPO is commonly employed in electric vehicles and other applications requiring a long lifecycle and high levels of safety [96].

The other type of Li-ion battery is lithium nickel manganese cobalt oxide (LNMCO) which also known as NMC battery. Basically, LNMCO is constructed of a variety of materials that are common in Li-ion battery including a mix of nickel, manganese, and cobalt to build its cathode. LNMCO battery, like other Li-ion battery types, can have a high specific energy density or power other than very low self-heating rate which most commonly used in power tools and automobile powertrains. 60% nickel, 20% manganese, and 20% cobalt are commonly the combination ratio for the cathode. This means that the cost of the raw material is lower than other Li-ion battery solutions, as cobalt may be fairly costly. In the future, the price of LNMCO battery may drop even more because some battery manufacturers plan to convert their battery chemistry to a larger percentage of nickel, in order to utilize less cobalt [94].

Lithium nickel cobalt aluminium oxide (LNCAO) is another type of Li-ion battery, which also called as NCA battery. Nowadays, this type of battery is becoming significantly increasing in grid storage and electric powertrains. To ensure driver safety, LNCAO battery must be accompanied by monitoring systems. Although LNCAO battery is not widely used in consumer electronics, it has a great potential in the automobile industry because it offers a high-energy choice with a long lifespan, however it is rather expensive and less safe than other Li-ion battery types. As a result, given the widespread use of LNCAO battery in electric vehicles, demand for this battery may increase as the number of electric vehicles grows [94].

Lastly, lithium titanate (LT) battery which familiar with Li-titanate is a type of battery that has a growing number of uses. According to Battery Space, the key benefit of LT battery is its very fast recharge time, which is due to sophisticated nanotechnology. LT battery is being employed by manufacturers of electric vehicles, and this type of battery has the potential to be used in electric buses for public transit. However, the disadvantage of LT battery is it has a lower intrinsic voltage or energy density than other Li-ion battery types, which might cause problems when it comes to efficiently powering automobiles. Despite this, LT battery still has a higher density compared to non-Li-ion battery, which is a benefit. Other than that, LT battery also could be used for military and aerospace applications, as well as storing wind and solar energy and constructing smart grids. Furthermore, according to Battery Space, LT battery could be employed in power system for system-critical backups [97].

In short, Li-ion battery exists in a variety number of shapes and sizes and can be used for various applications. As a result, some Li-ion battery in the market currently is more suited to specific applications than others. The most important point is to select the battery that is most appropriate for the task required. Thus, it is crucial to determine the best polymer used to construct a membrane to be incorporated in the Li-ion battery. It is also noteworthy that the Li-ion battery market is always evolving where companies and scientists all over the world are developing novel batteries to supply or replace Li-ion battery. It will be interesting to perceive which of this new battery rise to the top as time goes on.

4.3   Electrically conductive membranes

Advancements in the area of materials science and membrane preparation processes significantly leading to novel possibilities for mitigating the issues of conventional membrane filtration by the usage of electrically conductive membrane (ECM). Over traditional technologies, there are numerous advantages of ECM such as high efficiency, easy to operate, low chemical consumption, and smaller footprint. Other than that, ECM also benefits in terms of the separation effect by the membrane and the electrical effect via conducting elements. Recently, various types of conducting elements had been used for ECM such as carbon nanotube, conducting polymers, and graphene. However, membrane technology still has some drawbacks, such as fouling, selectivity in pressure-driven membranes for uncharged contaminants as well as energy consumption [98].

Basically, ECM is studied in terms of synthesis/fabrication, mechanism of action, and application. Mode of ECM's operations are usually divided into three major types, thermal-driven, osmotically-driven, and pressure-driven. Currently, several types of ECM were synthesized with promising properties, cost-effective, and stability in conducting electricity. For instance, stacked graphene oxide, carbon nanotube, Cu-nanowires conductive, polyurethane, polypyrrole, poly(ether sulfone), poly(vinylidene fluoride) (PVDF), polyaniline-poly(2-acrylamido-2-methyl-1-propanesulfonic acid) (PANI-PAMPSA), porous conductive diamond, and membrane in dilute electrolyte solutions [98].

Currently, numerous polymers are used for ECM such as polyacetylene (PAc), polyaniline (PANI), polypyrrole (PPy), polyetheretherketone (PEEK), polythiophene (PTh), poly(p-phenylene) (PPP), polyazulene (PAZ), polyfuran (PFu), polyisoprene (PIP) and polybutadiene (PBD) in various applications include biosensors (food, medicine, environment), gas sensors, electrochemical sensors, corrosion inhibitors and functionalized biomaterials [99-101].

4.4   Thin film for optoelectronic

Optoelectronic is a wide application for electronic devices to detect and control light includes visible light and invisible radiation. Lighting, sensor, and energy harvesting devices made of thin-film optoelectronics are commonly used in medical devices, semiconductor lasers, consumer electronics and aerospace. Therefore, the study of important classes of materials such as colloidal quantum dots and perovskites is being done to improve device's knowledge and performance as industrial, military and consumer demands are constantly expanding [102].

Enhanced properties such as hardness, transmission, reflectivity and damage threshold of optoelectronic devices require thin film coating because of the sensitivity of these core devices. This thin film coating also can protect the optoelectronic devices from external factors such as heat, dust and abrasion. kFor instance, conjugated and crystalline semiconductor thin film of poly(o-phenylene diamine-co-m-phenylene diamine) copolymer [PoPmP]TF is used for polymer solar cell. Other than that, the properties of zinc oxide (ZnO) which are high chemical stability, high binding energy, wide band gap, and transparency make its thin film and nanostructures promising to be applied for sensor, thin film transistor, and dye-sensitized solar cell industries. While, indium monoselenide (InSe) is considered as an emerging two-dimensional semiconductor with excellent optical and electrical properties that potential for high performance optoelectronics application [103].

In optoelectronic, the whole device can be twisted, bent, or stretched to achieve the requirements for numerous applications, in which not applicable via conventional devices that can twist or bend only. Moreover, rapid development in optoelectronic devices has been benefited in the advancements of science and engineering of photon detection and manipulation, image sensing, high efficiency and power density light emission, displays, communications and renewable energy harvesting. Particularly, promising material class for optoelectronics is colloidal nanomaterials because of their functionality, cost effectiveness, low dimensionality, and solution processability which reduces the time and cost required to fabricate thin film devices. As a result, it provides broad compatibility with existing materials interfaces and device structures [104].

4.4.1 Classifications

Currently, optoelectronics can be classified in several types include photodiode (PD), solar cell (SC), light emitting diode (LED), optical fiber (OF), and laser diode (LD). PD is a light sensor which basically made up of an active P-N junction that is biased in the opposite direction. It works in three different modes, forward bias (LED), reverse bias (photo detector) and photovoltaic (SC) by producing current or voltage when light falls on the junction. PD usually used in safety, medical and industrial equipment as well as cameras. While, SC is a device that transforms sunlight directly into electricity. When photons from the sun collide with SC, it produces a current and voltage, which generates electricity. Typically, SC is applied for rural electrifications, communications and nautical navigations [105].

On the other hand, LED is a semiconductor diode which usually used in household appliances, computer components, medical equipment and instrument panels. Electroluminescence (EL) is the effect of the photon produced via the recombination of electrons and holes in LED. The advantages of LED include uses less energy, emits less heat and long-lasting compared to incandescent lamp [106]. Whereas, the reasons of OF is preferable compared to electrical cable are mainly due to very low degradation, higher bandwidth, not vulnerable to EMI/EMC interference and can deliver different wavelengths of signals without interfering one another. OF is commonly used in sensors, biomedicals, lasers and telecommunications to transmit information via modulated light [107]. While, LD resembles LED in appearance and operation, which basically converts electrical energy into light, that typically found in telecommunications, surgical and military applications.

The advantages of LD include more efficient and reliable compared to conventional LED, and reduces signal losses due to higher coherence light and single optical wavelength [108].

4.5   Challenges and future direction of functional polymer membranes for emerging technologies

Several challenges encountered for functional polymer membranes in emerging technologies. In fuel cell, the membrane must enable only the suitable ions to move between anode and cathode, thus, it is a vital component in fuel cell technology, because, if additional chemicals went through the electrolyte of the fuel cell, the chemical process would be disrupted. While membrane in Li-ion battery acts as a barrier between positive and negative electrodes, preventing electrical short connections as well as serves as an electrolyte reservoir for ions transport. Therefore, the membrane must have strong ionic conductivity with excellent thermal and mechanical stability. Whereas, in the electrically conductive membrane, challenges such as fouling, selectivity in pressure-driven membranes for uncharged contaminants, as well as energy consumption are quite significant. On the other hand, thin film coating can be used to protect the devices from external factors such as heat, dust and abrasion for optoelectronic membrane application. Tremendous studies have been performed in developing and innovating functional polymer membranes for emerging technologies to be extended to large-scale production and industrial applications to improve the environment and quality of life. Apart from that, it is also noteworthy that the market is always evolving over the world as researchers are developing novel membrane for the emerging technologies which will be fascinating to observe which of the new membrane technology escalate to the top as the time pass by. Functional polymeric membranes have shown very significant evolution in many applications including these new emerging technologies applications.

Conclusion

The effectiveness of the membrane selectivity, permeability and separation process is determined by the polymer properties. Regardless, the functional polymer selection is crucial in the development of membrane technology for the water/gas separation, medical, and emerging technologies (fuel cells, Li-ion batteries, electric conductive, optoelectronic) owing to its manufacturability, cost, mechanism, type, and resources. Other factors such as challenges and future direction also must be considered in determining the best polymers to be used for the development of the membrane. The objective of the specific tasks requires a critical consideration in determining the best membrane technology suits the applications where it is performed, therefore, the advancement of the membrane used can be benefited thoroughly. It is also remarkable that the membrane technology is always evolving especially in the functional polymer membrane development in various applications, so that the quality of life is improved.

References

[1] T.A. Saleh, V.K. Gupta, An Overview of Membrane Science and Technology, Nanomater. Polym. Membr. 1 (2016) 1-23. https://doi.org/10.1016/B978-0-12-804703-3.00001-2

[2] V. Abetz, T. Brinkmann, M. Sözbilir, Fabrication and function of polymer membranes, Chem. Teach. Int. 3 (2021) 141-154. https://doi.org/10.1515/cti-2020-0023

[3] E.O. Ezugbe, S. Rathilal, Membrane Technologies in Wastewater Treatment: A Review, Membr. 10 (2020) 89. https://doi.org/10.3390/membranes10050089

[4] Z. Jiang, L. Chu, X. Wu, Z. Wang, X. Jiang, X. Ju, X. Ruan, G. He, Membrane-based separation technologies: from polymeric materials to novel process: an outlook from China, Rev. Chem. Eng. 36 (2020) 67-105. https://doi.org/10.1515/revce-2017-0066

[5] M. Ulbricht, Advanced functional polymer membranes, Polym. 47(7) (2006) 2217-2262. https://doi.org/10.1016/j.polymer.2006.01.084

[6] J.T.E. Goh, A.R. Abdul Rahim, M.S. Masdar, L.K. Shyuan, Enhanced Performance of Polymer Electrolyte Membranes via Modification with Ionic Liquids for Fuel Cell Applications, Membr. 11 (2021) 395. https://doi.org/10.3390/membranes11060395

[7] K. Buruga, H. Song, J. Shang, N. Bolan, T.K. Jagannathan, K.H. Kim, A review on functional polymer-clay based nanocomposite membranes for treatment of water, J. Hazard. Mater. 379 (2019) 120584. https://doi.org/10.1016/j.jhazmat.2019.04.067

[8] M.B. Hägg, Gas Permeation: Permeability, Permeance, and Separation Factor, Encycl. Membr. (2016) 1-4. https://doi.org/10.1007/978-3-642-40872-4_2215-1

[9] A. Lee, J.W. Elam, S.B. Darling, Membrane materials for water purification: design, development, and application, Environ. Sci.: Water Res. Technol. 2(1) (2016) 17-42. https://doi.org/10.1039/C5EW00159E

[10] M. Galizia, W.S. Chi, Z.P. Smith, T.C. Merkel, R.W. Baker, B.D. Freeman, Polymers and Mixed Matrix Membranes for Gas and Vapor Separation: A Review and Prospective Opportunities, Macromol. 50 (2017) 7809-7843. https://doi.org/10.1021/acs.macromol.7b01718

[11] P. Wang, T.S. Chung, Recent advances in membrane distillation processes: Membrane development, configuration design and application exploring, J. Membr. Sci. 474 (2015) 39-56. https://doi.org/10.1016/j.memsci.2014.09.016

[12] D.F. Sanders, Z.P. Smith, R. Guo, L.M. Robeson, J.E. McGrath, D.R. Paul, B.D. Freeman, Energy-efficient polymeric gas separation membranes for a sustainable future: A review, Polym. 54 (2013) 4729-4761. https://doi.org/10.1016/j.polymer.2013.05.075

[13] P. Hao, G.G. Lipscomb, The Effect of Sweep Uniformity on Gas Dehydration Module Performance, Membr. Gas Sep. (2010) 333-353. https://doi.org/10.1002/9780470665626.ch16

[14] V. Abetz, T. Brinkmann, M. Dijkstra, K. Ebert, D. Fritsch, K. Ohlrogge, D. Paul, K.V. Peinemann, S. Pereira-Nunes, N. Scharnagl, Developments in membrane research: from material via process design to industrial application, Adv. Eng. Mater. 8 (2006) 328-358. https://doi.org/10.1002/adem.200600032

[15] F.Y.C. Huang, A. Arning, Performance Comparison between Polyvinylidene Fluoride and Polytetrafluoroethylene Hollow Fiber Membranes for Direct Contact Membrane Distillation, Membr. 9 (2019) 52. https://doi.org/10.3390/membranes9040052

[16] A. Mollahosseini, A. Abdelrasoul, A. Shoker, A critical review of recent advances in hemodialysis membranes hemocompatibility and guidelines for future development, Mater. Chem. Phys. 248 (2020) 122911. https://doi.org/10.1016/j.matchemphys.2020.122911

[17] B. Krause, M. Storr, T. Ertl, R. Buck, H. Hildwein, R. Deppisch, H. Gohl, Polymeric membranes for medical applications, Chem. Ing. Tech. 75 (2003) 1725-1732. https://doi.org/10.1002/cite.200306149

[18] J.S. Eswari, S. Naik, A critical analysis on various technologies and functionalized materials for manufacturing dialysis membranes, Mater. Sci. Energy Technol. 3 (2020) 116-126. https://doi.org/10.1016/j.mset.2019.10.011

[19] V. Sirolli, S. Di Stante, S. Stuard, L. Di Liberato, L. Amoroso, P. Cappelli, M. Bonomini, Biocompatibility and functional performance of a polyethylene glycol acid-grafted cellulosic membrane for hemodialysis, Int. J. Artif. Org. 23 (2000) 356-364. https://doi.org/10.1177/039139880002300603

[20] N.J. Kaleekkal, A. Thanigaivelan, M. Tarun, D. Mohan, A functional PES membrane for hemodialysis - Preparation, Characterization and Biocompatibility, Chin. J. Chem. Eng. 23 (2015) 1236–1244. https://doi.org/10.1016/j.cjche.2015.04.009

[21] M.N. Zainol Abidin, P.S. Goh, A.F. Ismail, M.H.D. Othman, H. Hasbullah, N. Said, S.H. Abdul Kadir, F. Kamal, M.S. Abdullah, B.C. Ng, Antifouling polyethersulfone hemodialysis membranes incorporated with poly (citric acid) polymerized multi-walled carbon nanotubes, Mater. Sci. Eng. C 68 (2016) 540–550. https://doi.org/10.1016/j.msec.2016.06.039

Advanced Functional Membranes   Materials Research Forum LLC
Materials Research Foundations **120** (2022) 72-110   https://doi.org/10.21741/9781644901816-3

[22] A.M.D. Santos, A.C. Habert, H.C. Ferraz, Development of functionalized polyetherimide/ polyvinylpyrrolidone membranes for application in hemodialysis, J. Mater. Sci.: Mater. Med. 28 (2017) 131. https://doi.org/10.1007/s10856-017-5946-z

[23] D. Kaigler, G. Avila, L. Wisner-Lynch, M.L. Nevins, M. Nevins, G. Rasperini, S.E. Lynch, W.V. Giannobile, Platelet-derived growth factor applications in periodontal and peri-implant bone regeneration, Expert Opin. Biol. Ther. 11 (2011) 375-385. https://doi.org/10.1517/14712598.2011.554814

[24] L. Saigoa, V. Kumar, Y. Liu, J. Lim, S.H. Teoh, B.T. Goh, A pilot study: Clinical efficacy of novel polycaprolactone-tricalciumphosphate membrane for guided bone regeneration in rabbit calvarial defect model, J. Oral Maxillofac. Surg. Med. Pathol. 30 (2018) 212–219. https://doi.org/10.1016/j.ajoms.2017.12.007

[25] P. Gentile, V. Chiono, C. Tonda-Turo, A.M. Ferreira, G. Ciardelli, Polymeric membranes for guided bone regeneration, Biotechnol. J. 6 (2011) 1187–1197. https://doi.org/10.1002/biot.201100294

[26] S. Liao, W. Wang, M. Uo, S. Ohkawa, T. Akasaka, K. Tamura, F. Cui, F. Watari, A three-layered nano-carbonated hydroxyapatite/collagen/PLGA composite membrane for guided tissue regeneration, Biomater. 26 (2005) 7564–7571. https://doi.org/10.1016/j.biomaterials.2005.05.050

[27] J.M. Carbonell, I.S. Martı́n, A. Santos, A. Pujol, J.D. Sanz-Moliner, J. Nart, High-density polytetrafluoroethylene membranes in guided bone and tissue regeneration procedures: a literature review, Int. J. Oral Maxillofac. Surg. 43 (2014) 75–84. https://doi.org/10.1016/j.ijom.2013.05.017

[28] K. Fujihara, M. Kotaki, S. Ramakrishna, Guided bone regeneration membrane made of polycaprolactone/calcium carbonate composite nano-fibers, Biomater. 26 (2005) 4139–4147. https://doi.org/10.1016/j.biomaterials.2004.09.014

[29] L. Fu, Z. Wang, S. Dong, Y. Cai, Y. Ni, T. Zhang, L. Wang, Y. Zhou, Bilayer Poly(Lactic-co-glycolic acid)/Nano-Hydroxyapatite Membrane with Barrier Function and Osteogenesis Promotion for Guided Bone Regeneration, Mater. 10 (2017) 257. https://doi.org/10.3390/ma10030257

[30] Z.S. Haidar, Bio-inspired-functional colloidal core-shell polymeric-based nanosystems: Technology promise in tissue engineering, bioimaging and nanomedicine, Polym. 2 (2010) 323–352. https://doi.org/10.3390/polym2030323

[31] K. Zhang, M. Zhao, L. Cai, Z.K. Wang, Y.F. Sun, Q.L. Hu, Preparation of chitosan/ hydroxyapatite guided membrane used for periodontal tissue regeneration, Chin. J. Polym. Sci. 28 (2010) 555–561. https://doi.org/10.1007/s10118-010-9087-9

[32] V.C. Dumont, A.A. Mansur, S.M. Carvalho, F.G.M. Borsagli, M.M. Pereira, H.S. Mansur, Chitosan and carboxymethyl-chitosan capping ligands: effects on the nucleation and growth of hydroxyapatite nanoparticles for producing biocomposite membranes, Mater. Sci. Eng. C Mater. Biol. Appl. 59 (2016) 265–277. https://doi.org/10.1016/j.msec.2015.10.018

[33] A.M. Dobos, M.D. Onofrei, S. Ioan, Cellulose acetate nanocomposites with antimicrobial properties, in: V.K. Thakur, M.K. Thakur (Eds.), Eco-friendly Polymer Nanocomposites, India, 2015. pp. 367–398. https://doi.org/10.1007/978-81-322-2470-9_12

[34] S. Torgbo, P. Sukyai, Bacterial cellulose-based scaffold materials for bone tissue engineering, Appl. Mater. Today 11 (2018) 34–49. https://doi.org/10.1016/j.apmt.2018.01.004

[35] H. Shin, S. Jo, A.G. Mikos, Biomimetic materials for tissue engineering, Biomater. 24 (2003) 4353–4364. https://doi.org/10.1016/S0142-9612(03)00339-9

[36] D.S. Thoma, U.W. Jung, J.Y. Park, S.P. Bienz, J. Hüsler, R.E. Jung, Bone augmentation at peri-implant dehiscence defects comparing a synthetic polyethylene glycol hydrogel matrix vs. standard guided bone regeneration techniques, Clin. Oral Implants Res. 28 (2017) 76–83. https://doi.org/10.1111/clr.12877

[37] G.I. Benic, D.S. Thoma, F. Muñoz, I.S. Martin, R.E. Jung, C.H. Hämmerle, Guided bone regeneration of peri-implant defects with particulated and block xenogenic bone substitutes, Clin. Oral Implants Res. 27 (2016) 567–576. https://doi.org/10.1111/clr.12625

[38] Z. Sheikh, S. Hasanpour, M. Glogauer, Bone Replacement Materials and Techniques Used for Achieving Vertical Alveolar Bone Augmentation, Mater. 8 (2015) 2953-2993. https://doi.org/10.3390/ma8062953

[39] S. Tamburaci, F. Tihminlioglu, Novel poss reinforced chitosan composite membranes for guided bone tissue regeneration, J. Mater. Sci.: Mater. Med. 29 (2018) 1-14. https://doi.org/10.1007/s10856-017-6005-5

[40] C.A. Dascalu, A. Maidaniuc, A.M. Pandele, S.I. Voicu, T. Machedon-Pisu, G.E. Stan, A. Cîmpean, V. Mitran, I.V. Antoniac, F. Miculescu, Synthesis and characterization of biocompatible polymer-ceramic film structures as favorable interface in guided bone regeneration, Appl. Surf. Sci. 494 (2019) 335-352. https://doi.org/10.1016/j.apsusc.2019.07.098

[41] H.W. Toh, D.W.Y. Toong, J.C.K. Ng, V. Ow, S. Lu, L.P. Tan, P.E.H. Wong, S. Venkatraman, Y. Huang, H.Y. Ang, Polymer blends and polymer composites for cardiovascular implants, Eur. Polym. J. 146 (2021) 1102. https://doi.org/10.1016/j.eurpolymj.2020.110249

[42] M. Castilho, D. Feyen, M. Flandes-Iparraguirre, G. Hochleitner, J. Groll, P.A.F. Doevendans, T. Vermonden, K. Ito, J.P.G. Sluijter, J. Malda, Melt Electrospinning Writing of Poly-Hydroxymethylglycolideco-ε-Caprolactone-Based Scaffolds for Cardiac Tissue Engineering, Adv. Healthc. Mater. 6 (2017) 1700311. https://doi.org/10.1002/adhm.201700311

[43] J. Nogic, L.M. McCormick, R. Francis, N. Nerlekar, C. Jaworski, N.E.J. West, A.J. Brown, Novel bioabsorbable polymer and polymer-free metallic drug-eluting stents, J. Cardiol. 71 (2018) 435–443. https://doi.org/10.1016/j.jjcc.2017.12.007

[44] V.S. Sivasankarapillai, S.S. Das, F. Sabir, M.A. Sundaramahalingam, J.C. Colmenares, S. Prasannakumar, M. Rajan, A. Rahdar, G.Z. Kyzas, Progress in natural polymer engineered biomaterials for transdermal drug delivery systems, Mater. Today Chem. 19 (2021) 100382. https://doi.org/10.1016/j.mtchem.2020.100382

[45] G. Jeon, S.Y. Yang, J.K. Kim, Functional nanoporous membranes for drug delivery, J. Mater. Chem. 22 (2012) 14814. https://doi.org/10.1039/c2jm32430j

[46] R. Grilloa, F.V. Dias, S.M. Querobino, C. Alberto-Silva, L.F. Fracetoc, E. de Paula, D.R. de Araujo, Influence of hybrid polymeric nanoparticle/thermosensitive hydrogel systems on formulation tracking and in vitro artificial membrane permeation: A promising system for skin drug-delivery, Coll. Surf. B: Biointerfaces 174 (2019) 56–62. https://doi.org/10.1016/j.colsurfb.2018.10.063

[47] J. Wang, L. Wang, Z. Zhou, L. Lai, P. Xu, L. Liao, J. Wei, Biodegradable Polymer Membranes Applied in Guided Bone/Tissue Regeneration: A Review, Polym. 8 (2016) 115. https://doi.org/10.3390/polym8040115

[48] K.T. Shalumon, G.J. Lai, C.H. Chen, J.P. Chen, Modulation of Bone-Specific Tissue Regeneration by Incorporating Bone Morphogenetic Protein and Controlling the Shell Thickness of Silk Fibroin/Chitosan/Nanohydroxyapatite Core−Shell Nanofibrous Membranes, ACS Appl. Mater. Interfaces 7 (2015) 21170−21181. https://doi.org/10.1021/acsami.5b04962

[49] S.N. Jayasinghe, Bio-electrosprays and cell electrospinning: Rapidly emerging physical protocols for potential life science applications, Biotechnol. J. 6 (2007) 43–51. https://doi.org/10.12665/J63.Jayasinghe

[50] S.N. Jayasinghe, S. Irvine, J.R. McEwan, Cell electrospinning highly concentrated cellular suspensions containing primary living organisms into cell-bearing threads and scaffolds, Nanomed. 2 (2007) 555–567. https://doi.org/10.2217/17435889.2.4.555

[51] P. Joly, N. Chavda, A. Eddaoudi, S.N. Jayasinghe, Bio-electrospraying and aerodynamically assisted bio-jetting whole human blood: interrogating cell surface marker integrity, Biomicrofluid. 4 (2010) 011101. https://doi.org/10.1063/1.3294083

[52] C.H. Gonzalez, S.N. Jayasinghe, P. Ferretti, Bio-electrosprayed human neural stem cells are viable and maintain their differentiation potential, F1000Res. 9 (2020) 267. https://doi.org/10.12688/f1000research.19901.1

[53] U. Siemann, Solvent cast technology – a versatile tool for thin film production, Scatt. Methods Prop. Polym. Mater. (2005) 1-14. https://doi.org/10.1007/b107336

[54] M.A. Rahman, M.A. Khan, S.M. Tareq, Preparation and characterization of polyethylene oxide (PEO)/gelatin blend for biomedical application: effect of gamma radiation, J. Appl. Polym. Sci. 117 (2010) 2075–2082. https://doi.org/10.1002/app.32034

[55] S.M. Lai, W.W. Sun, T.M. Don, Preparation and characterization of biodegradable polymer blends from poly(3-hydroxybutyrate)/poly(vinyl acetate)-modified corn starch, Polym. Eng. Sci. 55 (2015) 1321–1329. https://doi.org/10.1002/pen.24071

[56] J. Mota, N. Yu, S.G. Caridade, G.M. Luz, M.E. Gomes, R.L. Reis, J.A. Jansen, X.F. Walboomers, J.F. Mano, Chitosan/bioactive glass nanoparticle composite membranes for periodontal regeneration, Acta Biomater. 8 (2012) 4173–4180. https://doi.org/10.1016/j.actbio.2012.06.040

[57] E.F. Hansen, M.R. Derrick, M.R. Schilling, The effects of solution application on some mechanical and physical properties of thermoplastic amorphous polymers used in conservation: poly(vinyl acetate)s, J. Am. Inst. Conserv. 30 (2013) 203–213. https://doi.org/10.1179/019713691806066764

[58] S. Gorgieva, T. Vuherer, V. Kokol, Autofluorescence-aided assessment of integration and μ-structuring in chitosan/gelatin bilayer membranes with rapidly mineralized interface in relevance to guided tissue regeneration, Mater. Sci. Eng. C 93 (2018) 226–241. https://doi.org/10.1016/j.msec.2018.07.077

[59] S.A. Altinkaya, Modeling of asymmetric membrane formation by a combination of dry/wet phase inversion processes, Desalin. 199 (2006) 459–460. https://doi.org/10.1016/j.desal.2006.03.200

[60] B. Khan, W. Zhan, C. Lina, Cellulose acetate (CA) hybrid membrane prepared by phase inversion method combined with chemical reaction with enhanced permeability and good anti-fouling property, J. Appl. Polym. Sci. (2020) 49556. https://doi.org/10.1002/app.49556

[61] C. Liu, W. Qiao, C. Wang, H. Wang, Y. Zhou, S. Gu, W. Xu, Y. Zhuang, J. Shi, H. Yang, Effect of poly (lactic acid) porous membrane prepared via phase inversion induced by water droplets on 3T3 cell behavior, Int. J. Biol. Macromol. 183 (2021) 2205–2214. https://doi.org/10.1016/j.ijbiomac.2021.05.197

Advanced Functional Membranes

Materials Research Foundations **120** (2022) 72-110

Materials Research Forum LLC

https://doi.org/10.21741/9781644901816-3

[62] Fuel cell, Encyclopedia Britannica. https://www.britannica.com/technology/fuel-cell, 2021 (accessed 29 November 2021).

[63] M.K. Mahapatra, P. Singh, Fuel Cells: Energy Conversion Technology, Future Energy 2 (2014) 511-547. https://doi.org/10.1016/B978-0-08-099424-6.00024-7

[64] I. Staffell, D. Scamman, A.V. Abad, P. Balcombe, P.E. Dodds, P. Ekins, N. Shah, K.R. Ward, The role of hydrogen and fuel cells in the global energy system, Energy Environ. Sci. 12 (2019) 463-491. https://doi.org/10.1039/C8EE01157E

[65] Y. DongHao, Z. Zhan, A review on the sealing structures of membrane electrode assembly of proton exchange membrane fuel cells, J. Power Sources 231 (2013) 285-292. https://doi.org/10.1016/j.jpowsour.2013.01.009

[66] S. Hossain, A.M. Abdalla, S.N. Jamain, J. Zaini, A.K. Azad, A review on proton conducting electrolytes for clean energy and intermediate temperature-solid oxide fuel cells, Renew. Sustain. Energ. Rev. 79 (2017) 750-764. https://doi.org/10.1016/j.rser.2017.05.147

[67] P. Toudret, J.F. Blachot, M. Heitzmann, P.A. Jacques, Impact of the Cathode Layer Printing Process on the Performance of MEA Integrating PGM Free Catalyst, Catal. 11 (2021) 669. https://doi.org/10.3390/catal11060669

[68] S.B. Park, Y. Park, Fabrication of gas diffusion layer (GDL) containing microporous layer using flourinated ethylene prophylene (FEP) for proton exchange membrane fuel cell (PEMFC), Int. J. Precis. Eng. Manuf. 13 (2012)1145-1151. https://doi.org/10.1007/s12541-012-0152-x

[69] S. Karimi, N. Fraser, B. Roberts, R. Foulkes, Review of Metallic Bipolar Plates for Proton Exchange Membrane Fuel Cells: Materials and Fabrication Methods, Adv. Mater. Sci. Eng. (2012). https://doi.org/10.1155/2012/828070

[70] B.K. Kakati, D. Deka, Effect of resin matrix precursor on the properties of graphite composite bipolar plate for PEM fuel cell, Energ. Fuels 21 (2007) 1681-1687. https://doi.org/10.1021/ef0603582

[71] F. Yalcinkaya, E. Boyraz, J. Maryska, K. Kucerova, A Review on Membrane Technology and Chemical Surface Modification for the Oily Wastewater Treatment, Mater. 13 (2020) 493. https://doi.org/10.3390/ma13020493

[72] S. Hube, J. Wang, L.N. Sim, T.H. Chong, B. Wu, Direct membrane filtration of municipal wastewater: Linking periodical physical cleaning with fouling mechanisms, Sep. Purif. Technol. 259 (2021) 118125. https://doi.org/10.1016/j.seppur.2020.118125

[73] A. Raza, S. Farrukh, A. Hussain, I. Khan, M.H.D. Othman, M. Ahsan, Performance Analysis of Blended Membranes of Cellulose Acetate with Variable Degree of

Acetylation for CO_2/CH_4 Separation, Membr. 11 (2021) 245.
https://doi.org/10.3390/membranes11040245

[74] X.M. Tan, D. Rodrigue, A Review on Porous Polymeric Membrane Preparation. Part I: Production Techniques with Polysulfone and Poly(Vinylidene Fluoride), Polym. 11 (2019) 1160. https://doi.org/10.3390/polym11071160

[75] J.E.K. Schawe, C. Wrana, Competition between Structural Relaxation and Crystallization in the Glass Transition Range of Random Copolymers, Polym. 12 (2020) 1778. https://doi.org/10.3390/polym12081778

[76] A. Kongkanand, W. Gu, M.F. Mathias, Proton-Exchange Membrane Fuel Cells with Low-Pt Content, in: R. Meyers (Eds.), Encyclopedia of Sustainability Science and Technology, New York, 2018. https://doi.org/10.1007/978-1-4939-7789-5_1022

[77] J. Thangavelautham, Degradation in PEM Fuel Cells and Mitigation Strategies Using System Design and Control, in: IntechOpen (Eds.), Proton Exchange Membrane Fuel Cell, Arizona, 2018, pp. 72208. https://doi.org/10.5772/intechopen.72208

[78] P. Joghee, J. Malik, S. Pylypenko, R. O'Hayre, A review on direct methanol fuel cells – In the perspective of energy and sustainability, MRS Energ. Sustain. 2 (2015) E3. https://doi.org/10.1557/mre.2015.4

[79] K. Kordesch, M. Cifrain, A comparison between the alkaline fuel cell (AFC) and the polymer electrolyte membrane (PEM) fuel cell, Fuel Cell Technol. Appl. (2010). https://doi.org/10.1002/9780470974001.f304065

[80] E.H. Yu, U. Krewer, K. Scott, Principles and Materials Aspects of Direct Alkaline Alcohol Fuel Cells, Energ. 3 (2010) 1499-1528. https://doi.org/10.3390/en3081499

[81] L. Giorgi, F. Leccese, Fuel Cells: Technologies and Applications, Open Fuel Cell J. 6 (2013) 1-20. https://doi.org/10.2174/1875932720130719001

[82] D. Akinyele, E. Olabode, A. Amole, Review of Fuel Cell Technologies and Applications for Sustainable Microgrid Systems, Invent. 5 (2020) 42. https://doi.org/10.3390/inventions5030042

[83] S.J. McPhail, Status and Challenges of Molten Carbonate Fuel Cells, Adv. Sci. Technol. 72 (2010) 283-290. https://doi.org/10.4028/www.scientific.net/AST.72.283

[84] J.H. Yu, C.W. Lee, Effect of Cell Size on the Performance and Temperature Distribution of Molten Carbonate Fuel Cells, Energ. 13 (2020) 1361. https://doi.org/10.3390/en13061361

[85] A.B. Stambouli, E. Travesa, Solid oxide fuel cells (SOFCs): a review of an environmentally clean and efficient source of energy, Renew. Sustain. Energ. Rev. 6 (2002) 433-455. https://doi.org/10.1016/S1364-0321(02)00014-X

[86] W. Li, Y. Wang, W. Liu, A review of solid oxide fuel cell application, IOP Conf. Ser.: Earth Environ. Sci. 619 (2020) 012012. https://doi.org/10.1088/1755-1315/619/1/012012

[87] V.N. Nguyen, L. Blum, Reversible fuel cells, Compend. Hydrog. Energ. 3 (2016)115-145. https://doi.org/10.1016/B978-1-78242-363-8.00005-0

[88] Y. Liang, C.Z. Zhao, H. Yuan, Y. Chen, W. Zhang, J.Q. Huang, D. Yu, Y. Liu, M.M. Titirici, Y.L. Chueh, H. Yu, Q. Zhang, A review of rechargeable batteries for portable electronic devices, InfoMat 1 (2019) 6-32. https://doi.org/10.1002/inf2.12000

[89] K. Xu, Nonaqueous liquid electrolytes for lithium-based rechargeable batteries, Chem. Rev. 104 (2004) 4303-4418. https://doi.org/10.1021/cr030203g

[90] Y. Chen, Y. Kang, Y. Zhao, L. Wang, J. Liu, Y. Li, Z. Liang, X. He, X. Li, N. Tavajohi, B. Li, A review of lithium-ion battery safety concerns: The issues, strategies, and testing standards, J. Energ. Chem. 59 (2021) 83-99. https://doi.org/10.1016/j.jechem.2020.10.017

[91] Z. Xue, D. He, X. Xie, Poly(ethylene oxide)-based electrolytes for lithium-ion batteries, J. Mater. Chem. A 3 (2015) 19218-19253. https://doi.org/10.1039/C5TA03471J

[92] H.R. Rezaie, H.B. Rizi, M.R. Khamseh, A. Öchsner, A Review on Dental Materials, Adv. Struct. Mater. 123 (2020).

[93] V.P.H. Huy, S. So, J. Hur, Inorganic Fillers in Composite Gel Polymer Electrolytes for High-Performance Lithium and Non-Lithium Polymer Batteries, Nanomater. 11 (2021) 614. https://doi.org/10.3390/nano11030614

[94] Y. Miao, P. Hynan, A. van Jouanne, A. Yokochi, Current Li-Ion Battery Technologies in Electric Vehicles and Opportunities for Advancements, Energ. 12 (2019) 1074. https://doi.org/10.3390/en12061074

[95] A.H. Marincas, P. Ilea, Enhancing Lithium Manganese Oxide Electrochemical Behavior by Doping and Surface Modifications, Coat. 11 (2021) 456. https://doi.org/10.3390/coatings11040456

[96] Y. Hato, H.C. Chien, T. Hirota, Y. Kamiya, Y. Daisho, S. Inami, Degradation Predictions of Lithium Iron Phosphate Battery, World Electr. Veh. J. 7 (2015) 25-31. https://doi.org/10.3390/wevj7010025

[97] R.C. Masse, E. Uchaker, G. Cao, Beyond Li-ion: electrode materials for sodium- and magnesium-ion batteries, Sci. Chin. Mater. 58 (2015) 715-766. https://doi.org/10.1007/s40843-015-0084-8

[98] N.H. Barbhuiya, U. Misra, S.P. Singh, Synthesis, fabrication, and mechanism of action of electrically conductive membranes: a review, Environ. Sci.: Water Res. Technol. 7 (2021) 671-705. https://doi.org/10.1039/D0EW01070G

[99] B. Lakard, Electrochemical Biosensors Based on Conducting Polymers: A Review, Appl. Sci. 10 (2020) 6614. https://doi.org/10.3390/app10186614

[100] J.H.T. Luong, T. Narayan, S. Solanki, B.D. Malhotra, Recent Advances of Conducting Polymers and Their Composites for Electrochemical Biosensing Applications, J. Funct. Biomater. 11 (2020) 71. https://doi.org/10.3390/jfb11040071

[101] S. Ramanavicius, A. Ramanavicius, Conducting Polymers in the Design of Biosensors and Biofuel Cells, Polym. 13 (2021) 49. https://doi.org/10.3390/polym13010049

[102] Z. Chen, S.N. Obaid, L. Lu, Recent advances in organic optoelectronic devices for biomedical applications, Opt. Mater. Expr. 9 (2019) 3843-3856. https://doi.org/10.1364/OME.9.003843

[103] D.K. Baisnab, S. Mukherjee, S. Das, A short review on inorganic thin films from device perspective, Chem. Solut. Synth. Mater. Des. Thin Film Device Appl. 1 (2021) 231-275. https://doi.org/10.1016/B978-0-12-819718-9.00007-8

[104] J. Zhao, Z. Chi, Z. Yang, X. Chen, M.S. Arnold, Y. Zhang, J. Xu, Z. Chi, M.P. Aldred, Recent developments of truly stretchable thin film electronic and optoelectronic devices, Nanoscale 10 (2018) 5764-5792. https://doi.org/10.1039/C7NR09472H

[105] M. Ahmadi, T. Wu, B. Hu, A Review on Organic–Inorganic Halide Perovskite Photodetectors: Device Engineering and Fundamental Physics, Adv. Mater. 29 (2017) 1605242. https://doi.org/10.1002/adma.201605242

[106] L. Cao, X. Liu, Z. Guo, L. Zhou, Surface/Interface Engineering for Constructing Advanced Nanostructured Light-Emitting Diodes with Improved Performance: A Brief Review, Micromach. 10 (2019) 821. https://doi.org/10.3390/mi10120821

[107] P.T. Huckabee, Optic Fiber Distributed Temperature for Fracture Stimulation Diagnostics and Well Performance Evaluation, SPE Hydraul. Fract. Technol. Conf. (2009). https://doi.org/10.2118/118831-MS

[108] M. Sciamanna, K.A. Shore, Physics and applications of laser diode chaos, Nat. Photonics 9 (2015) 151-16. https://doi.org/10.1038/nphoton.2014.326

Advanced Functional Membranes
Materials Research Foundations **120** (2022) 111-150

Materials Research Forum LLC
https://doi.org/10.21741/9781644901816-4

Chapter 4

Sustainable Membranes and its Applications

L. Oliveira[1], A.N. Módenes[1], C.C. Triques[1], M.L. Fiorese[1], L.C. Silva[2], V. Slusarski-Santana[3*], L.D. Fiorentin-Ferrari[3]

[1]Postgraduate Program in Chemical Engineering, West Paraná State University, Rua da Faculdade 645, Toledo, 85903-000 PR, Brazil

[2]Postgraduate Program in Chemical, West Paraná State University, Rua da Faculdade 645, Toledo, 85903-000 PR, Brazil

[3]Graduation Program in Chemical Engineering, West Paraná State University, Rua da Faculdade 645, Toledo, 85903-000 PR, Brazil

veronice.santana@unioeste.br

Abstract

The utilization of sustainable and high-performance technologies is a growing tendency in industrial processes. The membrane separation process is a clean technology alternative that can be used to separate, concentrate, and/or purify substances. But there is still a great interest in turning this process increasingly sustainable by the structural modification of the membranes using biopolymers and green solvents, and the functionalization with organic and inorganic materials, also improving their performance and anti-fouling characteristics. Thus, this review chapter brings the main concepts about membranes and their main structural modifications to obtain a process that meets the concepts involved in Green Technology.

Keywords

Membrane Separation Process, Polymers, Biopolymers, Green Solvents, Membrane Functionalization, Green Technology

Contents

1.   Introduction

The membrane separation process is used in food, pharmaceutical, biotechnology, steel, paper, cellulose, and petrochemical industries as a technique to separate, concentrate, and/or purify, and also to decontaminate wastewater, especially those containing emerging pollutants such as antibiotics and hormones, dyes, heavy metals, persistent organic pollutants, in addition to water desalination [1-14].

Due to the operational characteristics, mainly the ones that include low energy consumption during the separation, concentration and/or purification of substances, high solute rejection capacity, no generation of secondary pollutants, and capacity of reusability for multiple cycles, the membrane separation process is included in the concepts of sustainable productive processes defined as Green Technology. In Green Technology, the utilization of ecologically correct alternatives is essential during the transformation of the raw material into a product, with control of the pollution generated and without causing damages to the environment, as is the case of the membrane separation process [11, 15-30, 31].

The membranes used industrially are, mostly, fabricated with ceramic materials that possess the capacity to operate in extreme conditions of pH and temperature, besides having a high lifespan. However, the ceramic membranes possess a high fabrication cost due to the raw material used in their preparation [32-34]. Due to the cost, the ceramic materials used in the membranes are being substituted by natural or synthetic polymeric materials. Depending on the structural characteristics of the polymer macromolecule, the membrane fabricated will present equal or higher mechanical, thermal, and chemical resistance than ceramic membranes. Besides, structural modification in the polymers and mixture of polymeric materials can produce membranes with higher fouling resistance [5, 7, 13, 14, 27].

The polymeric membranes currently commercialized are fabricated with synthetic polymers, mainly from petroleum, and thus, non-biodegradable. The polymeric residues from the processes of fabrication and utilization of the membranes contribute to the increase of residual water pollution, mainly by microplastics [37, 38]. Besides, during the fabrication of the membranes, reprotoxic organic solvents are used, which are harmful to the environment. The substitution of synthetic polymers by biopolymers, such as cellulose acetate and chitosan, which are natural, biodegradable polymers from renewable sources, widely available and at low cost, become a promising alternative. The utilization of green solvents, such as water, ethyl lactate, ionic liquids, and supercritical CO_2 is another possibility to replace commonly used solvents for fabrication membranes [27, 28, 37-39].

Another problem associated with the filtration process with ceramic or polymeric membranes is the fouling development from the accumulation of organic, inorganic, and biological macromolecules that can occur in both the surface and in the membrane pores. Fouling causes a reduction in the selectivity and the permeate flow, and consequently the decrease of the membrane lifespan [9, 40-42, 49]. Several factors affect the fouling development during filtration, however, the superficial characteristics of the membrane, such as high hydrophilicity and low roughness, can contribute to reducing the deposition of foulants agents [5, 40, 44-47].

The morphological and structural characteristics of the membrane can be changed during its fabrication process by its functionalization with materials and/or nanomaterials that present desirable characteristics to the membrane, such as, high surface area, hydrophilicity and mechanical, thermal and physical resistance. The incorporation of organic or inorganic nanomaterials or materials can be performed in the solution or the membrane surface during the fabrication. Due to the physical and chemical characteristics of the incorporated material or nanomaterial, functionalized membranes can present adsorptive, photocatalytic, catalytic, antioxidant, and antimicrobial characteristics. Still, the structural modification by functionalization allows the obtention of membranes with anti-fouling property [19, 48- 55].

In general, the structural modification during membrane fabrication with ceramic or polymeric biomaterials and solvents that are less pollutant, as well as the functionalization with organic or inorganic materials, allow that the membrane separation process meets the sustainable principles becoming a technique closer to the concepts of Green Technology [26, 28, 48, 56]. Thus, this review chapter aims to present general aspects of the conventional membrane separation process and ways of turning the process more sustainable by using biopolymers and green solvents and functionalizing the membranes with organic and inorganic additives. This structural modification, besides keeping the industrially desired selectivity, influences the anti-fouling properties, promoting an increase in the membrane lifespan.

2.   Membrane separation processes

2.1   Morphological classification of the membranes

The membrane separation process is known since 1748 when Abbé Nolet observed that water diffused through a semi-permeable membrane from a dilute solution to a more concentrated one. However, between the 18th and 19th centuries, membranes had no industrial and commercial use and were used exclusively for laboratory applications. Natural membranes were made from the intestine and stomach of animals, or bladders from pigs, cattle, or fish [57-60]. However, a great advance in the membrane separation process occurred in the early 1960s, when Loeb and Sourirajan performed the annealing of commercial acetate membranes and obtained anisotropic reverse osmosis membranes with smaller pore sizes and ten times greater flux than previously existing membranes. In this period, the utilization of reverse osmosis membranes to desalinate water started [2, 18, 61]. After the discovery of Loeb and Sourirajan, many others happened in the following years, such as the fabrication of membranes using ceramic and polymeric materials, which is performed up to now [1-3].

The membrane filtration can be defined as the separation of two or more components through a semi-permeable barrier, based on the particle size difference, in which the permeate is what passes through the membrane, and what is retained in the membrane is the retentate [51, 60, 61]. The membrane separation process is commercially attractive by its technical and economic characteristics, being mainly based on the sieving and electrostatics principles [18]. Often, the membranes are considered a more efficient technology than conventional methods of gravity separation, chemical precipitation, adsorption, and ion exchange [11, 60, 62].

The conventional separation methods that are consolidated in several industrial processes possess unfavorable characteristics, especially concerning the utilization of large physical areas, the necessity of toxic compounds, and generation of secondary pollutants that need adequate treatment [11, 18, 60, 62, 63]. The chemical precipitation method, for example, requires the utilization of a large quantity of chemical products to adjust the pH so that the foulant precipitate, while the gravity separation needs a wide physical area and considerable time to separate the foulant agent. Adsorption is presented as the more viable alternative among the conventional and consolidated techniques, due to the high removal capacity of several types of solutes/pollutants. However, adsorption has as its challenge, be abel to reuse of the adsorbent for several cycles keeping efficiency [11, 18, 63]. Thus, when the characteristics of the membrane separation process are evaluated and compared to the conventional and consolidated methods used up to now, membranes are verified to be able to be used in separation, concentration, and/or purification of several types of substances and several industries.

Membranes can be classified concerning their origin, morphology, structure, fabrication method, and material used [60]. The first classification refers to the nature or origin, which can be biological or synthetic in the liquid or solid form, according to Fig. 1 [40]. Dense and porous polymeric synthetic membranes can be classified into symmetrical and asymmetrical according to the morphology and pore distribution [40, 59, 64]. According to the distribution and size of the pores, dense and porous membranes can be isotropic symmetric when the pores present similar size, and anisotropic asymmetric when the structure is not uniform with pores distributed in different sizes in the membrane [2, 62].

In the dense morphology, the pores are not visible and the solute moves by diffusion through the membrane. The separation of the solutes from a solution is determined by the diffusivity and solubility in the membrane material, thus being able to perform the separation of particles with similar sizes [59]. The microporous morphology aims to reject all the solutes with a size higher than the pore sizes of the membrane. Thus, solutes or particles with similar pore sizes to the membrane can cause the pore blocking, clogging it up, and decreasing the filtration process performance [2, 61].

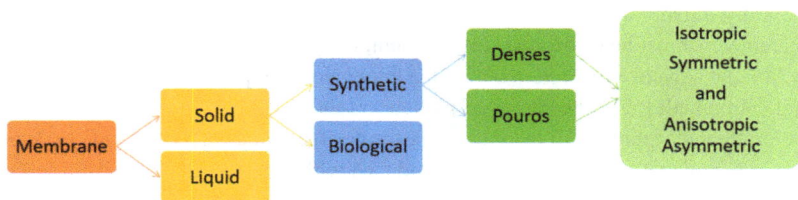

Figure 1 Classification of the membranes concerning origin, morphology, and structure.
Source: author

Asymmetric membranes have a structure consisting of a very thin polymer top layer (0.1 to 1 μm) supported by a highly porous 100 to 200 μm sublayer that acts only as a support. Thus, the pore size and the material nature of the top layer are responsible for the membrane separation characteristics, while the permeate flow is determined by the thickness and sublayer of the membrane [40, 60, 64]. Thus, asymmetrical membranes possess high flows when compared to the symmetrical ones, and present lower fouling, because the highest solute rejection happens in the membrane surface, making the internal structure of the pores present low clogging [61,65].

One factor that influences the morphology and structure of the membrane, polymeric or ceramic, is the fabrication technique used. For symmetrical polymeric membranes, the most used methods are sintering, stretching, and track-etching, while the asymmetrical membranes are produced mainly by the phase inversion method from a sole polymer or by the blend of polymeric materials [19, 60, 65].

The phase inversion method basically consists in dissolving a polymer into an adequate solvent and transforming the polymeric solution into a liquid state to a solid state, thus forming the membrane [66, 67]. For such transformation, after the polymer dissolution, the dissolved polymeric solution is spread under inert support as a thin liquid film and the phase inversion of the liquid film can be induced by the contact with a non-solvent or vapor, thermal contact, or solvent evaporation [19, 68]. During the phase inversion, the highest polymer concentration forms the membrane surface with lower pores, while the pores in the sublayer of the membrane are constituted of lower polymeric concentration [40, 64, 68, 69]. Due to the simplicity of the method and the fabrication cost, the phase separation induced by a non-solvent, also known as phase inversion via immersion precipitation, is the most used and reported in the literature fabrication method of asymmetrical membranes [19, 49, 67, 70, 71].

To improve the membrane performance concerning the permeate flow and selectivity, several parameters must be controlled, among them, the polymeric concentration, relative humidity, time of solvent evaporation before the immersion, temperature, and composition of the immersion bath [69]. The choice of parameters that will be controlled will directly influence the final morphology of the membranes, since the process of fabrication must be performed in a way that the membranes are produced with adequate characteristics to the kind of process (separation, purification, or concentration) that they will be industrially subjected. Thus, the property of solute transport through solid membranes can be performed from the application of concentration, pressure, temperature, or electric potential gradient [40, 64, 72].

In general, the lower the pore sizes, the higher the hydrodynamic resistance and the higher the necessary driving force so that permeate flow is not reduced. Thus, the pressure-driven membranes are classified into microfiltration, ultrafiltration, nanofiltration, and reverse osmosis. Table 1 shows the main characteristics of these membranes [2, 40, 61, 73].

Advanced Functional Membranes

Materials Research Forum LLC

Materials Research Foundations **120** (2022) 111-150

https://doi.org/10.21741/9781644901816-4

Table 1 Characteristics of the pressure-driven membranes according to the pores' sizes and rejected material

Membrane	Pressure [bar]	Pore Size [μm]	Retentate
Microfiltration	0.1 - 2	Macroporous pore 0.1 - 10	Bacteria, viruses, pathogens, and suspended particles
Ultrafiltration	1 - 5	Mesoporous pore 0.001 - 0.1	Macromolecules
Nanofiltration	5 - 20	Microporous pore 0.001 - 0.01	Sugars and divalent ions
Reverse Osmosis	10 - 100	Non-porous 0.0001 - 0.001	Monovalent ions

Source: author

Microfiltration membranes are commonly applied in the pharmaceutical, food and beverage, and semiconductor industries to remove bacteria, viruses, pathogens, in addition to the filtration of suspended particles, proteins, and emulsion separation. The microfiltration membranes use the principle of physical separation to reject particles with sizes higher than the range between 0.1 to 10 μm. Compared to the other kinds, the microfiltration membrane presents more open pores and consequently lower operating pressure than the other classifications [3, 31, 59, 60, 74]. The separation of solutes in the ultrafiltration membrane occurs by the pore size difference. This kind of membrane presents high product yield, relative easiness of scale-up, cleaning and equipment sanitation easiness, being mainly applied in the biotechnological and food area, where they are largely used to purify, concentrate, and fractionate solutes such as colloids, microorganisms, proteins, and lipids [3, 51, 75, 76].

The nanofiltration membrane possesses pore sizes between 0.001 and 0.01 μm, being responsible for the separation of particles with molecular weight cut from 200 to 1000 Dalton. Both the nanofiltration and the reverse osmosis membranes require high pressure during the filtration and present a high solute rejection rate, despite the nanofiltration membrane presenting low rejection of monovalent ions, such as chlorides. Due to the high rejection rate of divalent ions and sugars, the nanofiltration membranes have several applications, being the main ones in the potable water, sewage treatment, pharmaceutical, biotech, and food engineering segments. While the reverse osmosis membranes possess pore size between 0.0001 and 0.001 μm and can be applied in the

Advanced Functional Membranes
Materials Research Foundations **120** (2022) 111-150

Materials Research Forum LLC
https://doi.org/10.21741/9781644901816-4

concentration of a feed flow, due to the rejection of water molecules, or even produce ultra-pure water in the permeate since it possesses a solute rejection higher than 99%. Ultrafiltration membranes are considered porous, while reverse osmosis ones are dense [51, 60, 67, 75, 77-81].

Another factor that influences the performance of the pressure-driven membranes (Table 1) is the kind of material used for the fabrication. The characteristics of the ceramic or polymeric material directly influence the obtention of a membrane that presents a high lifespan, besides thermal, mechanical, and physical stability [2]. The organic membranes are usually made of polymers such as polysulfone, polyvinylidenefluoride, polyethersulfone, polyacrylonitrile, polyvinyl alcohol, polyvinyl chloride, polyethylene, polypropylene, polyamide, chitosan, cellulose acetate, and cellulose [3, 19, 31, 75]. In the fabrication of inorganic membranes are commonly used materials ceramics, zeolites, carbon, glass, or metal [32, 60, 82].

When compared to polymeric membranes, the inorganic membranes possess higher mechanical resistance, thermal, and chemical stability, being thus able to operate in extreme conditions. Besides, ceramic membranes possess easier cleaning and longer lifespan, being able to be used industrially for more than 8 years, while the polymeric membranes need to be replaced every 2 years [2, 32-34]. However, the inorganic membranes do not present good performance during the microfiltration and ultrafiltration processes, besides being too fragile and being easily damaged by improper vibration or fall. Besides, the biggest disadvantage of inorganic membranes is the high initial investment cost [62, 83], while the polymeric membranes are widely used in industrial processes, due to their low cost, easiness of fabrication, and they are available at a wide range of pore sizes [60, 62]. However, a great limitation in the use and commercialization of polymeric membranes occurs due to the fouling development [19, 84, 85].

2.2 Concentration polarization and fouling

Fouling can be defined as the accumulation of particles, colloids, salts, and macromolecules such as proteins and polysaccharides on the membrane surface, inside and/or in the pore walls. The deposition of macromolecules in the membrane surface begins in the first minutes of filtration and over time it causes an additional resistance to mass transfer, causing the phenomenon known as concentration polarization [40,59]. The main consequence of concentration polarization is the drop in permeate flow. Especially in porous membranes, the continuous drop of the permeate flow is commonly observed during the filtration, with a decrease in the solute rejection. At this moment, another phenomenon known as fouling is developed, which is responsible for the lifespan decrease of the

Advanced Functional Membranes   Materials Research Forum LLC

Materials Research Foundations **120** (2022) 111-150   https://doi.org/10.21741/9781644901816-4

membranes, increasing operating costs of the membrane separation process [19, 31, 51, 86, 87].

Fouling can be reversible when the solutes are retained in the membrane surface forming a gel-cake layer, and irreversible when the membrane pores become obstructed by the solutes [88, 89]. The factors that influence the most in the fouling are the flow velocity, feed composition, concentration of the main constituents, transmembrane pressure, process temperature, feed pH, size and shape of the fouling material, besides the membrane properties such as hydrophilicity/hydrophobicity, roughness, size, and type of pores [31, 89]. Another factor that influences fouling development is the kind of filtration used, dead-end or cross-flow filtration. In the cross-flow kind of filtration, the flow parallel to the membrane surface limits the accumulation of retained material, delaying the development of concentration polarization and fouling [40]. Fouling can occur from the clogging by organic, inorganic, colloidal, and biofouling substances (Fig. 2) [31, 84, 89-93].

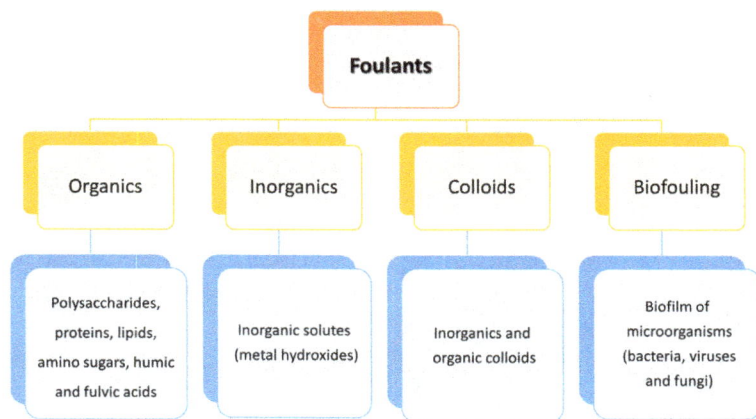

Figure 2 Types of foulants that can occur on the membrane during dead-end or cross-flow filtration. Source: author

With the development of fouling and consequent selectivity loss, after a certain filtration time, it is necessary to stop the process and perform a membrane cleaning. This cleaning can be performed physically, chemically, or by backwash, enabling to remove especially the reversible incrustations. The cleaning inefficiency causes a loss in the membrane performance, decreasing the lifespan and leading to a limited condition in which the membrane needs to be discarded. Besides, to perform the cleaning, the membrane needs to possess mechanical and chemical resistance, and tolerance to chlorine, and different pH ranges [2, 19, 31, 40, 94, 95]. Thus, the choice of the material and technique used to fabricate the membrane together with the knowledge of the incrustation mechanism and kind, as well as the optimization of several cleaning cycles with increase of the lifespan are important topics to still be studied and developed to improve the performance of the membrane separation process.

3   Fabrication of sustainable membranes

3.1   Biopolymers, green solvents, and functionalization

Currently, there is a great search for sustainability in industrial processes due to the accentuated growth of chemical industries, and it is undeniable that the environment has been suffering aggression by this sector. In addition, the requirements from society and environmental protection agencies with their respective legislation have been growing. Thus, the improvement of conventional industrial production processes with green and sustainable strategies, maintaining the quality and productivity required by the consumer market is essential [21, 26, 27, 39, 96-100].

In general, the membrane separation process presents as its main advantages: mild operating conditions with low energy consumption, without the use of additives to promote the separation, concentration and/or purification, operating flexibility, combination with other techniques causing a reduction in the industrial process stages, easiness of scale-up, lower residues generation, and great selectivity during separation, which can occur by the molecular size, or by the physicochemical behavior of the compounds involved. These characteristics make the membrane separation process able to be included in the current concepts of sustainable processes that combine care for the environment and maintenance of productivity. However, much still can be done and improved concerning the processing conditions of the membrane separation process, as well as structural modifications in the membranes to make them more sustainable [15, 16, 40, 101-104].

Normally, membranes are considered sustainable when they use renewable and clean sources in their fabrication, such as the biopolymers and green solvents, or even, when a morphological modification based on synthetic polymers or organic and inorganic additives is performed to increase their performance, anti-fouling properties, and lifespan of the membranes, and consequently, the membrane separation process [10, 22, 28, 36, 37, 41, 105-107]. Among this sustainable tendency, a problem found in the membrane is the polymers used for its fabrication. Polymers are predominantly synthetic, formed by non-biodegradable hydrocarbons, resistant to chemical and biological attack, ensuring longevity, and other properties maintained for long periods [35, 36]. In this context, the environmental impact caused by the use of synthetic materials in the form of plastics has stimulated the development of membranes from biodegradable biopolymers, produced from raw materials of renewable sources, such as corn, sugar cane, cellulose, and chitin [39,108].

Among the biopolymers, the polysaccharides cellulose and chitin, which are the most abundant natural polymers found in nature, have been the aim of several studies in the last years [22, 37-39, 105, 108-112]. Cellulose is a renewable polymer available in abundant quantity, biodegradable, with low cost, being, thus, another ecologically correct option to be used in the membranes' fabrication [37, 38, 113]. Through the deacetylation of chitin, chitosan is obtained, a biorenewable chemical product that is low-cost, with good stability characteristics, antibacterial activity, chelating properties, hydrophilicity, water solubility (at acidic pH), and is also biocompatible, biodegradable, and non-toxic, contributing to a cleaner and more environmentally correct fabrication process [15, 39, 114-116]. Chitosan can be used in the modification and development of membranes with higher hydrophilicity and performance for the reverse osmosis, ultrafiltration, and nanofiltration processes. Membranes based on chitosan have good separation efficiency, high adsorption capacity, fast kinetics, and present reuse capacity for several cycles, besides anti-fouling characteristics. These features enable chitosan membranes to be used mainly in the treatment of water and effluents containing heavy metals and dyes [114, 117-125].

When the use of synthetic and non-biodegradable polymers is inevitable, solvents less harmful to the environment can be used in an attempt to make the membrane with more sustainable characteristics. Usually, the fabrication of polymeric membranes is based on the phase separation induced by a non-solvent, which used a great quantity of reprotoxic organic solvents such as N,N-dimethylformamide, N-Methyl-2-pyrrolidone, and N,N-dimethylacetamide [126, 127]. In the literature, studies that use green solvents such as ethyl lactate, ionic liquids, supercritical CO_2, and even water are found [27, 28, 106, 126]. Besides, studies point that the membrane fabrication processes that performed the complete substitution of toxic solvents by more ecological solvents reached performances comparable to the ones produced by the classical approach [26, 28, 106].

In order to enable solvent substitution, modifications in the polymer dissolution need to be performed. For the separation of the aqueous phase without the utilization of organic solvents, three alternatives are mentioned in the literature. In the first one, the weak pH-responsive polyelectrolyte in its charged state is dissolved in water and then precipitated using a pH interrupter [27, 100]. In the second alternative, two polyelectrolytes, one weak and one strong with opposed charges, are mixed in a pH, where the weak polyelectrolyte is discharged, and then, a change in the pH induces the phase separation, due to the complexation between both polymers [99]. Lastly, in the third alternative, the

Advanced Functional Membranes   Materials Research Forum LLC
Materials Research Foundations **120** (2022) 111-150   https://doi.org/10.21741/9781644901816-4

complexation of the polymers occurs due to the mixture of two strong polyelectrolytes of opposed charge at a high salt concentration [128].

A promising alternative to turn the membranes more sustainable is the functionalization of the membrane from structural modification. This structural modification can be performed from the incorporation of organic or inorganic compounds/additives, providing antimicrobial, antioxidant, adsorptive, or catalytic characteristics to the membrane, depending on the kind of functional material employed, besides improving the anti-fouling properties and increasing the lifespan of the membranes [9, 18, 19, 129]. Thus, the structural functionalization allows the fabrication of membranes with morphologies that are more and more adequate to the specific demands of each industrial process, be it separation, purification, and/or concentration. This functionalization with sustainable materials still adds more characteristics associated with the Green Technology concepts to the separation membrane process.

Among the several functionalization techniques of polymeric porous membranes fabricated by phase inversion, the morphological modification from the insertion of organic and inorganic additives in the polymeric solution [89, 130] and under the liquid film in the form of a surface coating of the membrane [18, 48, 51, 89, 125] can be performed. Still, modifications of the macromolecules during the synthesis of the polymer that will be used in the fabrication of the membranes can be done [19, 41, 73, 89]. Thus, physical and/or chemical methods can be used to enable the modification of the membrane surface (Fig. 3).

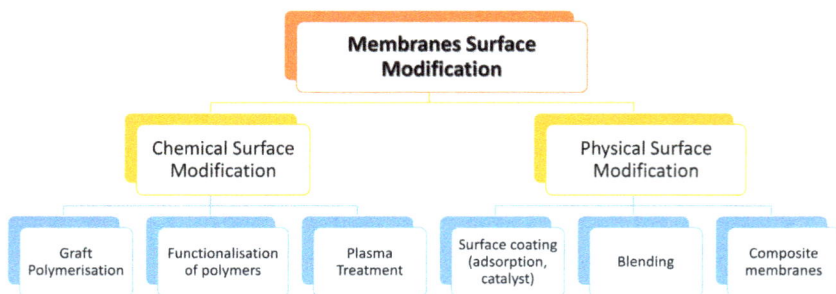

Figure 3 Functionalization techniques for porous polymeric membranes surface. Source: author

In the physical functionalization, the additive binds to the membrane surface through van der Waals bonds, electrostatic interaction, or hydrogen bonds which has a weaker interaction than the others and may not be stable in the long term. Thus, the membrane is coated by a very thin functional surface layer that provides anti-fouling properties, keeping high performance. Physical surface functionalization is the most used method to modify membranes. In addition, the surface coating technique can be applied to the fabrication of membranes layer-by-layer or double-membrane/in sandwich. However, the disadvantage of this method is the weak interaction between the coated layer and the membrane, which can allow that this layer suffers alteration and be removed during the chemical/physical cleaning and even during the operating process [131-136]. While in the surface adsorption technique, the membrane is modified by the insertion of compounds such as charged polyelectrolytes and surfactants in its surface, able to turn it into hydrophilic and consequently, improve the anti-fouling capacity [134, 137-139]. In the surface modification by blend, it is customary to mix organic and/or inorganic materials in order to transfer the main individual characteristics of each material to the membrane, such as hydrophilicity, thermal, physical, and chemical resistance [71, 140].

In chemical functionalization, the additive is fixed to the membrane surface by covalent bonds that have better chemical and structural stability. The functionalization with polymers, plasma treatment, and graft polymerization are examples of chemical functionalization techniques [18, 89, 134, 138, 141]. The plasma treatment has been widely used to modify the surface of polymeric membranes, due to the quick reaction time, high versatility, and being a process without secondary waste generation. In this method, the plasma species interact with the polymer surface, causing an electronically excited state in atoms with the formation of free radicals and unsaturated bonds. Thus, the reactions with the water and oxygen molecules in the surface of the polymer are eased and consequently, the surface of the membrane becomes with hydrophilic characteristics and higher fouling resistance [89, 133, 135, 142, 143]. Graft polymerization is the most used chemical method to modify the surface, which occurs through a covalent bond between the membrane surface and a polymer with the desired characteristic. This method is attractive due to its simplicity, and for not requiring an extra stage of chemical modification of the polymers during the membranes fabrication [18, 89, 130, 144, 145], which allows its utilization with a wide range of monomers that can be polymerized and grafted onto the porous membrane in a simple way, giving it several physicochemical surface properties [134, 135, 137, 138, 146].

In both chemical and physical functionalization, the use of nanomaterials is a widely used alternative in the membrane separation process. The nanomaterials possess as their main characteristics the high surface area that favors the adsorption, unique surface chemistry, and the photocatalytic, antimicrobial electrical, and optical properties that are desired in the membranes [7, 13, 21, 53, 107, 147, 148]. Several modifying agents are studied for the membrane functionalization and improvements of the anti-fouling properties, such as the metallic silver (Ag) and copper (Cu) nanoparticles, metal oxides (Fe_2O_3, ZrO_2, Al_2O_3, SiO_2, ZnO), in addition to various forms of carbon nanofillers, zeolites, $Mg(OH)_2$, $CaCO_3$, and MnO_2 [19, 107]. Comparing all these materials, it is worth mentioning that the carbonaceous materials are ecologically correct, of easy preparation, and abundant availability, besides presenting excellent biocompatibility, and low toxicity [53, 149]. The carbon nanomaterials have unique characteristics, being the main ones the possibility to reach the required pore size, high surface area, and surface functionality, which collaborate to increase thermomechanical stability, water permeability, hydrophilicity and consequently improve the anti-fouling properties of the membrane [52, 150]. Besides, the functionalizing nanomaterials can even provide the membranes antibacterial and photocatalytic characteristics [125, 151-153].

Currently, several studies are approaching the membranes' functionalization using graphene that is a material based on carbon, especially because of the efficiency that the graphene presents in environmental applications, involving the adsorption, catalysis, and photocatalysis techniques [49, 53, 98, 151, 152]. These several studies showed that membranes functionalized with graphene have easy permeate flow recovery after cleaning, chemical and physical stability of the nanomaterials added, and performance improvement concerning the permeate flow and solute rejection. Besides, due to lightness, flexibility, high conductivity, and resistance, the functionalization with graphene gives the membrane desirable characteristics that are closer to the concepts of sustainability [52, 98, 154].

Following the Green Technology principle, the functionalization with agricultural, lignocellulosic, civil construction, wood production, fruit peel, and general disposal residues is a promising and sustainable alternative for the fabrication of composites with minimization of environmental pollution [54, 155-157]. For example, there is eggshell, which is a residue consisting basically of minerals, such as calcium and magnesium carbonate, and lignocellulosic materials extracted from bamboo or other residues, which can be used in functionalization, providing catalytic, adsorptive, and anti-fouling characteristics to the polymeric membranes [50, 157-161].

The possibility of morphological and structural modification of membranes using biopolymers, green solvents, and the functionalization of materials and/or nanomaterials make the membrane separation process to be a technique closer to the sustainability principles, being highlighted as a Green Technology.

Advanced Functional Membranes   Materials Research Forum LLC
Materials Research Foundations **120** (2022) 111-150   https://doi.org/10.21741/9781644901816-4

3.2   Applications of the functionalized membranes

The membrane separation process is widely used in pharmaceutical, chemical, and food industries, in medical areas, and the treatment of domestic and industrial water. Some examples of applications can be mentioned, such as water treatment; residual water treatment; recuperation/purification of active pharmaceutical ingredients and valuable catalysts; recuperation of proteins and metallic ions; standardization, concentration, and fractionation of proteins; enzymes' purification; concentration of milk, cheese whey, and fruits' juice; whitening and desalcoholization of wines and beers; desalinization and demineralization of waters; as artificial kidneys and lungs in hemodialysis and oxygenators, respectively [3, 38, 66, 67, 112, 162].

In the literature, several relevant studies are found that used synthetic polymers and the combination with biopolymers, green solvents, and functionalization of the membranes with organic and/or inorganic additives to fabricate the membranes in order to make them more sustainable without losing the process efficiency. Rassol et al [28] fabricated nanofiltration membranes by the phase inversion technique using the biopolymer cellulose acetate dissolved in methyl lactate solvent, biodegradable and non-toxic, and in the green co-solvent 2-methyl tetrahydrofuran also from renewable resources. In this study, the cellulose acetate concentration in the casting solution, the addition of co-solvent, and the solvent evaporation time before the coagulation bath were evaluated. The performance of the membrane was evaluated with aqueous solutions of rose bengal or $MgSO_4$. The authors observed for the rose bengal, that the increase in the concentration of cellulose acetate from 8 to 20% (wt%) caused an increase in the rejection from 31 to 99.5%, and a decrease in the permeability from 31.8 to 2.4 L m^{-2} h^{-1} bar^{-1}, respectively. The addition of 10% (wt%) of co-solvent in the membranes with 10% (wt%) of cellulose acetate promoted rejection of the rose bengal solution higher than 92.0%, and mean permeability of 3.5 L m^{-2} h^{-1} bar^{-1}. Concerning the solvent evaporation time, in the membranes with 20% (wt%) of cellulose acetate, the $MgSO_4$ solution rejection was approximately 93.0% without evaporation and 96.5% with evaporation with a mean flow of 1.3 L m^{-2} h^{-1} bar^{-1}. With this study, the authors evidenced that the utilization of biopolymers associated with solvents that are less pollutant is a promising alternative to fabricate sustainable and ecologically correct nanofiltration membranes.

Marino et al [24] employed for the first time the Cyrene™ green bio-derived solvent to fabricate two membranes, being one of polyethersulfone and other polyvinylidene fluoride. Both membranes were fabricated by phase inversion induced by vapor and also non-solvent under controlled temperature and relative air humidity. The polyethersulfone membrane presented a structure formed by asymmetrical macrovoids and hydrophilic character with a contact angle lower than 90°. The polyvinylidene fluoride membrane had morphology with bicontinuous structure and contact angle between 110° and 120°, being, thus, more hydrophobic. The authors ranged the exposition time during phase inversion with vapor and observed that for the polyethersulfone membrane at lower exposition times, the membranes were in the ultrafiltration range, while for longer times, in the microfiltration range. For the polyvinylidene fluoride membrane, the phase inversion with vapor as non-solvent promoted the formation of membranes with pores of 0.03 µm and 0.55 µm, respectively. Thus, from these results, the viability of utilization of the green solvent Cyrene™ to fabricate more sustainable and ecologically correct membranes was verified.

Nielen et al [27] investigated an aqueous separation system based on pH change to produce the membranes. The authors chose to use the copolymer polystyrene-alt-maleic acid for having responsive monomers in its structure, which are necessary for the phase separation, and non-responsive hydrophobic monomers, which provide mechanical stability to the resulting membranes. The authors also tested the green solvent acetic acid and water to dilute the copolymer. Porous and dense membranes were fabricated in the microfiltration and nanofiltration range, which presented a rejection higher than 98% to oily waters. However, the membranes with dense and thin layers in the nanofiltration range showed rejections higher than 92% to small micropollutants. Thus, it is possible to observe that nanofiltration membranes can be fabricated to present excellent performance and sustainability characteristics through the separation method by the aqueous phase.

Aiming to improve the anti-fouling ability of the membranes and in the search for the development of sustainable membranes from the functionalization with natural polymers and nanoparticles of biological basis, Goetz et al [37] incorporated chitin nanocrystals in the surface of a membrane of cellulose acetate fabricated by the electrospinning technique. The authors tested the membrane to purify washing water from the process of a food industry containing biological and organic contaminants. The incorporation of chitin nanocrystals to the membrane increased 131% the mechanical resistance, 340% the rigidity, and the flow at 0.5 bar for $14217 \, L \, m^{-2} \, h^{-1}$ and decreased the membrane deformation. To complement the excellent results already obtained, the authors observed that the acetate functionalization, which presents a hydrophobic characteristic with a contact angle of 132°, produced a highly hydrophilic membrane, with a contact angle near zero. This alteration contributed to biofouling reduction in the membrane surface, as well as provided better resistance to fouling after the permeation of bovine serum albumin and humic acid solutions.

Most studies found in the literature about the application of the membrane separation process used synthetic polymers of hard biodegradation to fabricate the membranes. On the other hand, several authors fabricated membranes from a blend of polymers and biopolymers and/or performed the functionalization of the membranes incorporating organic and inorganic additives to make them more biodegradable. Among the organic additives, the chitosan biopolymer was the most used one to fabricate porous membranes, with more than 13,000 research articles published up to 2016 [105]. Another biopolymer that has been widely used is the cellulose acetate for having a low cost, good biocompatibility, and hydrophilicity [38, 108].

Al-Gharabli et al [112] performed the covalent anchorage of chitosan to the hydrophobic surface of polyvinylidene fluoride membranes in order to improve membrane permeability and anti-fouling properties. The membranes with and without chitosan were evaluated concerning the rejection of volatile organic compounds, desalination, and juice concentration. The membranes functionalized with chitosan, called hybrid, presented higher thermal stability, transport performance, selectivity, and fouling decrease when compared to the pure polyvinylidene fluoride membrane. The separation factor increased from 2.3 to 11.1 after the chitosan fixation to the surface of the membrane and the separation index of the process was 6.8 and 90 Kg m^{-2} h^{-1} for the membrane without and with chitosan, respectively. In juice concentration, the membrane without chitosan had a higher flow decrease (50%) than the one with chitosan (20%). The authors also evaluated that the membranes fabricated presented the potential to be used to separate oil in water and residual waters treatment.

Ji et al [109] also fabricated a membrane of a mixed matrix of chitosan with the copolymers 2-ethacryloyloxy ethyl trimethylammonium chloride and 2-hydroxyethyl acrylate. For the membranes with a concentration of 25% of co-polymers, the mechanical properties of elongation at break and tensile strength increased 2.4 and 2 times, reaching maximum values of 4.3% and 61.2 MPa, respectively. The authors observed high hydraulic permeability, salt selectivity, and that the membrane hydrophilicity increased with the concentration of copolymers, causing an increase in the anti-fouling characteristic of the membrane. For the concentration of 50% of the copolymer, the hydraulic permeability was 20.6 L m^{-2} h^{-1}, and the rejection of the inorganic salts $ZnCl_2$ and NaCl was 97.5 and 57.1%, respectively. Thus, this study showed that the mixture of chitosan with the copolymers is a promising strategy to prepare membranes positively charged with mechanical properties and great nanofiltration performances.

Zhou et al [38] fabricated, by the phase inversion technique, ultrafiltration membranes from the blend the polymers cellulose acetate and polyvinylpyrrolidone and the functionalization with cellulose nanocrystals. The results demonstrated that, after the incorporation of the cellulose nanocrystals, the membrane became more porous and the mechanical property of tensile strength increased 47%. However, only for concentrations lower than 0.5% (wt%) of cellulose nanocrystals, the water flow increased and high rejection of bovine serum albumin (>90%) was verified. The authors also observed that the membrane functionalization reduced the fouling, with a recovery of 68% of the flow.

The functionalization can also be performed by combining organic and inorganic additives. El Reash et al [105] fabricated membranes using the polymeric blend of chitosan and polyacrylamide and functionalized them with silver nanoparticles. The membranes were used for the rejection of Cu (II) ions in aqueous solutions. The composed membranes were efficient in the rejection of the Cu (II) ions, being that the temperature increase from 20 to 40°C accelerated the mass transfer of Cu (II) ions to the membrane surface. The functionalization of the membranes with silver nanoparticles was also verified to have provided antimicrobial properties against Gram-negative, Gram-positive, and Candida sp. microorganisms. The authors concluded that the functionalization allowed the lifespan of the membrane to be extended, due to the inability of bacteria to form a biofilm that can cause membrane fouling. Also, with the aim of increasing the anti-fouling property of the membrane and consequently its lifespan, Panda et al [163] fabricated a membrane of a mixed matrix using amine stabilized iron oxide (Fe_3O_4) nanoparticles and polyacrylonitrile (PAN) coated with chitosan. The chitosan coating increased the hydrophilicity of the membrane, however, the impregnation of Fe_3O_4 caused a slight decrease of this characteristic. Besides, the mechanical resistance and the zeta potential of the membrane increased with the concentration of Fe_3O_4 and the membrane became denser and with reduced roughness. The adsorption of humic acid and the fouling characteristic increased with the quantity of amine stabilized magnetic nanoparticles, having a higher value in the concentration of 0.4% (wt%), in which the maximum adsorption capacity of humic acid was reached (70 mg g^{-1}) with flow recovery of 95%.

Alhalili et al [164] used membranes fabricated with the biopolymer cellulose acetate and functionalized with titanium dioxide (TiO_2) and zinc oxide (ZnO) nanoparticles in the rejection of synthetic thiol (dithioterethiol). The TiO_2 and ZnO nanoparticles decreased the contact angle of the cellulose acetate membrane, turning it into more hydrophilic, and, thus, with a greater anti-fouling property. In addition, the permeability that was 8.82 L m^{-2} h^{-1} bar^{-1} with the cellulose acetate membrane, increased to 20.77 and 21.96 L m^{-2} h^{-1} bar^{-1} when functionalized with TiO_2 and ZnO, respectively. The rejection of dithioterethiol at 2 mM of concentration and pH 10 with filtration under the pressure of 2 bar was 99.67% for the membrane functionalized with ZnO. The bond ZnO-Sulfur caused a significant reduction of the mobility of sulfur diffusion, while the TiO_2 did not present great effects.

A mixed matrix membrane based on cellulose acetate with the incorporation of mixed metal oxides (Fe - Al - Mn) and chitosan was fabricated by Chaudhary and Maiti [165] with the aim of removing fluoride contamination in groundwater. To achieve that, several parameters were optimized, reaching the optimum conditions of transmembrane pressure of 6 to 8 bar, pH between 6 and 9, and adsorbent mass of 8%. At these conditions, the membrane with an area of 1 m^2 was capable of treating 4000 litters of fluoride contaminated water. The fluorine rejection occurred initially by adsorption and posteriorly by electrostatic repulsion, due to the ions adsorbed in the membrane that presented negative surface charges. Besides, the studied membrane presented antimicrobial potential in an aqueous medium, being less susceptible to microbial attack compared to the non-functionalized cellulose acetate membrane.

In Table 2, a summary of several studies that performed the functionalization of membranes to make them more sustainable and improve their performance, especially regarding their anti-fouling capacity, is presented. Among these studies, the incorporation of organic and/or inorganic materials and nanomaterials in the membrane surface is highlighted, which is becoming a promising alternative and able to change the hydrophilicity, mechanical, chemical, and thermal resistance, permeability, and selectivity of the membranes. This combination of organic and inorganic materials and the polymers/biopolymers can also provide the membrane characteristics of adsorption, catalysis, photocatalysis, and antimicrobial. Thus, the membranes are fabricated with characteristics and performance that are desirable to the industrial process in a more biodegradable and sustainable way.

Table 2 Recent studies of the functionalization of porous polymeric membranes.

Polymer	Functionalization	Property	Application	Permeability*	Anti-fouling capacity*	Ref.
Polyvinylidene fluoride	Titanium dioxide (TiO$_2$), and zinc oxide (ZnO)	Anti-fouling and Photocatalytic	Wastewater treatment	Increase of 33.5% of the permeability for the membrane with TiO$_2$/ZnO 3:1	Increase of 73% in the anti-fouling capacity for the membrane with TiO$_2$/ZnO 1:3 and under visible light	[21]
Polysulfone modified with polyamide thin film	Natural Clinoptilolite	Anti-fouling	Water desalination	Increase of 33% in the water flow	Increase of 88% in the anti-fouling property	[41]
Polyethersulfone and alumina	Poly(styrene sulfonate) and poly (allylamine hydrochloride)	High rejection of ionic liquids	Recovery of ionic liquid	Permeability of 2.5 L m^{-2} h^{-1} bar^{-1} for the polyethersulfone membrane and 4.8 L m^{-2} h^{-1} bar^{-1} for the alumina membrane. Permeability reduction according to the	Increase in the rejection of the sugars Cellobiose, Glucose, Xylose, Fructose, 1-ethyl-3-methylimidazolium acetate, and 1-butyl-3-methylimidazolium chloride according to the increase in the layers of	[166]

				increase in the layers of polymer	polymer for both membranes	
Polyethersulfone modified with polyamide thin film	Acyl chlorided graphene oxide	Anti-fouling and high ions rejection	Water desalination	Increase in the flow from 11.6 to 22.6 L m^{-2} h^{-1}	Rejection of Na$_2$SO$_4$ increased from 95.0% to 97.1%	[167]
Polyethersulfone	Atmospheric pressure argon jet plasma treatment	Anti-fouling	Concentration of whey proteins	Decrease of 86% in the hydraulic permeability of the functionalized membrane	Increase in the hydrophilicity of the membrane and decrease the surface roughness resulting in a membrane with better anti-fouling characteristics. Increase in the flow recovery rate of 70% and high capacity of rejection of the protein	[5]
Polysulfone	Polyamide activated with low pressure nitrogen plasma activated	Anti-fouling	Concentration of fruit juice	Increase in the membrane hydrophilicity and hydraulic permeability (3 times higher)	Decrease of 30% in the time of juice concentration by osmotic distillation for the functionalized membranes. The concentration of the juice from 17.6 °Brix and 18.8 °Brix, reaching up to 60.9 °Brix and 60.4 °Brix in 430 and 340 min, respectively	[168]
Polyamide	Titanium dioxide	Anti-fouling and Photocatalytic	Wastewater treatment	Permeability increase from 6.15 to 7.28 L h^{-1} m^{-2} bar^{-1}	Improvement in the anti-fouling properties due to the increase of hydrophilicity and promotion of photocatalytic activity	[147]
Polyethersulfone	Polyethylenimine and graphene oxide	Anti-fouling	Wastewater treatment	Reduction in the water permeability, being 99.4 L m^{-2} h^{-1} bar^{-1} for the best rejection result	Rejection of the Blue Corazol dye of 97.8% and flow recovery rate higher than 80%	[8]

Polyethersulfone	Cellulose nanofibers and nanocrystals	Anti-fouling	Wastewater treatment	Highest water flow for the membranes with cellulose nanocrystals, followed by the membranes with cellulose nanofibers	Reduction of fouling and highest rejection for bovine serum albumine and humic acid	[43]
Polyethersulfone	Copper-modified titanate nanotubes	Anti-fouling and antimicrobial	Permeation of bovine serum albumin protein	Highest flow in the membrane with 15% (wt%) of polyethersulfone and highest copper content	Improvement of the anti-fouling property and highest rejection of albumine for the membrane with lower flow	[107]
Polyethersulfone	TiO$_2$@Ni particles	Anti-fouling	Permeation of bovine serum albumin protein	5-times increase of the flow	Rejection of 95.85% of albumine, and flow recovery rate higher than 75%	[148]
Polyethersulfone	Zeolitic imidazolate framework-L nanoflakes and carboxylated graphene oxide	Anti-fouling	Removal of pharmaceutical compounds	Increase in the hydrophilicity of the membrane and increase in the water permeability (203.5 L m^{-2} h^{-1} bar^{-1}) in the membrane with 0.25% (wt%) of graphene and 0.25% (wt%) of zeolitic.	Improvement in the anti-fouling properties and flow recovery rate of 95.1%. Amoxicillin rejection of 98.9% in laboratory-produced waters	[9]
Polyethersulfone modified with polyamide thin film	Functionalized boron nitride nanosheets	Anti-fouling and high ions rejection	Water desalination	Increase of 25% in the water flow	96.4% rejection of the salt NaCl, and 96% of flow recovery	[10]
Polysulfone	Polyvinylpyrrolidone and silicon dioxide nanoparticles	Anti-fouling and high rejection of amoxicillin	Removal of antibiotics from the aqueous medium	Increase in the flow from 6.6 L m^{-2} h^{-1} bar^{-1} (pure membrane) to 42.28 L m^{-2} h^{-1} bar^{-1} (membrane functionalized with 3% of silicon dioxide)	Increase in the separation performance of the amoxicillin from 66.52% to 89.81% in the membrane with 4% (wt%) of silicon dioxide. Improvement in the hydrophilicity and the anti-fouling	[7]

					property of the membrane. The relative flow reduction ratio was 34.14% for the modified membrane	
Polyethersulfone	2-mercaptoethanol capped zinc sulfide quantum dots	Anti-fouling and high rejection of dye	Rejection of dye	Increase in the hydrophilicity of the membrane and the flow with water 249.1 L m^{-2} h^{-1} b ar^{-1} for 1% of added nanocomposite	Improvement in the anti-fouling capacity and increase in the flow recovery rate from 52.6% to 87.9% with the addition of the nanocomposite. Increase in the rejection of the Reactive Red 195 dye from 91.3% to 96.1%.	[12]
Polyvinylidene fluoride	Ionic liquid triocty (dodecyl) phosphonium chloride	High recovery of platinum group metals	Separation of metals	The isotropic dense membrane achieved lower values of initial flow and transport kinetics slower than the porous anisotropic membrane	Recovery of Pt and Pd higher than 90% for both membranes and excellent chemical resistance property during exposition of 10 days to aggressive chemical products	[14]
Polysulfone and polytetrafluoroethylene	Graphene oxide	Anti-fouling	Removal of lactose and concentration of fats and proteins	Improvement in the anti-fouling capacity and the hydrophilicity of the membranes. Flow recovery of 89% for the PS membrane and 43% for the PTFE membrane	Higher lactose concentration in the permeate for the functionalized membranes	[13]

* compared to the membrane without additive. *Source: author*

4.   Final Considerations

This review approached the possibility to improve the membrane separation process from the use of membranes fabricated in a more sustainable way and with similar or higher mechanical, chemical, and thermal resistance characteristics when compared to membranes fabricated with synthetic polymers. Basic concepts that involve the structural and morphological classification of the membranes, as well as problems related to the

Advanced Functional Membranes
Materials Research Foundations **120** (2022) 111-150

Materials Research Forum LLC
https://doi.org/10.21741/9781644901816-4

fouling development during the filtration process were also presented. Concerning the sustainability concepts, the membrane separation process was observed to be able to be considered as a Green Technology due to the low energy consumption, the non-generation of pollutants, and the possibility to reuse the membranes for several cycles without decreasing its performance regarding the permeate flow and selectivity. However, the studies found in the literature showed that it is possible to improve even more the sustainability concept in the membrane separation process. This improvement is achieved through structural modification of the membranes using biopolymers, solvents that are less harmful to human health, and functionalization of the membranes with organic and/or inorganic additives. Among the biopolymers, chitosan and cellulose acetate are the most used ones to fabricate membranes individually, but mainly as a blend of synthetic polymers, because the biopolymers dot not always present adequate chemical, physical, and thermal resistance to be used industrially. Thus, the blend between the polymers and biopolymers is a possibility to have a more sustainable membrane and with desirable characteristics. Concerning the green solvents, water, ethyl lactate, and ionic liquids would be the most indicated. This review also evidences the great volume of studies that were developed in the last 10 years regarding the functionalization of membranes, which was preferably performed inserting organic and/or inorganic materials and nanomaterials in the polymeric solution or the membrane surface. The functionalization of membranes becomes a great alternative to improve the morphology, increase the anti-fouling property, and lifespan of the membranes because it allows associating to the membranes adsorptive, catalytic, photocatalytic, and antimicrobial characteristics. Thus, this review demonstrated that, with the adequate structural modifications in the membranes, the environmentally sustainable Green Technology characteristic that does not harm the environment nor future generations can be associated with the membrane separation process, besides the separation, purification, and concentration efficiency, already consolidated in industrial processes.

References

[1]   J.G. Crespo, K.W. Boddeker, Membrane Processes in Separation and Purification, Vol 272, Springer Science+Business Media Dordrecht, 1994.

[2]   A.Y. Tamime, Membrane Processing: Dairy and Beverage Applications, first ed., 2013. https://doi.org/10.1002/9781118457009

[3]   K. Hu, J.M. Dickson, Membrane Processing for Dairy Ingredient Separation, first ed., Wiley Blackwell, 2015.

Materials Research Forum LLC
https://doi.org/10.21741/9781644901816-4

[4]   M. Omidvar, M. Soltanieh, S.M. Mousavi, E. Saljoughi, A. Moarefian, H. Saffaran, Preparation of hydrophilic nanofiltration membranes for removal of pharmaceuticals from water, J. Environ. Heal. Sci. Eng. 13 (2015) 1–9. https://doi.org/10.1186/s40201-015-0157-3

[5]   I. Damar Huner, H.A. Gulec, Fouling behavior of poly(ether)sulfone ultrafiltration membrane during concentration of whey proteins: Effect of hydrophilic modification using atmospheric pressure argon jet plasma, Colloids Surfaces B: Biointerfaces. 160 (2017) 510–519. https://doi.org/10.1016/j.colsurfb.2017.10.003

[6]   S.F. Zakeritabar, M. Jahanshahi, M. Peyravi, Photocatalytic Behavior of Induced Membrane by ZrO_2–SnO_2 Nanocomposite for Pharmaceutical Wastewater Treatment, Catal. Letters. 148 (2018) 882–893. https://doi.org/10.1007/s10562-018-2303-x

[7]   M. Shakak, R. Rezaee, A. Maleki, A. Jafari, M. Safari, B. Shahmoradi, H. Daraei, S.M. Lee, Synthesis and characterization of nanocomposite ultrafiltration membrane (PSF/PVP/SiO_2) and performance evaluation for the removal of amoxicillin from aqueous solutions, Environ. Technol. Innov. 17 (2020). https://doi.org/10.1016/j.eti.2019.100529

[8]   N.C. Homem, N. de C.L. Beluci, S. Amorim, R. Reis, A.M.S. Vieira, M.F. Vieira, R. Bergamasco, M.T.P. Amorim, Surface modification of a polyethersulfone microfiltration membrane with graphene oxide for reactive dyes removal, Appl. Surf. Sci. 486 (2019) 499–507. https://doi.org/10.1016/j.apsusc.2019.04.276

[9]   A. Modi, J. Bellare, Amoxicillin removal using polyethersulfone hollow fiber membranes blended with ZIF-L nanoflakes and cGO nanosheets: Improved flux and fouling-resistance, J. Environ. Chem. Eng. 8 (2020) 103973. https://doi.org/10.1016/j.jece.2020.103973

[10]   R. Wang, Z.X. Low, S. Liu, Y. Wang, S. Murthy, W. Shen, H. Wang, Thin-film composite polyamide membrane modified by embedding functionalized boron nitride nanosheets for reverse osmosis, J. Memb. Sci. 611 (2020). https://doi.org/10.1016/j.memsci.2020.118389

[11]   S.A. Gokulakrishnan, G. Arthanareeswaran, Z. László, G. Veréb, S. Kertész, J. Kweon, Recent development of photocatalytic nanomaterials in mixed matrix membrane for emerging pollutants and fouling control, membrane cleaning process, Chemosphere. 281 (2021). https://doi.org/10.1016/j.chemosphere.2021.130891

[12]   M.R. Ganjali, M.A. Al-Naqshabandi, B. Larijani, A. Badiei, V. Vatanpour, H.R. Rajabi, H. Rezania, S. Paziresh, G. Mahmodi, S.J. Kim, M.R. Saeb, Improvement of dye and protein filtration efficiency using modified PES membrane with 2-

mercaptoethanol capped zinc sulfide quantum dots, Chem. Eng. Res. Des. 168 (2021) 109–121. https://doi.org/10.1016/j.cherd.2020.12.026

[13]   A. Morelos-Gomez, S. Terashima, A. Yamanaka, R. Cruz-Silva, J. Ortiz-Medina, R. Sánchez-Salas, J.L. Fajardo-Díaz, E. Muñoz-Sandoval, F. López-Urías, K. Takeuchi, S. Tejima, M. Terrones, M. Endo, Graphene oxide membranes for lactose-free milk, Carbon. 181 (2021) 118–129. https://doi.org/10.1016/j.carbon.2021.05.005

[14]   A.T.N. Fajar, T. Hanada, M. Goto, Recovery of platinum group metals from a spent automotive catalyst using polymer inclusion membranes containing an ionic liquid carrier, J. Memb. Sci. 629 (2021). https://doi.org/10.1016/j.memsci.2021.119296

[15]   W.S. Wan Ngah, L.C. Teong, M.A.K.M. Hanafiah, Adsorption of dyes and heavy metal ions by chitosan composites: A review, Carbohydr. Polym. 83 (2011) 1446–1456. https://doi.org/10.1016/j.carbpol.2010.11.004

[16]   E. Drioli, A.I. Stankiewicz, F. Macedonio, Membrane engineering in process intensification-An overview, J. Memb. Sci. 380 (2011) 1–8. https://doi.org/10.1016/j.memsci.2011.06.043

[17]   Y. Medina-Gonzalez, P. Aimar, J.F. Lahitte, J.C. Remigy, Towards green membranes: Preparation of cellulose acetate ultrafiltration membranes using methyl lactate as a biosolvent, Int. J. Sustain. Eng. 4 (2011) 75–83. https://doi.org/10.1080/19397038.2010.497230

[18]   M. Padaki, R. Surya Murali, M.S. Abdullah, N. Misdan, A. Moslehyani, M.A. Kassim, N. Hilal, A.F. Ismail, Membrane technology enhancement in oil-water separation. A review, Desalination. 357 (2015) 197–207. https://doi.org/10.1016/j.desal.2014.11.023

[19]   J.H. Jhaveri, Z.V.P. Murthy, A comprehensive review on anti-fouling nanocomposite membranes for pressure driven membrane separation processes, Desalination. 379 (2016) 137–154. https://doi.org/10.1016/j.desal.2015.11.009

[20]   C. Santhosh, V. Velmurugan, G. Jacob, S.K. Jeong, A.N. Grace, A. Bhatnagar, Role of nanomaterials in water treatment applications: A review, Chem. Eng. J 306 (2016) 1116–1137. https://doi.org/10.1016/j.cej.2016.08.053

[21]   N. Li, Y. Tian, J. Zhang, Z. Sun, J. Zhao, J. Zhang, W. Zuo, Precisely-controlled modification of PVDF membranes with 3D TiO_2/ZnO nanolayer: enhanced anti-fouling performance by changing hydrophilicity and photocatalysis under visible light

irradiation, J. Memb. Sci. 528 (2017) 359–368.
https://doi.org/10.1016/j.memsci.2017.01.048

[22]   S. Beisl, S. Monteiro, R. Santos, A.S. Figueiredo, M.G. Sánchez-Loredo, M.A. Lemos, F. Lemos, M. Minhalma, M.N. de Pinho, Synthesis and bactericide activity of nanofiltration composite membranes - Cellulose acetate/silver nanoparticles and cellulose acetate/silver ion exchanged zeolites, Water Res. 149 (2019) 225–231. https://doi.org/10.1016/j.watres.2018.10.096

[23]   Z. Huang, B. Gong, C. Huang, S. Pan, P. Wu, Z. Dang, Performance evaluation of integrated adsorption-nanofiltration system for emerging compounds removal: Exemplified by caffeine, diclofenac and octylphenol, J. Environ. Manage., 231 (2019) 121–128. https://doi.org/10.1016/j.jenvman.2018.09.092

[24]   T. Marino, F. Galiano, A. Molino, A. Figoli, New frontiers in sustainable membrane preparation : Cyrene^{TM} as green bioderived solvent, J. Membr. Sci., 580 (2019) 224–234. https://doi.org/10.1016/j.memsci.2019.03.034

[25]   S. Samsami, M. Mohamadizaniani, M. Sarrafzadeh, E.R. Rene, M. Firoozbahr, Recent advances in the treatment of dye-containing wastewater from textile industries : Overview and perspectives, Process Saf. Environ. Prot. 143 (2020) 138–163. https://doi.org/10.1016/j.psep.2020.05.034

[26]   W.M. Nielen, J.D. Willott, Z.M. Esguerra, W.M. de Vos, Ion specific effects on aqueous phase separation of responsive copolymers for sustainable membranes, J. Colloid Interface Sci. 576 (2020) 186–194. https://doi.org/10.1016/j.jcis.2020.04.125

[27]   W.M. Nielen, J.D. Willott, W.M. De Vos, Aqueous Phase Separation of Responsive Copolymers for Sustainable and Mechanically Stable Membranes, Appl. Polym. Mater. 2 (2020) 1702–1710. https://doi.org/10.1021/acsapm.0c00119

[28]   M.A. Rasool, C. Van Goethem, I.F.J. Vankelecom, Green preparation process using methyl lactate for cellulose-acetate-based nanofiltration membranes, Sep. Purif. Technol. 232 (2020). https://doi.org/10.1016/j.seppur.2019.115903

[29]   T. Shindhal, P. Rakholiya, S. Varjani, A. Pandey, H.H. Ngo, W. Guo, H.Y. Ng, M.J. Taherzadeh, A critical review on advances in the practices and perspectives for the treatment of dye industry wastewater, Bioengineered. 12 (2021) 70–87. https://doi.org/10.1080/21655979.2020.1863034

[30]   E.K. Tetteh, S. Rathilal, D. Asante-Sackey, M.N. Chollom, Prospects of synthesized magnetic TiO$_2$-based membranes for wastewater treatment: A review, Materials. 14 (2021) 11–15. https://doi.org/10.3390/ma14133524

[31]   A. Gul, J. Hruza, F. Yalcinkaya, Fouling and chemical cleaning of microfiltration membranes: A mini-review, Polymers. 13 (2021). https://doi.org/10.3390/polym13060846

[32]   W. Kujawski, J. Kujawa, E. Wierzbowska, S. Cerneaux, M. Bryjak, J. Kujawski, Influence of hydrophobization conditions and ceramic membranes pore size on their properties in vacuum membrane distillation of water-organic solvent mixtures, J. Memb. Sci. 499 (2016) 442–451. https://doi.org/10.1016/j.memsci.2015.10.067

[33]   E. Bet-moushoul, Y. Mansourpanah, K. Farhadi, M. Tabatabaei, TiO_2 nanocomposite based polymeric membranes : A review on performance improvement for various applications in chemical engineering processes, Chem. Eng. J. 283 (2016) 29–46. https://doi.org/10.1016/j.cej.2015.06.124

[34]   C. Li, W. Sun, Z. Lu, X. Ao, S. Li, Ceramic nanocomposite membranes and membrane fouling: A review, Water Res. 175 (2020). https://doi.org/10.1016/j.watres.2020.115674

[35]   M. Aider, Chitosan application for active bio-based films production and potential in the food industry: Review, LWT - Food Sci. Technol. 43 (2010) 837–842. https://doi.org/10.1016/j.lwt.2010.01.021

[36]   F. Galiano, K. Briceño, T. Marino, A. Molino, K.V. Christensen, A. Figoli, Advances in biopolymer-based membrane preparation and applications, J. Memb. Sci. 564 (2018) 562–586. https://doi.org/10.1016/j.memsci.2018.07.059

[37]   L.A. Goetz, B. Jalvo, R. Rosal, A.P. Mathew, Superhydrophilic anti-fouling electrospun cellulose acetate membranes coated with chitin nanocrystals for water filtration, J. Memb. Sci. 510 (2016) 238–248. https://doi.org/10.1016/j.memsci.2016.02.069

[38]   J. Zhou, J. Chen, M. He, J. Yao, Cellulose acetate ultrafiltration membranes reinforced by cellulose nanocrystals: Preparation and characterization, J. Appl. Polym. Sci. 133 (2016) 1–7. https://doi.org/10.1002/app.43946

[39]   F. Russo, F. Galiano, A. Iulianelli, A. Basile, A. Figoli, Biopolymers for sustainable membranes in CO_2 separation: a review, Fuel Process. Technol. 213 (2021). https://doi.org/10.1016/j.fuproc.2020.106643

[40]   M. Mulder, Basic Principles of Membrane Technology, second ed., Kluwer Academic Publisher, 1996. https://doi.org/10.1007/978-94-009-1766-8

[41]   M. Safarpour, V. Vatanpour, A. Khataee, H. Zarrabi. P. Gholami, M. E. High flux and fouling resistant reverse osmosis membrane modified with plasma treated natural zeolite, Desalination. 411 (2017) 89–100. https://doi.org/10.1016/j.desal.2017.02.012

[42]   M.T. Alresheedi, B. Barbeau, O.D. Basu, Comparisons of NOM fouling and cleaning of ceramic and polymeric membranes during water treatment, Sep. Purif. Technol. 209 (2019) 452–460. https://doi.org/10.1016/j.seppur.2018.07.070

[43]   L. Bai, Y. Liu, A. Ding, N. Ren, G. Li, H. Liang, Surface coating of UF membranes to improve antifouling properties: A comparison study between cellulose nanocrystals (CNCs) and cellulose nano fibrils (CNFs), Chemosphere. 217 (2019) 76–84. https://doi.org/10.1016/j.chemosphere.2018.10.219

[44]   A. Rahimpour, S.S. Madaeni, Polyethersulfone (PES)/ cellulose acetate phthalate (CAP) blend ultrafiltration membranes: Preparation, morphology, performance and antifouling properties, J. Memb. Sci. 305 (2007) 299–312. https://doi.org/10.1016/j.memsci.2007.08.030

[45]   M.Z. Yunos, Z. Harun, H. Basri, A.F. Ismail, Studies on fouling by natural organic matter (NOM) on polysulfone membranes: Effect of polyethylene glycol (PEG), Desalination. 333 (2014) 36–44. https://doi.org/10.1016/j.desal.2013.11.019

[46]   R. Krishnamoorthy, V. Sagadevan, Polyethylene glycol and iron oxide nanoparticles blended polyethersulfone ultrafiltration membrane for enhanced performance in dye removal studies, E-Polymers. 15 (2015) 151–159. https://doi.org/10.1515/epoly-2014-0214

[47]   L.D. Fiorentin-Ferrari, K.M. Celant, B.C. Gonçalves, S.M. Teixeira, V. Slusarski-Santana, A.N. Módenes, Fabrication and characterization of polysulfone and polyethersulfone membranes applied in the treatment of fish skin tanning effluent, J. Clean. Prod. 294 (2021). https://doi.org/10.1016/j.jclepro.2021.126127

[48]   N. Nady, M.C.R. Franssen, H. Zuilhof, M.S.M. Eldin, R. Boom, K. Schroën, Modification methods for poly(arylsulfone) membranes: A mini-review focusing on surface modification, Desalination. 275 (2011) 1–9. https://doi.org/10.1016/j.desal.2011.03.010

[49]   L. Zhang, B. Chen, A. Ghaffar, X. Zhu, Nanocomposite Membrane with Polyethylenimine-Grafted Graphene Oxide as a Novel Additive to Enhance Pollutant Filtration Performance, Environ. Sci. Technol. 52 (2018) 5920–5930. https://doi.org/10.1021/acs.est.8b00524

[50] G.D. Değermenci, N. Değermenci, V. Ayvaoğlu, E. Durmaz, D. Çakır, E. Akan, Adsorption of reactive dyes on lignocellulosic waste; characterization, equilibrium, kinetic and thermodynamic studies, J. Clean. Prod. 225 (2019) 1220–1229. https://doi.org/10.1016/j.jclepro.2019.03.260

[51] F. Yalcinkaya, E. Boyraz, J. Maryska, K. Kucerova, A review on membrane technology and chemical surface modification for the oily wastewater treatment, Materials. 13 (2020). https://doi.org/10.3390/ma13020493

[52] H. Saleem, S.J. Zaidi, Nanoparticles in reverse osmosis membranes for desalination: A state of the art review, Desalination. 475 (2020). https://doi.org/10.1016/j.desal.2019.114171

[53] M. Jani, J.A. Arcos-Pareja, M. Ni, Engineered Zero-Dimensional Fullerene/Carbon Dots-Polymer Based Nanocomposite Membranes for Wastewater Treatment, Molecules. 25 (2020) 1–28. https://doi.org/10.3390/molecules25214934

[54] A.A. Amusa, A.L. Ahmad, J.K. Adewole, Study on lignin-free lignocellulosic biomass and PSF-PEG membrane compatibility, BioResources. 16 (2020) 1063–1075. https://doi.org/10.15376/biores.16.1.1063-1075

[55] O.S. Serbanescu, S.I. Voicu, V.K. Thakur, Polysulfone functionalized membranes: Properties and challenges, Mater. Today Chem. 17 (2020). https://doi.org/10.1016/j.mtchem.2020.100302

[56] J. Ayyavoo, T.P.N. Nguyen, B.M. Jun, I.C. Kim, Y.N. Kwon, Protection of polymeric membranes with antifouling surfacing via surface modifications, Colloids Surfaces A Physicochem. Eng. Asp. 506 (2016) 190–201. https://doi.org/10.1016/j.colsurfa.2016.06.026

[57] H.K. Lonsdale, The growth of membrane technology, J. Memb. Sci. 10 (1982) 81–181. https://doi.org/10.1016/S0376-7388(00)81408-8

[58] K.W. Böddeker, Commentary: Tracing membrane science, J. Memb. Sci. 100 (1995) 65–68. https://doi.org/10.1016/0376-7388(94)00223-L

[59] R.W. Baker, Membrane technologies and applications, second ed., John Wiley & Sons Ltd, 2004.

[60] R. Field, E. Bekassy-Molnar, F. Lipnizki, G. Vatai, Engineering aspects of membrane separation and application in food processing, first ed., Taylor & Francis Group, 2017. https://doi.org/10.4324/9781315374901

[61] M. Cheryan, Ultrafiltration and Microfiltration Handbook, second ed., Taylor & Francis Routledge, 1998. https://doi.org/10.1201/9781482278743

[62]   Z.F. Cui, H.S. Muralidhara, Membrane technology - A Practical Guide to Membrane Technology and Applications in Food and Bioprocessing, first ed., Elsevier Ltd, 2010.

[63]   G. Lofrano, M. Carotenuto, G. Libralato, R.F. Domingos, A. Markus, L. Dini, R.K. Gautam, D. Baldantoni, M. Rossi, S.K. Sharma, M.C. Chattopadhyaya, M. Giugni, S. Meric, Polymer functionalized nanocomposites for metals removal from water and wastewater: An overview, Water Res. 92 (2016) 22–37. https://doi.org/10.1016/j.watres.2016.01.033

[64]   H. Strathmann, Ion-Exchange membrane separation processes, first ed., Elsevier B.V., 2004. https://doi.org/10.1002/14356007.a16_187.pub2

[65]   H. Strathmann, Membrane separation processes, J. Memb. Sci. 9 (1981) 121–189. https://doi.org/10.1016/S0376-7388(00)85121-2

[66]   K. Mohanty, M.K. Purkait, Membrane technologies and applications, first ed., Taylor & Francis Group, 2012. https://doi.org/10.1201/b11416

[67]   L.W. Jye, A.F. Ismail, Nanofiltration membranes - Synthesis, Characterization, and Applications, first ed., Taylor & Francis Group, 2016. https://doi.org/10.1201/9781315181479

[68]   S. Mei, C. Xiao, X. Hu, Preparation of Porous PVC Membrane via a Phase Inversion Method from PVC/DMAc/Water/Additives, J. Appl. Polym. Sci. 120 (2010) 557–562. https://doi.org/10.1002/app.33219

[69]   K. Scott, Handbook of Industrial Membranes, second ed., Elsevier Science, 1995.

[70]   M. Kumar, Z. Gholamvand, A. Morrissey, K. Nolan, M. Ulbricht, J. Lawler, Preparation and characterization of low fouling novel hybrid ultrafiltration membranes based on the blends of GO-TiO$_2$ nanocomposite and polysulfone for humic acid removal, J. Memb. Sci. 506 (2016) 38–49. https://doi.org/10.1016/j.memsci.2016.02.005

[71]   S. Benkhaya, S. M'rabet, R. Hsissou, A El Harfi, Synthesis of new low-cost organic ultrafiltration membrane made from Polysulfone/Polyetherimide blends and its application for soluble azoic dyes removal, J. Mater. Res. Technol. 9 (2020) 4763–4772. https://doi.org/10.1016/j.jmrt.2020.02.102

[72]   M.T. Ravanchi, T. Kaghazchi, A. Kargari, Application of membrane separation processes in petrochemical industry: a review, Desalination. 235 (2009) 199–244. https://doi.org/10.1016/j.desal.2007.10.042

[73]   M. Ulbricht, Advanced functional polymer membranes, Polymer. 47 (2006) 2217–2262. https://doi.org/10.1016/j.polymer.2006.01.084

[74]   J.K. Shethji, S.M.C. Ritchie, Microfiltration membranes functionalized with multiple styrenic homopolymer and block copolymer grafts, J. Appl. Polym. Sci. 132 (2015) 1–11. https://doi.org/10.1002/app.42501

[75]   W. Ho, K.K. Sirkar, Membrane Handbook, Springer Science Business Media New York, 1992. https://doi.org/10.1007/978-1-4615-3548-5

[76]   A.K. Pabby, S.S.H. Rizvi, A.M. Sastre, Handbook of Membrane Separations - Chemical, Pharmaceutical, Food, and Biotechnological Applications, first ed., Taylor & Francis Group, 2008. https://doi.org/10.1201/9781420009484

[77]   T.J. Britz, R.K. Robinson, Advanced Dairy Science and Technology, first ed., Blackwell Publishing Ltd, 2008. https://doi.org/10.1002/9780470697634

[78]   C. Charcosset, Membrane Processes in Biotechnology and Pharmaceutics, first ed., Elsevier B.V., 2012. https://doi.org/10.1016/B978-0-444-56334-7.00007-1

[79]   A.W. Mohammad, Y.H. Teow, W.L. Ang, Y.T. Chung, D.L. Oatley-Radcliffe, N. Hilal, Nanofiltration membranes review: Recent advances and future prospects, Desalination. 356 (2015) 226–254. https://doi.org/10.1016/j.desal.2014.10.043

[80]   M. Paul, S.D. Jons, Chemistry and fabrication of polymeric nanofiltration membranes: A review, Polymer. 103 (2016) 417–456. https://doi.org/10.1016/j.polymer.2016.07.085

[81]   M.K. Selatile, S.S. Ray, V. Ojijo, R. Sadiku, Recent developments in polymeric electrospun nanofibrous membranes for seawater desalination, RSC Adv. 8 (2018) 37915–37938. https://doi.org/10.1039/C8RA07489E

[82]   N.N. Li, A.G. Fane, W.S.W. Ho, T. Matsuura, Advanced membrane technology and applications, John Wiley & Sons, 2008.

[83]   S.K. Hubadillah, M.H.D. Othman, T. Matsuura, A.F. Ismail, M.A. Rahman, Z. Harun, J. Jaafar, M. Nomura, Fabrications and applications of low cost ceramic membrane from kaolin: A comprehensive review, Ceram. Int. 44 (2018) 4538–4560. https://doi.org/10.1016/j.ceramint.2017.12.215

[84]   W. Zhang, L. Ding, J. Luo, M.Y. Jaffrin, B. Tang, Membrane fouling in photocatalytic membrane reactors (PMRs) for water and wastewater treatment: A critical review, Chem. Eng. J. 302 (2016) 446–458. https://doi.org/10.1016/j.cej.2016.05.071

[85]   S. Meng, W. Fan, X. Li, Y. Liu, D. Liang, X. Liu, Intermolecular interactions of polysaccharides in membrane fouling during microfiltration, Water Res. 143 (2018) 38–46. https://doi.org/10.1016/j.watres.2018.06.027

[86]   A. Basile, A. Cassano, N. Rastogi, Advances in Membrane Technologies for Water Treatment - Materials, Processes and applications, Elsevier Ltd, 2015.

[87]   B. Huang, Z. Wu, H. Zhou, J. Li, C. Zhou, Z. Xiong, Z. Pan, G. Yao, B. Lai, Recent advances in single-atom catalysts for advanced oxidation processes in water purification, J. Hazard. Mater. 412 (2021). https://doi.org/10.1016/j.jhazmat.2021.125253

[88]   A.W. Zularisam, A.F. Ismail, R. Salim, Behaviours of natural organic matter in membrane filtration for surface water treatment – a review, Desalination. 194 (2006) 211–231. https://doi.org/10.1016/j.desal.2005.10.030

[89]   Y. Subasi, B. Cicek, Recent advances in hydrophilic modification of PVDF ultrafiltration membranes – a review: part I, Membr. Technol. (2017) 7–12. https://doi.org/10.1016/S0958-2118(17)30191-X

[90]   F. Zhao, K. Xu, H. Ren, L. Ding, J. Geng, Y. Zhang, Combined effects of organic matter and calcium on biofouling of nanofiltration membranes, J. Memb. Sci. 486 (2015) 177–188. https://doi.org/10.1016/j.memsci.2015.03.032

[91]   L. Weinrich, M. Lechevallier, C.N. Haas, Contribution of assimilable organic carbon to biological fouling in seawater reverse osmosis membrane treatment, Water Res. 101 (2016) 203–213. https://doi.org/10.1016/j.watres.2016.05.075

[92]   J.N. Hakizimana, B. Gourich, C. Vial, P. Drogui, A. Oumani, J. Naja, L. Hilali, Assessment of hardness, microorganism and organic matter removal from seawater by electrocoagulation as a pretreatment of desalination by reverse osmosis, Desalination. 393 (2016) 90–101. https://doi.org/10.1016/j.desal.2015.12.025

[93]   S. Jiang, Y. Li, B.P. Ladewig, A review of reverse osmosis membrane fouling and control strategies, Sci. Total Environ. 595 (2017) 567–583. https://doi.org/10.1016/j.scitotenv.2017.03.235

[94]   H. Chang, H. Liang, F. Qu, B. Liu, H. Yu, X. Du, G. Li, S.A. Snyder, Hydraulic backwashing for low-pressure membranes in drinking water treatment : A review, J. Memb. Sci. 540 (2017) 362–380. https://doi.org/10.1016/j.memsci.2017.06.077

[95]   N. V Thombre, A.P. Gadhekar, A. V Patwardhan, P.R. Gogate, Ultrasound induced cleaning of polymeric nanofiltration membranes, J. Ultrason. Sonochemistry. 62 (2020). https://doi.org/10.1016/j.ultsonch.2019.104891

[96]   K. De Sitter, C. Dotremont, I. Genné, L. Stoops, The use of nanoparticles as alternative pore former for the production of more sustainable polyethersulfone ultrafiltration membranes, J. Memb. Sci. 471 (2014) 168–178. https://doi.org/10.1016/j.memsci.2014.06.061

[97]   P.S. Goh, A.F. Ismail, N. Hilal, Nano-enabled membranes technology: Sustainable and revolutionary solutions for membrane desalination?, Desalination. 380 (2016) 100–104. https://doi.org/10.1016/j.desal.2015.06.002

[98]   X. Zhu, K. Yang, B. Chen, Membranes prepared from graphene-based nanomaterials for sustainable applications: a review, Environ. Sci. Nano. 4 (2017) 2267–2285. https://doi.org/10.1039/C7EN00548B

[99]   M.I. Baig, E.N. Durmaz, J.D. Willott, W.M. De Vos, Sustainable Membrane Production through Polyelectrolyte Complexation Induced Aqueous Phase Separation, Adv. Funct. Mater. 30 (2020). https://doi.org/10.1002/adfm.201907344

[100] J.D. Willott, W.M. Nielen, W.M. De Vos, Stimuli-Responsive Membranes through Sustainable Aqueous Phase Separation, ACS Appl. Polym. Mater. 2 (2020) 659–667. https://doi.org/10.1021/acsapm.9b01006

[101] M. Mondal, S. De, Characterization and antifouling properties of polyethylene glycol doped PAN-CAP blend membrane, RSC Adv. (2015). https://doi.org/10.1039/C5RA02889B

[102] A. Said, T.J. Daou, L. Limousy, J. Bikai, J. Halwani, J. Toufaily, T. Hamieh, P. Dutournié, Surface energy modification of a Na-mordenite thin layer treated by an alkaline solution, Mater. Express. 5 (2015) 451–456. https://doi.org/10.1166/mex.2015.1253

[103] V.K. Thakur, S.I. Voicu, Recent Advances in Cellulose and Chitosan Based Membranes for Water Purification: A Concise Review, Carbohydr. Polym. (2016) 148-165. https://doi.org/10.1016/j.carbpol.2016.03.030

[104] S. Mondal, Carbon Nanomaterials Based Membranes, J. Membr. Sci. Technol. 6 (2017) 18–20.

[105] Y.G.A. El-reash, A.M. Abdelghany, A.A. Elrazak, Removal and separation of Cu (II) from aqueous solutions using nano-silver chitosan/polyacrylamide membranes, Int. J. Biol. Macromol. 86 (2016) 789–798. https://doi.org/10.1016/j.ijbiomac.2016.01.101

[106] T. Marino, E. Blasi, S. Tornaghi, E. Di Nicolo, A. Figoli, Polyethersulfone membranes prepared with Rhodiasolv® Polarclean as water soluble green solvent, J. Membr. Sci. J. 549 (2018) 192–204. https://doi.org/10.1016/j.memsci.2017.12.007

[107] K. Szymanski, D. Darowna, P. Sienkiewicz, M. Jose, K. Szymanska, M. Zgrzebnicki, S. Mozia, Novel polyethersulfone ultrafiltration membranes modified with Cu/titanate nanotubes, J. Water Process Eng. 33 (2020). https://doi.org/10.1016/j.jwpe.2019.101098

[108] M.N. Subramanian, Polymer Blends and Composites: Chemistry and Technology, John Wiley & Sons, 2017. https://doi.org/10.1002/9781119383581

[109] Y. Ji, Q. An, F. Zhao, C. Gao, Fabrication of chitosan / PDMCHEA blend positively charged membranes with improved mechanical properties and high nanofiltration performances, Desalination. 357 (2015) 8–15. https://doi.org/10.1016/j.desal.2014.11.005

[110] Z. Xu, G. Liu, H. Ye, W. Jin, Z. Cui, Two-dimensional MXene incorporated chitosan mixed-matrix membranes for efficient solvent dehydration, J. Membr. Sci. 563 (2018) 625–632. https://doi.org/10.1016/j.memsci.2018.05.044

[111] C.C. De Faria, M. Favero, M.M.M. Caetano, A.H. Rosa, P.S. Tonello, Application of chitosan film as a binding phase in the diffusive gradients in thin films technique (DGT) for measurement of metal ions in aqueous solution, Anal. Bioanal. Chem. 412 (2020) 703–714. https://doi.org/10.1007/s00216-019-02281-4

[112] S. Al-gharabli, B. Al-omari, W. Kujawski, J. Kujawa, Biomimetic hybrid membranes with covalently anchored chitosan – Material design, transport and separation, Desalination. 491 (2020). https://doi.org/10.1016/j.desal.2020.114550

[113] H. Mahdavi, T. Shahalizade, Preparation, characterization and performance study of cellulose acetate membranes modified by aliphatic hyperbranched polyester, J. Membr. Sci. 473 (2015) 256–266. https://doi.org/10.1016/j.memsci.2014.09.013

[114] A. Ghaee, M. Shariaty-niassar, J. Barzin, T. Matsuura, Effects of chitosan membrane morphology on copper ion adsorption, J. Chem. Eng. 165 (2010) 46–55. https://doi.org/10.1016/j.cej.2010.08.051

[115] S.T. Koev, P.H. Dykstra, X. Luo, G.W. Rubloff, W.E. Bentley, G.F. Payne, R. Ghodssi, Chitosan : an integrative biomaterial for lab-on-a-chip devices, Lab Chip. 10 (2010) 3026–3042. https://doi.org/10.1039/c0lc00047g

[116] J. Wang, B. Wu, S. Yang, Y. Liu, A.G. Fane, J.W. Chew, Characterizing the scouring efficiency of Granular Activated Carbon (GAC) particles in membrane fouling mitigation via wavelet decomposition of accelerometer signals, J. Memb. Sci. 498 (2016) 105–115. https://doi.org/10.1016/j.memsci.2015.09.061

[117] R. Kumar, A.M. Isloor, A.F. Ismail, T. Matsuura, Performance improvement of polysulfone ultrafiltration membrane using N-succinyl chitosan as additive, Desalination. 318 (2013) 1–8. https://doi.org/10.1016/j.desal.2013.03.003

[118] A. Akbari, Z. Derikvandi, S. M. M. Rostami, Influence of chitosan coating on the separation performance, morphology and anti-fouling properties of the polyamide nanofiltration membranes, J. Ind. Eng. Chem. 28 (2015) 268–273. https://doi.org/10.1016/j.jiec.2015.03.002

[119] E. Salehi, P. Daraei, A.A. Shamsabadi, A review on chitosan-based adsorptive membranes, Carbohydr. Polym. 152 (2016) 419–432. https://doi.org/10.1016/j.carbpol.2016.07.033

[120] P.O. Osifo, H.W.J.P. Neomagus, H. Van Der Merwe, D.J. Branken, Transport properties of chitosan membranes for zinc (II) removal from aqueous systems, Sep. Purif. Technol. 179 (2017) 428–437. https://doi.org/10.1016/j.seppur.2017.02.030

[121] C.N.B. Elizalde, S. Al-Gharabli, J. Kujawa, M. Mavukkandy, S.W. Hasan, H.A. Arafat, Fabrication of Blend Polyvinylidene fluoride/Chitosan Membranes for Enhanced Flux and Fouling Resistance, Sep. Purif. Technol. 190 (2018) 68–76. https://doi.org/10.1016/j.seppur.2017.08.053

[122] A. Shakeri, H. Salehi, M. Rastgar, Chitosan-based thin active layer membrane for forward osmosis desalination, Carbohydr. Polym. 174 (2017) 658–668. https://doi.org/10.1016/j.carbpol.2017.06.104

[123] F.F. Ghiggi, L.D. Pollo, N.S.M. Cardozo, I.C. Tessaro, Preparation and characterization of polyethersulfone/N-phthaloyl-chitosan ultrafiltration membrane with antifouling property, Eur. Polym. J. 92 (2017) 61–70. https://doi.org/10.1016/j.eurpolymj.2017.04.030

[124] E. Bagheripour, A.R. Moghadassi, S.M. Hosseini, M.B. Ray, F. Parvizian, B. Van Der Bruggen, Highly hydrophilic and antifouling nanofiltration membrane incorporated with water-dispersible composite activated carbon/chitosan nanoparticles, Chem. Eng. Res. Des. 132 (2018) 812–821. https://doi.org/10.1016/j.cherd.2018.02.027

[125] N. Li, Y. Tian, J. Zhao, J. Zhang, L. Kong, J. Zhang, W. Zuo, Static adsorption of protein-polysaccharide hybrids on hydrophilic modified membranes based on atomic layer: Anti-fouling performance and mechanism insight, J. Memb. Sci. 548 (2018) 470–480. https://doi.org/10.1016/j.memsci.2017.11.063

[126] A. Figoli, T. Marino, S. Simone, E. Di Nicolò, X.M. Li, T. He, S. Tornaghi, E. Drioli, Towards non-toxic solvents for membrane: a review, Green Chem. 16 (2014) 4034–4059. https://doi.org/10.1039/C4GC00613E

[127] M. Razali, J.F. Kim, M. Attfield, P.M. Budd, E. Drioli, Y.M. Lee, G. Szekely, Sustainable wastewater treatment and recycling in membrane manufacturing, Green Chem. 17 (2015) 5196–5205. https://doi.org/10.1039/C5GC01937K

[128] K. Sadman, D.E. Delgado, Y. Won, Q. Wang, K.A. Gray, K.R. Shull, Versatile and High-Throughput Polyelectrolyte Complex Membranes via Phase Inversion, ACS Appl. Mater. Interfaces. 11 (2019) 16018–16026. https://doi.org/10.1021/acsami.9b02115

[129] A. Bera, J.S. Trivedi, S. Binod, A.K. Singh, S. Haldar, S.K. Jewrajka, Anti-organic fouling and anti-biofouling poly(piperazineamide) thin film nanocomposite membranes for low pressure removal of heavy metal ions, J. Hazard. Mater. 343 (2018) 86–97. https://doi.org/10.1016/j.jhazmat.2017.09.016

[130] H. Susanto, M. Ulbricht, Characteristics, performance and stability of polyethersulfone ultrafiltration membranes prepared by phase separation method using different macromolecular additives, J. Membr. Sci. 327 (2009) 125–135. https://doi.org/10.1016/j.memsci.2008.11.025

[131] D. Zhao, S. Yu, A review of recent advance in fouling mitigation of NF/RO membranes in water treatment: pretreatment, membrane modification, and chemical cleaning, Desalin. Water Treat. 55 (2015) 870–891. https://doi.org/10.1080/19443994.2014.928804

[132] J. Sun, L. Zhu, Z. Wang, F. Hu, P. Zhang, B. Zhu, Improved chlorine resistance of polyamide thin-film composite membranes with a terpolymer coating, Sep. Purif. Technol. 157 (2016) 112–119. https://doi.org/10.1016/j.seppur.2015.11.034

[133] D.J. Miller, D.R. Dreyer, C.W. Bielawski, D.R. Paul, B.D. Freeman, Surface Modification of Water Purification Membranes: a review, Angew. Chemie. Int. Ed. 56 (2017) 4662–4711. https://doi.org/10.1002/anie.201601509

[134] M. Asadollahi, D. Bastani, S.A. Musavi, Enhancement of surface properties and performance of reverse osmosis membranes after surface modification: A review, Desalination. 420 (2017) 330–383. https://doi.org/10.1016/j.desal.2017.05.027

[135] M.A.A. El-ghaffar, H.A. Tieama, A Review of Membranes Classifications, Configurations, Surface Modifications, Characteristics and Its Applications in Water Purification, Chem. Biomol. Eng. 2 (2017) 57–82.

[136] N.D. Suzaimi, P.S. Goh, A.F. Ismail, S.C. Mamah, N.A.N.N. Malek, J.W. Lim, K.C. Wong, N. Hilal, Strategies in Forward Osmosis Membrane Substrate Fabrication and Modification: A Review, Membranes. 10 (2020) 332. https://doi.org/10.3390/membranes10110332

[137] Y. Zhou, S. Yu, C. Gao, X. Feng, Surface modification of thin film composite polyamide membranes by electrostatic self deposition of polycations for improved fouling resistance, Sep. Purif. Technol. 66 (2009) 287–294. https://doi.org/10.1016/j.seppur.2008.12.021

[138] M. Rezakazemi, A. Dashti, H.R. Harami, N. Hajilari, Inamuddin, Fouling - resistant membranes for water reuse, Environ. Chem. Lett. 16 (2018) 715–763. https://doi.org/10.1007/s10311-018-0717-8

[139] R.H. Hailemariam, Y.C. Woo, M.M. Damtie, B.C. Kim, K. Park, J. Choi, Reverse osmosis membrane fabrication and modification technologies and future trends: A review, Adv. Colloid Interface Sci. 276 (2020) 102100. https://doi.org/10.1016/j.cis.2019.102100

[140] A. Soleimany, S.S. Hosseini, F. Gallucci, Recent progress in developments of membrane materials and modification techniques for high performance helium separation and recovery: A review, Chem. Eng. Process.: Process Intensificatio. 122 (2017) 296–318. https://doi.org/10.1016/j.cep.2017.06.001

[141] D. Li, Y. Yan, H. Wang, Recent advances in polymer and polymer composite membranes for reverse and forward osmosis processes, Prog. Polym. Sci. 61 (2016) 104–155. https://doi.org/10.1016/j.progpolymsci.2016.03.003

[142] R. Reis, L.F. Dumée, L. He, F. She, J.D. Orbell, B. Winther-jensen, M.C. Duke, Amine Enrichment of Thin-Film Composite Membranes via Low Pressure Plasma Polymerization for Antimicrobial Adhesion, ACS Appl. Mater. Interfaces. 7 (2015) 14644–14653. https://doi.org/10.1021/acsami.5b01603

[143] J. Wang, X. Chen, R. Reis, Z. Chen, N. Milne, B. Winther-jensen, L. Kong, L.F. Dumée, Plasma Modification and Synthesis of Membrane Materials - A Mechanistic Review, Membranes. 8 (2018) 56. https://doi.org/10.3390/membranes8030056

[144] B.K. Chaturvedi, A.K. Ghosh, V. Ramachandhran, M.K. Trivedi, M.S. Hanra, B.M. Misra, Preparation, characterization and performance of polyethersulfone ultrafiltration membranes, Desalination. 133 (2001) 31–40. https://doi.org/10.1016/S0011-9164(01)00080-7

[145] E.M. V Hoek, A.K. Ghosh, X. Huang, M. Liong, J.I. Zink, Physical–chemical properties, separation performance, and fouling resistance of mixed-matrix ultrafiltration membranes, Desalination. 283 (2011) 89–99. https://doi.org/10.1016/j.desal.2011.04.008

[146] A. Bhattacharya, B.N. Misra, Grafting: a versatile means to modify polymers: techniques, factors and applications, Prog. Polym. Sci. 29 (2004) 767–814. https://doi.org/10.1016/j.progpolymsci.2004.05.002

[147] R. Bergamasco, P.F. Coldebella, F.P. Camacho, D. Rezende, N.U. Yamaguchi, M.R.F. Klen, C.J.M. Tavares, M.T.S.P. Amorim, Self-assembly modification of polyamide membrane by coating titanium dioxide nanoparticles for water treatment applications, Rev. Ambient. Água. 14 (2019). https://doi.org/10.4136/ambi-agua.2297

[148] T. Sun, Y. Liu, L. Shen, Y. Xu, R. Li, L. Huang, H. Lin, Magnetic field assisted arrangement of photocatalytic TiO_2 particles on membrane surface to enhance membrane antifouling performance for water treatment, J. Colloid Interface Sci. 570 (2020) 273–285. https://doi.org/10.1016/j.jcis.2020.03.008

[149] O.A. Williams, Nanodiamond, Royal Society of Chemistry, 2014. https://doi.org/10.1039/9781849737616

[150] B.S. Al-anzi, O.C. Siang, Recent developments of carbon based nanomaterials and membranes for oily wastewater treatment, RSC Adv. 7 (2017) 20981–20994. https://doi.org/10.1039/C7RA02501G

[151] C.D. Williams, P. Carbone, Selective Removal of Technetium from Water Using Graphene Oxide Membranes, Environ. Sci. Technol. 50 (2016) 3875–3881. https://doi.org/10.1021/acs.est.5b06032

[152] D.H. Seo, S. Pineda, Y.C. Woo, M. Xie, A.T. Murdock, E.Y.M. Ang, Y. Jiao, M.J. Park, S. Il Lim, M. Lawn, F.F. Borghi, Z.J. Han, S. Gray, G. Millar, A. Du, H.K. Shon, T.Y. Ng, K.K. Ostrikov, Anti-fouling graphene-based membranes for effective water desalination, Nat. Commun. 9 (2018). https://doi.org/10.1038/s41467-018-02871-3

[153] H. Yang, R.Z. Waldman, Z. Chen, S.B. Darling, Atomic layer deposition for membrane interface engineering, Nanoscale. 10 (2018) 20505–20513. https://doi.org/10.1039/C8NR08114J

[154] E.F.D. Januário, N. de C.L. Beluci, T.B. Vidovix, M.F. Vieira, R. Bergamasco, A.M.S. Vieira, Functionalization of membrane surface by layer-by-layer self-assembly method for dyes removal, Process Saf. Environ. Prot. 134 (2020) 140–148. https://doi.org/10.1016/j.psep.2019.11.030

[155] J.H. Lora, W.G. Glasser, Recent Industrial Applications of Lignin: A Sustainable Alternative to Nonrenewable Materials, J. Polym. Environ. 10 (2002) 39-48. https://doi.org/10.1023/A:1021070006895

[156] J.F. Kadla, S. Kubo, R.A. Venditti, R.D. Gilbert, A.L. Compere, W. Griffith, Lignin-based carbon fibers for composite fiber applications, Carbon. 40 (2002) 2913–2920. https://doi.org/10.1016/S0008-6223(02)00248-8

[157] S. Yang, T. Wang, R. Tang, Q. Yan, W. Tian, L. Zhang, Enhanced permeability, mechanical and antibacterial properties of cellulose acetate ultrafiltration membranes incorporated with lignocellulose nanofibrils, Int. J. Biol. Macromol. 151 (2020) 159–167. https://doi.org/10.1016/j.ijbiomac.2020.02.124

[158] M. Nasrollahzadeh, S.M. Sajadi, A. Hatamifard, Waste chicken eggshell as a natural valuable resource and environmentally benign support for biosynthesis of catalytically active Cu/eggshell, Fe_3O_4/eggshell and Cu/Fe_3O_4/eggshell nanocomposites, Appl. Catal. B: Environ. 191 (2016) 209–227. https://doi.org/10.1016/j.apcatb.2016.02.042

[159] X. Meng, D. Deng, Trash to Treasure: Waste Eggshells as Chemical Reactors for the Synthesis of Amorphous $Co(OH)_2$ Nanorod Arrays on Various Substrates for Applications in Rechargeable Alkaline Batteries and Electrocatalysis, ACS Appl. Mater. Interfaces. 9 (2017) 5244–5253. https://doi.org/10.1021/acsami.6b14053

[160] X. Zhang, X. He, Z. Kang, M. Cui, D.-P. Yang, R. Luque, Waste eggshell-derived dual-functional CuO/ZnO/Eggshell nanocomposites: Photocatalytic reduction and bacterial inactivation, ACS Sustain. Chem. Eng. 7 (2019) 15762–15771. https://doi.org/10.1021/acssuschemeng.9b04083

[161] C. Senthil, K. Vediappan, M. Nanthagopal, H. S. Kang, P. Santhoshkumara, R. Gnanamuthub, C. W. Lee, Thermochemical conversion of eggshell as biological waste and its application as a functional material for lithium-ion batteries, Chem. Eng. J. 372 (2019) 765–773. https://doi.org/10.1016/j.cej.2019.04.171

[162] K. V. Peinemann, S.P. Nunes, L. Giorno, Membrane Technology: Membrane for Food Applications, Vol. 3, Wiley-VCH, 2010. https://doi.org/10.1002/9783527631384

[163] S.R. Panda, M. Mukherjee, S. De, Preparation, characterization and humic acid removal capacity of chitosan coated iron-oxide-polyacrylonitrile mixed matrix membrane, J. Water Process Eng. 6 (2015) 93–104. https://doi.org/10.1016/j.jwpe.2015.03.007

[164] Z. Alhalili, C. Romdhani, H. Chemingui, M. Smiri, Removal of dithioterethiol (DTT) from water by membranes of cellulose acetate (AC) and AC doped ZnO and TiO_2 nanoparticles, J. Saudi Chem. Soc. 25 (2021) 101282. https://doi.org/10.1016/j.jscs.2021.101282

[165] M. Chaudhary, A. Maiti, Fe–Al–Mn@chitosan based metal oxides blended cellulose acetate mixed matrix membrane for fluoride decontamination from water: Removal mechanisms and antibacterial behavior, J. Membr. Sci. 611 (2020) 118372. https://doi.org/10.1016/j.memsci.2020.118372

[166] A.M. Avram, P. Ahmadiannamini, A. Vu, X. Qian, A. Sengupta, S.R. Wickramasinghe, Polyelectrolyte multilayer modified nanofiltration membranes for the recovery of ionic liquid from dilute aqueous solutions, J. Appl. Polym. Sci. 134 (2017) 45349. https://doi.org/10.1002/app.45349

[167] P. Wen, Y. Chen, X. Hu, B. Cheng, D. Liu, Y. Zhang, S. Nair, Polyamide thin film composite nanofiltration membrane modified with acyl chlorided graphene oxide. J. Memb. Sci. 535 (2017) 208–220. https://doi.org/10.1016/j.memsci.2017.04.043

[168] P.O. Bagci, M. Akbas, H.A. Gulec, U. Bagci, Coupling reverse osmosis and osmotic distillation for clarified pomegranate juice concentration: Use of plasma modified reverse osmosis membranes for improved performance, Innov. Food Sci. Emerg. Technol. 52 (2019) 213–220. https://doi.org/10.1016/j.ifset.2018.12.013

Advanced Functional Membranes
Materials Research Foundations **120** (2022) 151-183

Materials Research Forum LLC
https://doi.org/10.21741/9781644901816-5

Chapter 5

Self-Assembled Membranes and their Applications

Bhushan Thipe and Ravi Kumar Pujala[*]

Soft and Active Matter group, Department of Physics, Indian Institute of Science Education and Research (IISER), Tirupati, India

[*]pujalaravikumar@iisertirupati.ac.in

Abstract

Self-assembled membranes have gained a lot of attention in the scientific community. They offer a simple and cheap way to form membranes that serves specific purposes like water filtration, drug delivery, anti-glare coatings, dielectrics, scaffold tissue engineering, etc. Polymers are the most popular choice for making self-assembled membranes, however there are other available options like proteins, small molecules, etc. Each new method/material offers certain advantages and disadvantages. Polymer membranes are strong and durable, but offer low porosity. Other materials like proteins degrade quickly. Hybrid materials are thus synthesized to combine the advantageous properties of materials and form self-assembled membranes. It is difficult to obtain membranes that are thin, sturdy and long lasting. New materials are being synthesized. However, materials that already exist can also serve well enough. A different perspective is needed to look into current materials and put them into good use. Self-assembled membranes offer a lot of potential uses and a lot of research is being pursued in this regard.

Keywords

Self-Assembled, Polymers, Copolymers, Proteins, Phospholipids, Porous, Non-Porous, Mesh

Contents

1.   Introduction

Membranes are defined as thin sheets that allow materials to pass through it. Simple as they may sound, membranes are what make life possible. They constitute very important parts of cellular life. They form the walls of the cell and other organelles that separate the interior of cells from its surroundings (fig. 1). They provide structural support by rendering the cell with shape and rigidness [1]. They protect the cell against pathogens. They help build osmotic pressure that allows for the movement of material into and out of the cell. They decide what goes through it, and what doesn't. They allow important materials like water and oxygen to pass through it, and reject materials that may be harmful for the cell [2]. Apart from being the regulators of life, membranes have been put into use in various other aspects of life as well.

The water filters that we use in our houses use these membranes to filter out unwanted materials like clay, salts, etc. The filtration that occurs there is based on a process called reverse osmosis (RO). Osmosis is a naturally occurring process in which a solvent moves spontaneously from a region of high solvent or low solute concentration to a region of low concentration of solvent or high solute through a semipermeable membrane [3], whereas in RO the opposite happens. The solvent is forced to move from a region of low concentration to high concentration, by applying pressure on it. Water containing impurities like chemicals, clay, etc. is moved across a semipermeable membrane by

Advanced Functional Membranes   Materials Research Forum LLC
Materials Research Foundations **120** (2022) 151-183   https://doi.org/10.21741/9781644901816-5

applying pressure on it. The membrane allows only water to pass through it and blocks out the impurities, thus filtering the water out (fig. 2) [4].

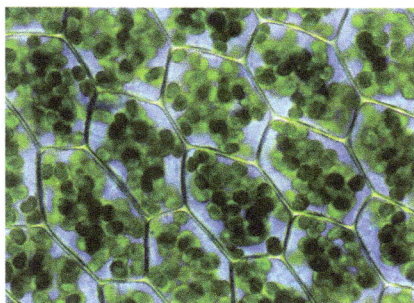

Figure 1 Cell wall: It separates the interior of the cell from its surroundings thus protecting it from damage from external agents. Image source: [5]

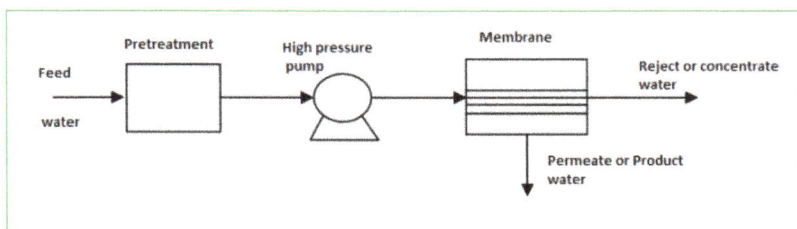

Figure 2 Working principle of reverse osmosis. Since it is not a naturally occurring phenomena, external pressure needs to be applied for RO to occur. Image obtained from ref.[7] with permission from IJARCS.

Of course, the water still has other impurities like microbes that can't be simply filtered out due to their small sizes. They are removed/killed using other processes like Ultra-violet filtration and nano-filtration [6].

The osmosis phenomenon is also used in medicinal purposes. The dialysis technique that is used to help people with dysfunctional kidneys is also based on osmosis, albeit a bit differently. In dialysis, instead of the solvent, the solute in blood (impurities, toxins, etc.) moves through a semipermeable membrane. The membrane doesn't allow the solvent (blood) to pass through it. Thus, the blood gets purified and is returned back to the body (fig. 3) [8].

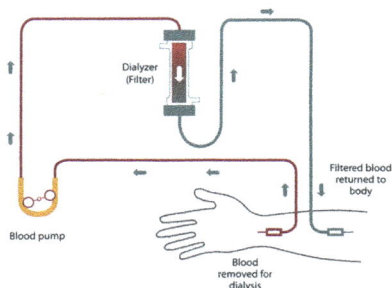

Figure 3 Working of dialysis. Blood moves through a membrane to filter out toxins. The filtered blood is returned back to the body. Image obtained with permission from ref.[9].

Dialysis is a life saver. All the people who are on a waitlist for kidney transplants undergo dialysis, since their own kidneys can't purify the blood. So, until their turn arrives, dialysis is what keeps them alive.

Another area where membranes are being used these days is the field of optics. Lenses used in spectacles are now being coated with certain material, like magnesium fluoride (MgF_2), that prevent the "glare" phenomenon (fig. 4). It is based on the idea that the reflectivity of the outer surface can be reduced if the transition of reflective index between the lens and surrounding is decreased [10]. Doing this vastly improves the efficiency of lenses since less light is lost due to reflection. This makes driving, especially during nights, much easier and safer. One of the latest advancements in this field is the development of a coating that reduces the amount of blue light that passes through the lens. Blue light is the major cause of eye strain and headaches and prolonged exposure can be very harmful [11]. These coatings are a boon to people who can't help but sit in front of electronic gadgets all day long.

Self-assembled membranes are defined as membranes that arrange themselves on their own, a cell wall made up of phospholipids being one of the many examples. Over the past few years, the interest in self-assembled membranes has risen exponentially, owing to their real-life applications, self-assembling properties, low-cost preparation, self-structure regulation and the opportunity to create tailor made membranes that serve specific purposes, like the anti-glare coatings.

Figure 4 Difference in images formed of a white object. Left side is coated with AS-AG plastic, and the right side is uncoated. The coated side shows significantly reduced reflection. Image obtained from ref.[12] with permission from Elsevier.

The scope for their applications is vast. They can be employed in almost every sphere of life. The current research is focused on:

- Developing membranes that offer separation based on size and purification

- Controlled release of materials

- Scaffold tissue engineering: 3-D porous materials are used to determine a region that later on gets filled by a developing tissue. The membrane provides mechanical support and also allows exchange of necessary materials, owing to its porous nature.

- Sensors and catalysts

- Low dielectric materials, to be used in microelectronics

- Anti-reflective coatings for lenses in spectacles and telescopes

- Proton-exchange in fuel cells: the membranes allow for protons to pass through it, while keeping the reactants separate.

Separation processes based on membranes need membranes that allow for enhanced chemical selectivity by predetermined pore sizes and shapes. They should be easily tunable by subjecting it to an external stimulus. However, the membranes need to be thin enough, preferably in the nano thickness range. It's difficult to achieve this. Even if the desired thickness is attained, their robustness remains questionable. A nano thin membrane that breaks down or degrades easily serves very little purpose. Also, the thinner the membranes get, the tougher it gets to tune them according to our needs [13].

In RO, solvents move through the membrane with application of a pressure in the range of 10 Bar and above. Nanofiltration membranes are used to remove particle sizes smaller than 2nm, with the applied pressure being in the range of 5 to 20 Bar. In ultrafiltration (UF), particle sizes between 2 nm and 100 nm can perfuse with pressure being 1 to 5 Bar. Particle sizez upto 500 nm can be filtered out using microfiltration. The external stimulus (pressure) allows us to fine tune to properties of the membrane and allow for selective separation [14].

Great strides towards development of better membranes have been made to date, and yet a lot remains to be done. More robust membranes that last long and don't degrade under chemical or physical pressure need to be synthesized. Over the years, synthetic self-assembled membranes based on block copolymer systems have gained a lot of popularity. Polymer membranes still remain the most researched, but various other materials have also been successfully synthesized. Every new method offers a new advantage, but also carries certain drawbacks. The next few sections focus on these newer approaches, their advantages and disadvantages.

2.   Self-assembled membranes based on Polymers

Polymer membranes are the most commonly used membranes right now. They are great substitutes for lipid membranes. They provide good mechanical support and they can also be combined with other materials like other polymers, proteins, etc. They are used in industries for separation of gas, reverse osmosis (RO), ultrafiltration (UF), nanofiltration (NF) and microfiltration (MF). They are also used in fuel cells [15]. They are generally composed of a single monomer type, but are often combined with other polymers to enhance and fine tune properties like strength, pore size and distribution, hydrophilicity, etc. The membranes created can be non-porous or porous. In the non-porous case, the membranes form a mesh-like structure (fig. 5). There aren't any predefined passages for materials to pass through them, and hence they diffuse through it. The density of the mesh can be adjusted as per need and thus the size of the particle that passes through it can be manipulated. Denser the mesh, smaller will be the size of particles that can pass through it. In the porous case, the membranes form designated pores and only through these pores can the materials pass through. By choosing suitable constituents for membranes, the pore sizes can be manipulated.

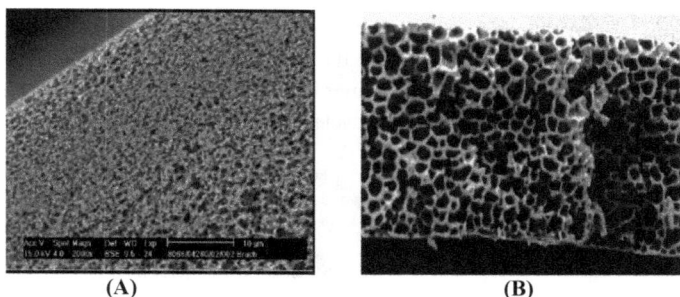

(A)   **(B)**

Figure 5 (A) Mesh structure (non-porous): There are no defined paths for the material to move through. It diffuses through the structure. (B) Porous structure: Cross section of porous polysulfone membrane. The material moves through these well-defined networks of pores. Image obtained with from ref.[16] and [17] with permission from ACS and Elsevier, respectively

2.1   Separation of phase in block-copolymers

When two or more different monomer units combine together and form the repeating unit of a membrane, then the membrane generated is called a block-copolymer. They are bonded covalently to each other. Since each monomer will have its own properties, combining them together to form a block-copolymer can give rise to various other interesting properties. The individual monomers have significantly different polarities, hydrophilicity and hydrophobicity. They remain separate on the nanoscale, but do not show bulk separation. They can give rise to various structures like the cylindrical, spherical, gyroidal and lamellar structures by changing the relative lengths of each block [18,19]. The molecular weight of the block-copolymer can also be easily maintained. When put in a suitable solvent at low concentration, the repeating units arrange themselves into a spherical micelle-like structure or thin lamellar membrane structure, much like the phospholipids [20,21].

2.1.1   Phase separation in thin films

As mentioned before, the phase separated regions lie in the nanometer range. This can be advantageous in the formation of thin film membranes. Since the two blocks have different properties, the best morphologies arise when the two different ends of the membranes are connected. Thus, gyroidal, cylindrical and lamellar morphologies arise [23]. These morphologies lead to formation of very distinct pores, like the shape of an upstanding cylinder (Fig. 6). The number of pores and their sizes can easily be tuned by using different blocks, thus giving us the flexibility to tune the properties of the membrane. But there's a problem. The pores aren't empty. The polymer itself occupies these pores. The pore being blocked effectively renders the membrane useless. The question to be asked is what good can a membrane with blocked pores do.

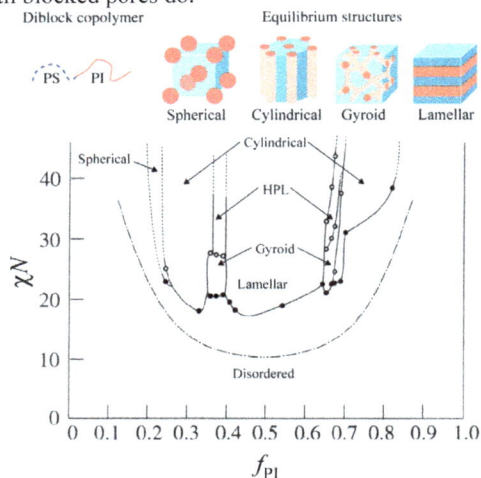

Figure 6 Change in morphology as a function of the volume fraction of the polymer present (This figure is for the case of PS-b-PI copolymer). HPL stands for hexagonally perforated layer structure. Image obtained from ref.[22] with permission from ACS.

Selective solvents are used for such cases (fig. 7). A solvent selective for cylindrical phase would cause it to swell up and expand. This creates a gel phase filled with solvent having a fixed mesh size in the case of a non-porous membrane. If only solvent can pass through the mesh, then this acts as solvent purification membrane. Choosing appropriate molecular weight and block ratio would create a highly entangled, dense and strong enough mesh structure. To open up the pore, solvents that affect different blocks differently are chosen. Since the blocks are different from each other, their reaction towards the solvent would also be different. The blocks would swell and shrink differently, this opening up the pore. For e.g. polystyrene-b-polyvinylpyridine (PS-b-P4VP) have blocks that have different selectivity for water. Water is more selective towards P4VP and less towards PS [25]. The PS phase hardens and solidifies, whereas the P4VP shrinks to a much lesser extent. This leads to opening up of cylindrical pores, while the morphology remains the same. The pore is still smaller than the expected size since the polymer still occupies it, but it is much more opened up as compared to before.

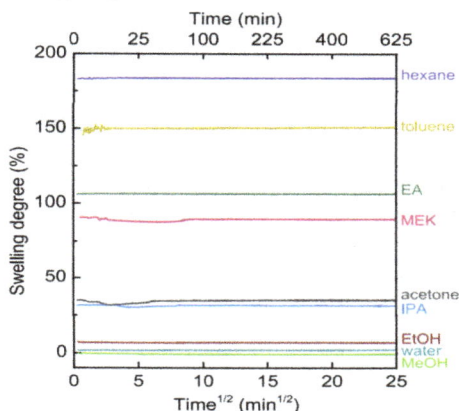

Figure 7 Swelling of PDMS in various selective solvents. A suitable solvent thus can be induced to induce different degrees of swelling. Image obtained from ref.[24] with permission from Elsevier.

This approach of using selective solvents is preferred because of the following advantages:

- No chemical reactions involved
- No weight loss
- Pore formation can be reversed
- High pore regularity

The problem isn't completely solved though. Overall porosity still remains low, around 20-30%. This is mainly because of the non-permeable components that provide structural stability to the membrane. Making more structurally sound membranes would involve the usage of rigid non permeable components, leading to low porosity. To attain high porosity, we would need to use less rigid and more permeable components, but then the membrane becomes structurally weak.

In the above two methods, the overall structure remained the same. However, it is also possible to use decomposable components as the minor structure. The minor phase can be selectively removed later, thus opening up the pores that were previously occupied by the block. For e.g., Degradation of polylactide using water or polydimethylsiloxane (PDMS) using tetrabutylammonium fluoride leads to formation of gyroidal, cylindrical and lamellar structures. The non-decomposing block can be additionally stabilized as well [26–28].

1,2-polybutadiene-*b*-polydimethylsiloxane has been used to make nano porous materials. 1,2-polybutadiene is cross-linked and then degraded using the polydimethylsiloxane. To attain different morphologies, the cross linking is performed at different temperatures. Although the structures are locally well-ordered, attaining large order still remains a challenge. The most common morphologies are cylindrical and gyroidal because of the phases that span the entire membrane. A spherical phase, although possible, does not offer a lot of possibilities. It was attained by Sivaniah et al though by forming PMMA spheres in the PS matrix using PS-*b*-PMMA. For PMMA, Acetic acid was used as the solvent, whereas PS was used to create the semipermeable blockage towards it. Dipping the entire film in acetic acid results in the systematic burst of the spheres, due to collective osmotic shock (COS), leading to formation of pillar-like structures [29].

2.1.2 Nanostructures based on solutions

Amphiphilic materials have both hydrophobic and hydrophilic ends. Using amphiphilic polymers, interesting morphologies like the ones mentioned below can be generated (fig. 8).

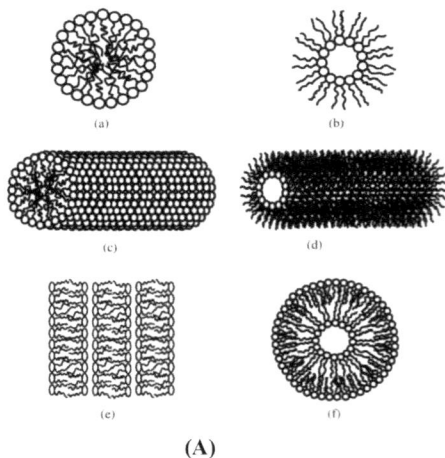

(A)

Figure 8 Different types of micelle morphologies (a) Spherical micelle (b) Inverted micelle (occurs in non-aqueous solutions) (c)Normal Hexagonal (d)Reverse Hexagonal (e)Lamellar (f)Vesicle. Image obtained from ref. [30] with permission from Elsevier.

Phospholipids are of interest to us, apart from the reason that they form the membranes for cells and organelles, because they can also arrange to form spherical vesicles. The interior of the vesicle remains shielded from the surroundings, owing to the structure of the vesicles. This can be extremely useful in drug delivery, where the drug needs to be delivered to designated sites [32]. This has been effectively used for delivery of poorly water-soluble anticancer drugs. For a long time, the treatment of cancer relied on water insoluble drugs. They had a short circulation time in the body, which made their use in chemotherapy very difficult. The hydrophobic end of the micelles attaches to the drug, whereas the hydrophilic ends protrude away from it. The drug remains protected from the effects of the environment. The drug remains in the bloodstream for a longer period of time, thus enhancing its effects on the tumour cells [33]. Paul Ehrlich won the Nobel Prize in 1908 for physiology and medicine for envisioning the concept of targeted drug delivery system, which he coined as the "Magic bullet", much akin to a bullet hitting its target [34].

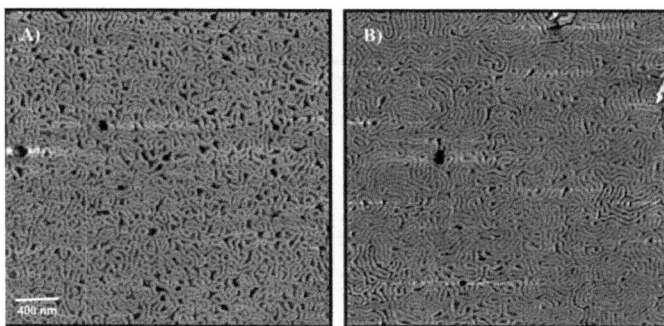

Figure 9 Phase difference in PS-b-PSL polymer when annealed for 72 hours at 100°C and 120°C. Thus, phases are highly dependent on the temperatures at which they are prepared. Image obtained from ref.[31] with permission from Elsevier.

Polymers are chosen as the preferred medium because they have many responsive properties like pH, temperature, ionic strength, etc. (fig. 9). They hold great importance in the field of medicine, pharmacy and bio-technology. They are used to form synthetic tiny hollow spheres, called polymersomes, that contain a solution, analogous to the micelle vesicles. Biodegradable polymers have been used for loading and delivery of drugs (fig. 10). They break down in acidic environments, thus releasing the drug on the designated target [35].

Advanced Functional Membranes | Materials Research Forum LLC

Materials Research Foundations **120** (2022) 151-183 | https://doi.org/10.21741/9781644901816-5

Different types of polymersomes have been synthesized using polymers bonded with ethylene glycol. Various responses like temperature, ultrasound, electrostatic forces, pH, redox potential, light and hydrogen bonding are used to break the polymersomes when they arrive at the target. Phospholipids can be used to form lamellar structures by placing them on a supporting material. Polymersomes, being extra stable, can pose a problem too. They don't break down and spread easily upon reaching the substrate. A high osmotic pressure is created using surface tethering techniques and a high concentration of salts to burst open polymersomes.

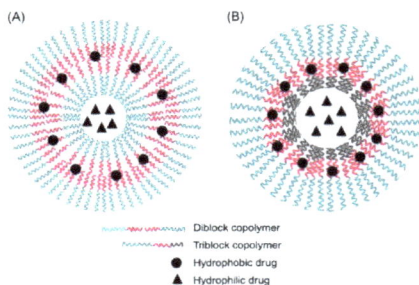

Figure 10 Drug delivery using polymersomes. Different regions of the polymersome can be used to protect different types of drugs from degradation. Image obtained from ref.[36] with permission from Elsevier.

Micelle membranes can also be formed by block copolymer. Thin films of PS-b-P4VP micelles were formed by Nunes et al. PS forms the core and is surrounded by P4VP to form the film. Transition metal ions are used as coordinating materials. They also provide additional stability for easy handling. The PS-b-P4VP structures and morphologies have strong susceptibility towards pH. Thus, in this case, pH is the stimulus to control the membrane structure (fig. 11).

2.1.3 Other approaches

Above-mentioned processes involved a templating approach in which one of the polymers could be removed or modified such that pores would open up. This method serves the purpose well, but there are other possibilities as well. A homopolymer can also be used in combination with other templates. The templates can be taken out later on since they are not bonded to the polymer. A variety of polymer and inorganic materials can be used. They don't necessarily need to be removed. Colloidal crystal lattices can be generated. These lattices show long range order and good porous structures. Cheng et al created one such

Advanced Functional Membranes
Materials Research Foundations **120** (2022) 151-183

Materials Research Forum LLC
https://doi.org/10.21741/9781644901816-5

colloidal structure. Polystyrene microspheres were used on a solid support in combination with PDMS or a polyurethane precursor to fill the voids (fig. 12). [38].

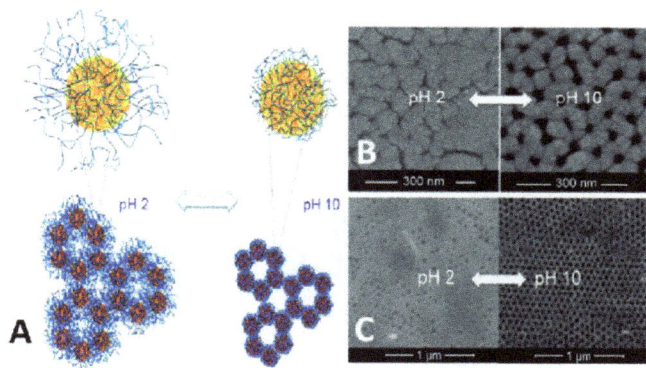

Figure 11 For PS-b-P4VP (A) pH dependent difference in size and assembly. (B) Image produced using Cryo-FESEM (C) ESEM analysis of membranes from solution in DMF– THF–Cu(acetate) 2, treated with highly acidic and basic solvents. Image obtained from ref.[37] with permission from ACS.

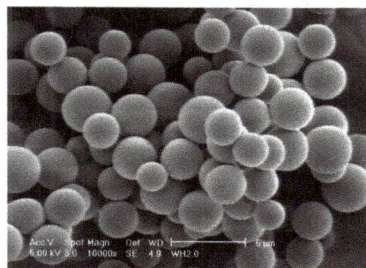

Figure 12 Polystyrene microspheres. Image obtained with permission from ref.[39].

A stable polymer matrix was made by suspending the substrate upside down. The microparticles expose to the surface depending on the annealing times and temperatures. The PS microspheres were later on removed by dissolving them into toluene. Pores with different sizes can be obtained depending on the exposure of the particles to the surface. A similar approach can be used in case of SiO_2. They can be dissolved in HF or other colloid susceptible solutions.

Solvent templates provide the advantage of just evaporating them and forming a stable matrix with defined voids. Another technique that has garnered attention is the "breath figure" technique. Water droplets are used as templates to form porous films. A polymer solution is cast on a substrate and treated with humid air (hence the name "breath") The polymer solution is evaporated, leaving behind a cooled surface. Water vapour droplets liquify on the surface and arrange themselves in a hexagonal array due to the capillary force. Hexagonally arranged pores are obtained after the total evaporation of the solvent (fig. 13) [40].

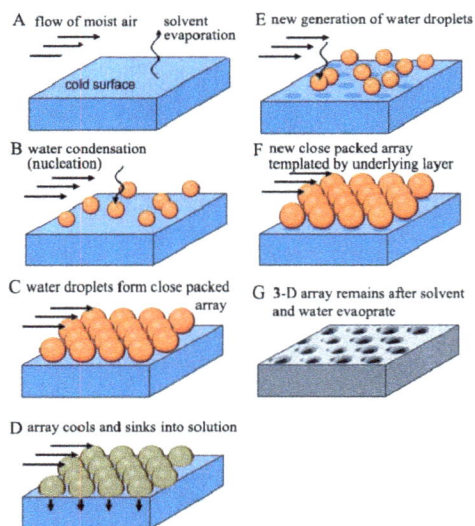

Figure 13 Working principle of the Breath Figure Technique. Image obtained from ref.[41] with permission from MDPI.

3.   Self-assembled membranes based on particles

A colloidal crystal is an ordered arrangement of colloidal particles, similar to solid crystals, that are suspended in a solvent. Their properties depend on sizes of constituent particles, the packing arrangement and the degree of regularity. An example of colloidal crystal is opal. Colloidal crystals can be used to form membranes by using organic and inorganic materials.   In the previous section we saw them being used as a template. However, colloidal crystals can themselves be used to form a fine porous structure. The voids between the colloidal crystals act like a well-connected network of pores (fig. 14). Nano and microparticles of silica have been used to form such types of structures. Silica is preferred because its surface can be easily functionalized using charged coatings, with cross-linking materials for extra stability. Such charged colloidal systems have been put into use to synthesize ion-transport membranes. These membranes are responsive to stimuli like pH and ionic-strength of the medium [42].

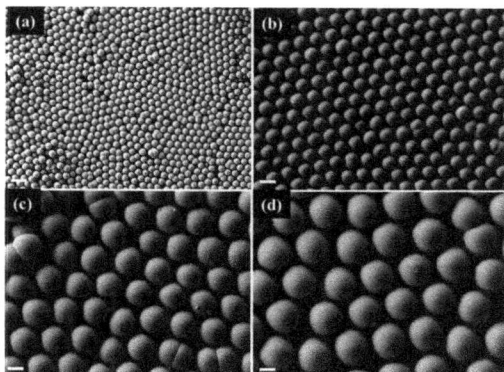

Figure 14 Colloidal crystal (a) Si 1 lm, (b) Si 2 lm, (c) Si 3.7 lm, and (d) Si 5 lm. The gaps between the arrangement are well connected and act like a passage for materials to pass through. Density of the arrangement determines the size of the particle that can move through the colloid. Image obtained from ref.[43] with permission from AIP publishing.

Proteins can also be considered as particles. As compared to inorganic particles that need coatings to enhance their reactivity, proteins already have exposed amino acids to provide the necessary reactivity. Protein based membranes that showed ultra-fast permeation of water through it have been synthesized. Glutaraldehyde, owing to its double aldehyde groups, was used to crosslink two proteins where lysine residues provided the accessible amine groups. Peng et al used ferritin to form such a membrane. Ferritin was first mixed with nano strands of cadmium hydroxide. They acted like confinement structures during the deposition [44]. The proteins, now immobilized, were then crosslinked using glutaraldehyde. The nano strands were removed after the crosslinking was finished. A membrane with channel diameters 2.2 nm and thickness between 25 to 4000 nm was generated, depending on the amount of deposition solution.

Viruses have also been used to form functional membranes. Viruses can also be treated as particles because when outside of a living body they barely show any living traits. They behave a lot like colloidal particles, by stabilizing the polar-non polar fluid surfaces. This was first put into effect by Pickering to stabilize emulsions, but the same concept can be extended towards structures containing protein (fig. 18). The particles get adsorbed on the surface, making water-particle and oil-particle interactions favourable than water-oil interactions and thereby minimizing the interfacial energy [45]. For e.g., Cowpea Mosaic virus particles were used to stabilize emulsions containing protein in them. Glutaraldehyde was then added to this emulsion to crosslink the proteins and form a membrane like structure surrounding the oil droplet [46]. Hollow virus particles offer a large range of possibilities since they can act as nano-containers.

Semiconductor nanoparticles and FePt-nanoparticles that can be crosslinked have also been used to form capsules and planar membranes. Terpyridine ligand has an affinity for Fe(ii). Dispersible particles of toluene and an aqueous solution of Fe(ii) tetrafluoroborate was used. Water was then added to the dispersion leading to the production of aqueous capsules. Using the bulk separated system of toluene and water yields planar membrane [47]. Additionally, FePt has magnetic properties that can be used to separate the capsules (fig. 15).

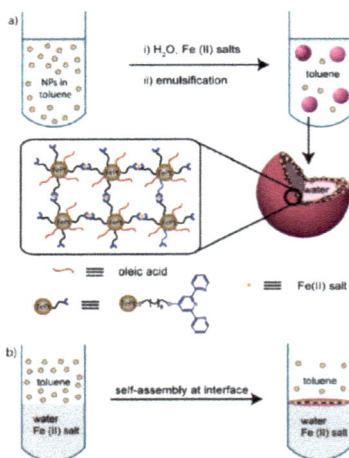

Figure 15 (a) Formation of Fe-Pt nanoparticles colloid at the water toluene interface. Cross-section shows nanoparticle arrangement formed by the ligand with Fe (II) tetrafluoroborate hexahydrate salt. (b) Formation of Fe-Pt membrane at the water-toluene interface. Image obtained from ref.[47] with permission from ACS.

Advanced Functional Membranes   Materials Research Forum LLC
Materials Research Foundations **120** (2022) 151-183   https://doi.org/10.21741/9781644901816-5

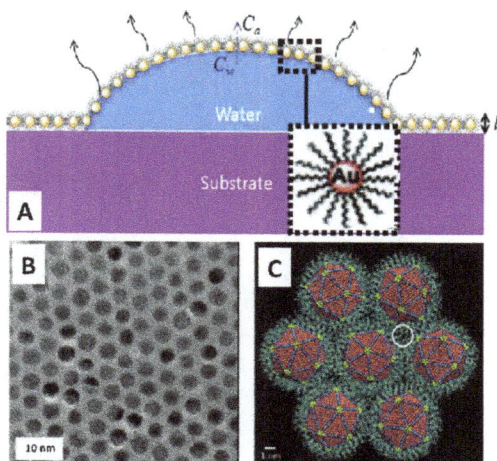

Figure 16 (A) Working of drying mediated assembly (B) TEM analysis shows a hexagonal array (C) White circle shows a small mesh where three particles join. Image obtained from ref.[49] with permission from ACS.

Recently, a self-assembly process mediated by drying was used to produce ultra-thin colloidal nanocrystal membrane [48]. Au, Fe/Fe_3O_4, CoO and capping ligands like oleic acid, oleylamine and dodecanethiol, were used to form closely packed monolayer membranes by placing a water droplet on porous Si_3N_4 and then covering it with an organic droplet containing the nanoparticles. The solvent was then evaporated, leading to formation of a densely filled coating at the interface. When the liquids evaporate, a single layer of nanoparticles covering the base is left behind (fig. 16).

Using particles provides an easy and straightforward way to form new membranes. As can be observed, a lot of the breakthroughs were discovered from materials that are already known to us. They were also not very complex particles. Creating new particles to form novel membranes isn't the only way to move ahead. We need to look at materials that already exist with a fresh vision. Who would've thought viruses could possibly be used in membrane synthesis? A different perspective to look at existing materials and a proper method to utilize them would surely be helpful in developing membrane science!

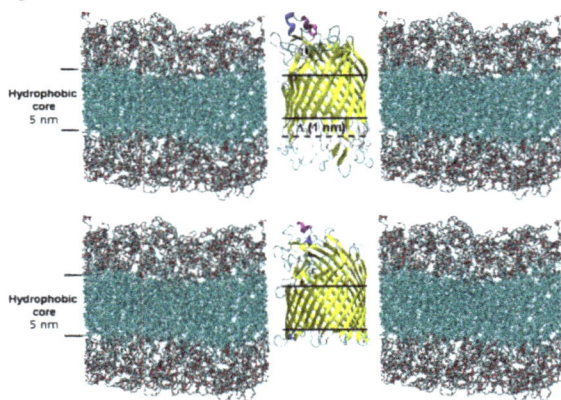

Figure 17 The top image shows the extension in length of the FhuA and the bottom image shows the non-extended FhuA within triblock copolymer PIB 1000 –PEG 6000 – PIB 1000 membranes. Image obtained from ref.[58] with permission from BioMed Central ltd.

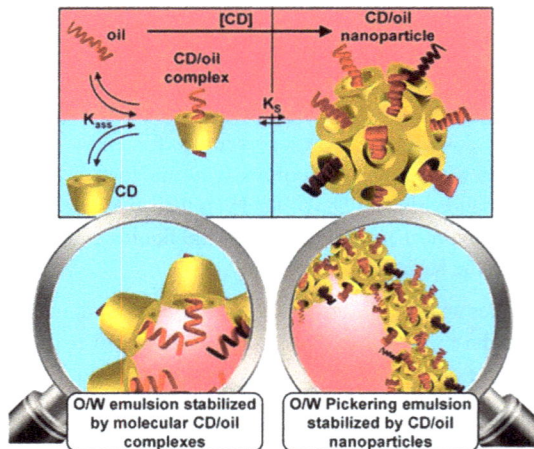

Figure 18 Illustration showing the formation of a Pickering emulsion, affected by the amount of cyclodextrin. K_{ass} and K_s are the binding and solubility constants, respectively. Image obtained with permission from ref.[65].

Advanced Functional Membranes   Materials Research Forum LLC
Materials Research Foundations **120** (2022) 151-183   https://doi.org/10.21741/9781644901816-5

4.   Self-assembled membranes based on small molecules

Molecules that exhibit various interactions like hydrophobicity, π stacking, hydrogen bonding, electrostatic interactions and weigh less than 1.5kD can self-assemble and form complex architectures, phospholipids being the most well-known example. Several other molecules have been explored and synthesized to form membranes. They not only form bilayer membranes, but can also be formed into a fibrous structure. When these fibrous structures are confined into a dense region, they get tangled up and form a mesh like structure. These structures can be used for filtration of nanoparticles.

Phospholipids have been an inspiration in the synthesis of novel membranes, cyclodextrin being one of the examples. Cyclodextrin is a cyclic molecule made up of single glucose moieties. The variants, α-CD, β-CD and γ-CD arise when 6-, 7-, 8-glucose moieties are joined in a ring-like structure. One side was substituted using chains of aliphatic nature and the remaining with hydrophilic nature. This structure gets self-assembled into a structure similar to phospholipids. CDs have generally been used to make artificial enzymatic structures. Since CD has hydrophobic ends in its inner space, it can easily bind with other hydrophobic structures like adamantane.

Azobenzene exists in the form of trans and cis isomers. Irradiating with UV light causes it go from the trans to cis configuration. Thus, it is like a molecular switch. Ravoo and coworkers used this property to reversibly bind the azobenzene structures. This type of binding nature has been used to prompt reversible processes like aggregation of inter-membrane, complexing of Proteins and DNA [50]. This system corresponds to a phospholipid very well and thus creating planar membranes using them should be possible. Such membranes would have reversible binding, where the CD molecules act like pores. Molecular gelator species have been designed using the amphiphilic character of aggregation. Two perylene units were connected using bipyridine bridges. These structures show hydrophobic and π bonding, and therefore undergo stacking [51]. Attaching an ethylene glycol chain on both sides led to increase in solubility and rendered the structure an amphiphilic nature. When it was put in a Water-THF mixture, the composition self-assembled itself into a fibrous arrangement. A gel having a mesh like structure was obtained. The obtained structure was mechanically robust as it was easy to transfer it onto a macroporous structure by simply filtering the solution on a syringe filter. The molecular membrane was used to filter an aqueous solution of nano gold particles. It allowed gold particles between the size of 1 to 5 nm to pass through it, and rejected the bigger sizes. Everything larger than 5 nm was stuck in the membrane. It was also very easy to obtain the larger particles from the membrane. The filter was washed with water and ethanol mixture. The particles and membrane matrix both got dissolved and later collected. A water dichloromethane partition can also be used to separate the particles and the gel. This leaves the gel unaffected and it can be used for another filtering process [52].

Complementary structures can also be combined and used to form fibrous structures. Membranes can be obtained from these structures by mixing two solutions containing the complementary structures or by using the solutions alternatingly, like in the case of layer-by-layer deposition of polyelectrolytes. If the aggregation is fast enough, it's not even necessary to have two different interfaces like oil and water, just water-water works fine. A fibrous structure was formed by mixing amphiphilic aqueous solution of small polypeptides with aqueous solution of hyaluronic acid. The membrane can be spherical or planar depending upon the incubation time or the concentration of the acid [53].

The membranes obtained in the previous cases produce thick membranes. Phospholipids and the CD case produce about twice as wide membranes as compared to the length of molecules. A-B-A membranes are one particle wide, however length of the polymer is quite long resulting in a membrane in the range of a few nanometers. Gölzhäuser and coworkers further stretched this limit and produced a 1 nm thick membrane. Using a gold substrate, they used the self-assembly of biphenyl containing thiol to produce a monolayer. Exposing it to electrons with energy in the range of 10^2 eV and 0.05C cm^{-2} crosslinked the monolayer. A PMMA resist was used to detach the membrane from the surface. The mica subphase was dissolved using HF to expose the gold layer. The gold layer was then removed by etching with I_2/KI. The PMMA resist was removed using acetone, thus producing the monolayer. Functionalization of both sides of the membrane with fluorescent dyes was performed using isothiocyanate and maleimide-functionalized dye [54].

The membranes formed in the above-mentioned cases are generally very thick. Thicker membranes are difficult to handle and isolating a free thin membrane from is a big problem. Moreover, higher flux and pressure need to be applied on thicker membranes to make them work.

5.   Membranes based on hybrid materials

Hybrid materials are composite structures formed by combining organic and inorganic constituents at the molecular or nanoscale. They offer an exotic way of forming novel functional materials. Inorganic, bio and polymer materials having suitable properties are covalently combined. In membranes obtained via hydrophobic interaction, as in the case of polymersomes, it is difficult to generate pores whereas in block-copolymers, defined cylindrical pores can be obtained and removed. Polymersomes are responsive towards temperature and pH when the hydrophobic phase is responsive and the membranes become less densely packed. However, doing this disrupts the membrane structure. The pore sizes get varied. Thus, a well-defined pore structure needs to be introduced. Nanometer proteins can be ideal candidates for these. They exist in various well-defined 3D structures, pore sizes and are very reactive towards stimuli [55]. Some well-known building blocks are viral particles (like CCMV, TMV), β-barrel proteins (like OmpF, FhuA, Tsx), cage-proteins (such as ferritin), and ion selective helix bundles. Protein structures suffer from certain drawbacks though. They are highly reactive and thus unstable. Their structures are not completely understood. They often suffer a mismatch in size when combined with other materials. Their template synthesis is neither easy nor very cost effective [56,57].

Channel proteins like FhuA, α-hemolysin, OmpF and aquaporin, are attractive candidates in membrane synthesis. They can be used to make pores that are selective towards size, specific ions and molecules. These pathways are very useful in maintaining cellular passages and are used in phospholipids for very specific purposes. As mentioned before, block copolymer membranes are very similar to phospholipid membranes. Incorporating channel proteins for translation will be a major development for applications in medicine and industry. Polymersomes are expected to be stable and impenetrable by smaller molecules so that the entire flow can be conducted through the pores while withstanding the osmotic pressure. The proteins also need to be compatible with polymer membranes even though their hydrophobic regions in biological membranes are a few nanometers thinner. This difference in thickness leads to the usage of functional proteins inefficiently. The hydrophobic part of the membrane thus needs to be thinned out, or the hydrophobic zone of the pore-protein should be thickened. Reduction of the membrane width would decrease its stability and robustness. Therefore, the size of the hydrophobic region of the pore protein seems like a viable option. This can be achieved by protein engineering and gene synthesis. For e.g. The hydrophobic region of FhuA has been stretched by 1 nm by adding 5 more amino groups in all of the 22 β-sheets that constitute the β-barrel protein [58]. Although the increase seems small, it accounts for a net 33% extension in length of the hydrophobic zone (fig. 15). This lengthening minimizes the incongruity between protein and the hydrophobic region of the polymer, leading to a better usage of the protein in the polymer membrane in comparison with naturally occurring FhuA channels. Another benefit is that the well-defined structure of proteins can be used to add or remove functional groups like NH2 and -SH inside the channel. The interior of the channel can thus be functionalized by introducing potential pore blockers that can be triggered by a stimulus. For e.g., FhuA has responsive triggers like chemical reduction or UV irradiation [59,60]. Adding DTT (dithio-threitol) cleaved the dithio-group linker by reduction and calcein dye was released. Further studies showed that responsive systems can be of great use in medicine and industries.

Materials Research Forum LLC

https://doi.org/10.21741/9781644901816-5

Kaucher et al used a method constituting multi-molecular ingredients. He used a determined dendritic molecular structure that self-assembled into a helical structure [61]. The boundaries of the structure were synthesized to have a hydrophobic nature to match with the hydrophobic nature of lipid membranes, whereas the inside of the structure was stabilized by pi stacking. Analysis showed that the channel selectively transports only protons and water and rejects other ions.

Although membranes that self-assemble are very lucrative, they are not always very practical. It is difficult to have total control over the pore size, pore location and their distribution. The inefficiency of such membranes also is not favorable. However, when we combine selective properties of proteins and polymers, the hybrid material thus formed offers great ways to form membranes. The protein polymer arrangement can be considered like a microgel particle. Interface stabilization or Pickering emulsion can also be used for the system (fig. 18). A membrane array is generated with proteins embedded in a polymer matrix [62]. As mentioned before, proteins are highly reactive and thus unstable. They get denatured upon exposure to conditions like high temperature, pH or chemicals that affect the hydrogen bond or disulfide bridges. Thus, they can be easily removed from the structure and leave behind a pore.

Inorganic-polymer membranes can also be generated by following similar approaches. Silica in combination with a polymer was used to develop elastic and nano porous membranes using the Langmuir–Blodgett techniques at the liquid-air interface [63]. The polymers then crosslinked and the silica was dissolved away by suitable agents, leaving behind the desired membrane. Pickering emulsion templating can also be used to form spherical membranes [64]. The same inorganic template method was employed to form a 3D membrane or microcapsules. The microcapsules are formed using adsorption of polyelectrolyte multilayers using alternating electrostatic adsorption. Most approaches use an inorganic template that is insoluble in water that gets removed after the formation of a capsule to form membranes that are hollow or planar. The process however suffers a major drawback. There is no control over the pore sizes. Removing the template also requires the usage of strong chemicals like HF. This means that large industrial production of such membranes isn't viable. Also, other delicate molecules can't be added to the mixture due to the presence of these strong chemicals.

Additionally, for convenient handling, the membranes need to be put on substructures. For solution-based systems where pressure is applied, a porous substrate of inert nature is needed for a sub-phase. Generally, a free-floating film is prepared and then the membrane is lifted and placed onto the support. Otherwise, the Langmuir Blodgett technique is followed where the substrate is vertically pulled through the water-air interface. For the previous case of the membrane-particle system, the water droplet on which the membrane was formed was gradually dried. As the water evaporates, the membrane gently gets placed on top of the substrate. Goedel and co-workers used a hierarchical build up technique. They first prepared a porous substrate by micro-sieve preparation. The polymer-nanoparticle membrane prepared in the water-oil interface was then lowered onto it, subsequently followed by removal of the water phase. A porous and mechanically stable membrane was thus obtained. Table 1 briefly summarizes the ideas presented in this chapter.

Table 1 A summary of all mentioned methods, their advantages and disadvantages.

Method	General idea	Advantages	Disadvantages
Thin film phase separation (polymer based)	(1) Two or more monomers are covalently bonded to each other, each having different properties. (2) Different morphologies arise due to difference in properties (3) Monomers remain separate on nanoscale, but don't show bulk separation.	(1) No chemical reactions involved (2) No membrane weight loss (3) Reversible pore formation (4) High pore regularity (5) Different morphologies are possible	(1) Pores created aren't empty. They need to be opened up by other techniques. (2) Overall low porosity, ~20-30%
Solution based nanostructure-es (polymer based)	(1) Amphiphilic materials (materials that have both hydrophobic and hydrophilic ends) are used to form different morphologies. (2) Particles of interest (e.g., drugs) are enclosed inside these morphologies, thus preventing its degradation.	(1) Useful in medical and biotech for drug delivery. (2) Responsive to many stimuli like pH, temperature, pressure, etc.	(1) Difficult to bust open the micelle structures at the target site. (2) High osmotic pressures are needed to break them.
Other approaches (polymer based)	The templating material need not be necessarily removed. Materials are chosen in a way such that, regardless of whether or not they get removed in the end, the membrane works perfectly fine.	(1) Involves lesser work since the template need not be necessarily removed. (2) If need be, the template can be removed using simple processes like evaporation.	(1) Polymers in general are non-biodegradable materials. They need to be carefully disposed of after use.

Self-assembled membranes from particles	(1) Colloidal systems are used to form membranes. (2) The gaps between the colloidal particles acts as the pores	(1) Particles can be organic or inorganic, like viruses, semiconductor nano-particles. (2) Possibility of creating new functional particles	(1) Developing new particles is not an easy task. (2) Conventional particles, like proteins, are susceptible to stimuli and degrade easily.
Self-assembled membranes from small molecules	(1) Molecules that weigh less than 1.5kD can combine by different interactions and form porous and non-porous structures. (2) Bilayer membranes or a fibrous network can be obtained and used for filtration purposes.	(1) Self-assembly by interactions like hydrogen bonding, π stacking, etc. (2) Binding can be reversible in some cases	(1) The membranes are generally very thick and obtaining a thin slice is difficult (2) High pressure and flux are needed to make them efficient
Membranes from hybrid materials	(1) Particles having different advantageous properties are combined to form membranes (2) A very exotic yet promising technique.	(1) Task specific membranes can be made, for e.g., membranes for proton and water transport.	(1) Difficult to control pore size, location and distribution (2) Low efficiency

6. Applications

Above discussed methods offer exciting new ways for membrane synthesis that can have potential applications. Being able to fine tune pore size, shapes and distribution offers a great way to form membranes that offer higher chemical selectivity. Higher chemical selectivity would enable it to filter and block out very specific materials, thus making use of membranes in niche cases possible. Their applications are also possible in the field of medical physics and biotechnology. Controlled release of drugs using external stimuli being one of the applications. As mentioned before, in chemotherapy target specific drugs can directly affect the tumour cells. They remain for a longer time in the bloodstream, and thus their efficiency increases. The membranes should ideally be thin so as to achieve a good amount of flux at lower applied pressures. In this regard, block copolymers offer a lot of possibilities. Their pore size and distribution are tunable [66]. Thin film membranes can be obtained using block copolymers. Thus, they show a lot of potential in water

Advanced Functional Membranes

Materials Research Foundations **120** (2022) 151-183

Materials Research Forum LLC

https://doi.org/10.21741/9781644901816-5

purification applications. However, to make them good candidates for use in industrial purposes, we need membranes that have a large surface area. Industries depend on the bulk production of goods. For bulk production to be profitable, it needs to be large, reusable and long lasting. Although some membranes have been developed with areas in the cm^2 regime, most lie in the mm^2 area regime. Producing membranes with large surface area still remains a challenge. Ultra-thin membranes often need a good porous support that doesn't interfere with the membrane structure. Thus, development of support materials is also important as well.

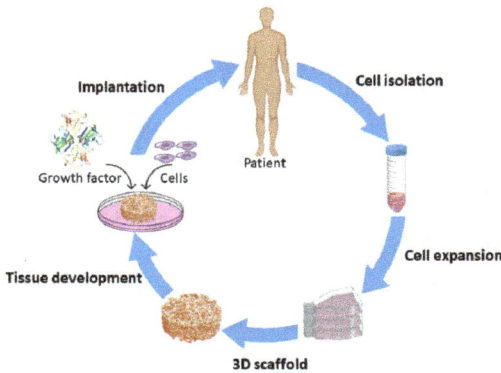

Figure 19 Flowchart showing scaffold tissue engineering. Image obtained from ref.[69] with permission from MDPI.

Industrial applications of self-assembled remains a far-fetched dream, but there are some areas where the small size membranes are useful. Drug delivery and biosensors are the fields that do not need large or bulk membranes. Engineering protein channels into lipids and polymers membranes seems to be the way to go ahead. It is possible that self-assembled membranes remain applicable only for small scale purposes. This isn't necessarily a bad thing. Households need water purifiers mostly to serve a single family. Medical science needs membranes to be small. The membrane needs to do a specific job and it being small would be advantageous in serving its purpose.

Biosensors can be synthesized using polymers. Biomolecules that react to certain stimuli are embedded in a polymer support. The stimuli may be a physical stimulus or another biomolecule. The response to the stimuli could be something like a change in fluorescence. The most common biosensing membranes have polymer bilayers with protein embedded in them. The protein plays a crucial role in the sensitivity and selectivity of the membranes.

For e.g., α-Hemolysin (α-HL) embedded in the membrane is used in DNA sequencing. α-Hemolysin (α-HL) has a very narrow (1.4 nm diameter) pore, and thus when a DNA strand passes through it, it can be detected by the change in ion current for that base. The pore being very small can't be used for larger sequences. The pore size can be altered by chemical modification to modify the sensitivity of such membranes. The channel can also be modified by using different channel proteins in the membrane. For e.g., OmpF has a large channel size and can be used in sensing devices. It is also resistant towards pH, temperature and solvents. The large channel size can be increases or decreased on demand using a pH switch [67].

Interconnect scaling requires the development of materials with dielectric constants lower than 2.5. Such materials have been developed by introducing porosity, reducing polarizability or both. Self-assembled organic polymers are used to form such materials. They effectively have no polar bonds. They exhibit low polarizability even if the porosity is low. Smaller pore sizes can easily be engineered. The membranes are resistant to process induced damages. They are also resistant to plasma damage owing to their mono component nature [68].

Fractures like cranial, oral and even large bone flaws are cured using auto and allograft procedures. They suffer from various limitations like auto-immune reaction, donor site morbidity and unavailability of a donor. Therefore, over the years the field of tissue engineering has acquired a lot of interest in the scientific community. Such purposes require a 3D.

It maintains the sites for the cells to connect, grow and function normally as the bone tissue grows in size. Molecular and structural biology is applied in combination with materials science to form such scaffolds. Bio scaffolds produced by self-assembly are preferred because they provide spatial and temporal resolution. Self-assembling peptides that are capable of gelation are attractive candidates as they offer a lot of possibilities with minimal surgical interventions in bone therapy [70].

7. Summary

Over the past few years, a lot has been achieved in the field of self-assembled membranes. Several membranes have been developed that serve the purpose in purification, drug delivery and confinement. But, the synthesizing stable, thin and durable self-assembled membranes with uniform pores still remains a challenge to overcome. The field is garnering a lot of interest because of the affordable preparation, self-structural maintenance, tunable properties and the ability to create compound materials. Polymers and nanoparticles appear to be the most promising candidates for self-assembled membranes.

However, composite materials also offer a lot of benefits. Polymers provide the required stability, flexibility and elasticity to the membrane. Combining with nano particles that provide high chemical selectivity, pore size control, catalytic activity, etc. Nanoparticles can be organic, inorganic or based on proteins. It is believed that most of the functional membranes that will be developed in the future will constitute of the above-mentioned materials. It isn't necessary to develop never seen before materials. Materials that exist right now could very well serve our purpose in the future. A renewed vision and better techniques need to be developed. Even with all the achievements and difficulties that exist, there is still a plethora of possibilities in membrane science.

Acknowledgement

Authors are grateful to IISER Tirupati, for providing the necessary funding and research facilities. R.K.P. acknowledges the Department of Science and Technology for an INSPIRE Faculty Award Grant [DST/INSPIRE/04/2016/002370] and a Core Research Grant (CRG/2020/006281, DST-SERB), Government of India for funding.

References

[1]   P.D. William Gahl, M.D., Cell Membrane (Plasma Membrane), (n.d.). https://www.genome.gov/genetics-glossary/Cell-Membrane (accessed July 12, 2021).

[2]   T.E. of E. Brittanica, Cell membrane, (n.d.). https://www.britannica.com/science/cell-membrane (accessed July 21, 2021).

[3]   J. Feher, 2.7 - Osmosis and Osmotic Pressure, in: J. Feher (Ed.), Quant. Hum. Physiol. (Second Ed., Second Edi, Academic Press, Boston, 2017: pp. 182–198. https://doi.org/https://doi.org/10.1016/B978-0-12-800883-6.00017-3.

[4]   What is Reverse Osmosis?, (n.d.). https://puretecwater.com/reverse-osmosis/what-is-reverse-osmosis (accessed July 21, 2021).

[5]   Plagiomnium affine laminazellen.jpeg, Wikimedia Commons. (n.d.). https://commons.wikimedia.org/w/index.php?curid=1350193 (accessed July 21, 2021).

[6]   Is UV Required With Reverse Osmosis (RO)?, (n.d.). https://waterpurificationguide.com/is-uv-required-with-reverse-osmosis-ro/.

[7]   A. Ahuchaogu, J. Chukwu, A. Obike, C. Igara, I. Nnorom, J. Bull, O. Echeme, Reverse Osmosis Technology, its Applications and Nano-Enabled Membrane, 5 (2018). https://doi.org/10.20431/2349-0403.0502005.

[8]   A. Green, Dialysis: principles and treatment options, (n.d.). https://pharmaceutical-journal.com/article/ld/dialysis-principles-and-treatment-options (accessed July 23, 2021).

[9]   National Institute of Diabetes and Digestive and Kidney Diseases, (n.d.). https://www.niddk.nih.gov/ (accessed July 23, 2021).

[10]   M. Keshavarz Hedayati, M. Elbahri, Antireflective Coatings: Conventional Stacking Layers and Ultrathin Plasmonic Metasurfaces, A Mini-Review, Materials (Basel). 9 (2016). https://doi.org/10.3390/ma9060497.

[11]   J. Lawrenson, C. Hull, L. Downie, The effect of blue-light blocking spectacle lenses on visual performance, macular health and the sleep-wake cycle: a systematic review of the literature, Ophthalmic Physiol. Opt. 37 (2017) 644–654. https://doi.org/10.1111/opo.12406.

[12]   N. Al-Dahoudi, H. Bisht, C. Göbbert, T. Krajewski, M. Aegerter, Transparent Conducting, Anti-static and Anti-static–Anti-glare Coatings on Plastic Substrates, Thin Solid Films. 392 (2001) 299–304. https://doi.org/10.1016/S0040-6090(01)01047-1.

[13]   K.K. Ho, W. S. W., & Sirkar, Membrane Handbook, Boston, MA: Springer US, 1992.

[14]   R.D.N.& S.A. Stern, Membrane Separations Technology:Principles and Applications, ELSEVIER, n.d.

[15]   S.P. Nunes, Block Copolymer Membranes for Aqueous Solution Applications, Macromolecules. 49 (2016) 2905–2916. https://doi.org/10.1021/acs.macromol.5b02579.

[16]   M. Ulbricht, Advanced functional polymer membranes, Polymer (Guildf). 47 (2006) 2217–2262. https://doi.org/https://doi.org/10.1016/j.polymer.2006.01.084.

[17]   I. Pinnau, B.D. Freeman, Formation and Modification of Polymeric Membranes: Overview, in: Membr. Form. Modif., American Chemical Society, 1999: p. 1. https://doi.org/doi:10.1021/bk-2000-0744.ch001.

[18]   H. Feng, X. Lu, W. Wang, N.-G. Kang, J.W. Mays, Block Copolymers: Synthesis, Self-Assembly, and Applications, Polymers (Basel). 9 (2017). https://doi.org/10.3390/polym9100494.

Advanced Functional Membranes   Materials Research Forum LLC
Materials Research Foundations **120** (2022) 151-183   https://doi.org/10.21741/9781644901816-5

[19]   Y. Wang, F. Li, An Emerging Pore-Making Strategy: Confined Swelling-Induced Pore Generation in Block Copolymer Materials, Adv. Mater. 23 (2011) 2134–2148. https://doi.org/https://doi.org/10.1002/adma.201004022.

[20]   M.W. Matsen, F.S. Bates, Unifying Weak- and Strong-Segregation Block Copolymer Theories, Macromolecules. 29 (1996) 1091–1098. https://doi.org/10.1021/ma951138i.

[21]   A. Mecke, C. Dittrich, W. Meier, Biomimetic membranes designed from amphiphilic block copolymers, Soft Matter. 2 (2006) 751–759. https://doi.org/10.1039/B605165K.

[22]   A.K. Khandpur, S. Foerster, F.S. Bates, I.W. Hamley, A.J. Ryan, W. Bras, K. Almdal, K. Mortensen, Polyisoprene-Polystyrene Diblock Copolymer Phase Diagram near the Order-Disorder Transition, Macromolecules. 28 (1995) 8796–8806. https://doi.org/10.1021/ma00130a012.

[23]   L.M. Pitet, M.A. Amendt, M.A. Hillmyer, Nanoporous Linear Polyethylene from a Block Polymer Precursor, J. Am. Chem. Soc. 132 (2010) 8230–8231. https://doi.org/10.1021/ja100985d.

[24]   E.J. Kappert, M.J.T. Raaijmakers, K. Tempelman, F.P. Cuperus, W. Ogieglo, N.E. Benes, Swelling of 9 polymers commonly employed for solvent-resistant nanofiltration membranes: A comprehensive dataset, J. Memb. Sci. 569 (2019) 177–199. https://doi.org/https://doi.org/10.1016/j.memsci.2018.09.059.

[25]   K.-V. Peinemann, V. Abetz, P.F.W. Simon, Asymmetric superstructure formed in a block copolymer via phase separation, Nat. Mater. 6 (2007) 992–996. https://doi.org/10.1038/nmat2038.

[26]   L. Schulte, A. Grydgaard, M.R. Jakobsen, P.P. Szewczykowski, F. Guo, M.E. Vigild, R.H. Berg, S. Ndoni, Nanoporous materials from stable and metastable structures of 1,2-PB-b-PDMS block copolymers, Polymer (Guildf). 52 (2011) 422–429. https://doi.org/https://doi.org/10.1016/j.polymer.2010.11.038.

[27]   L. Li, L. Schulte, L.D. Clausen, K.M. Hansen, G.E. Jonsson, S. Ndoni, Gyroid Nanoporous Membranes with Tunable Permeability, ACS Nano. 5 (2011) 7754–7766. https://doi.org/10.1021/nn200610r.

[28]   H.-Y. Hsueh, H.-Y. Chen, M.-S. She, C.-K. Chen, R.-M. Ho, S. Gwo, H. Hasegawa, E.L. Thomas, Inorganic Gyroid with Exceptionally Low Refractive Index from Block Copolymer Templating, Nano Lett. 10 (2010) 4994–5000. https://doi.org/10.1021/nl103104w.

Materials Research Forum LLC
https://doi.org/10.21741/9781644901816-5

[29]   P. Zavala-Rivera, K. Channon, V. Nguyen, E. Sivaniah, D. Kabra, R.H. Friend, S.K. Nataraj, S.A. Al-Muhtaseb, A. Hexemer, M.E. Calvo, H. Miguez, Collective osmotic shock in ordered materials, Nat. Mater. 11 (2012) 53–57. https://doi.org/10.1038/nmat3179.

[30]   S.M. Peker, Ş.Ş. Helvacı, H.B. Yener, B. İkizler, A. Alparslan, eds., 1 - The Particulate Phase: A Voyage from the Molecule to the Granule, in: Solid-Liquid Two Phase Flow, Elsevier, Amsterdam, 2008: pp. 1–70. https://doi.org/https://doi.org/10.1016/B978-044452237-5.50003-5.

[31]   I. Barandiaran, A. Cappelletti, M. Strumia, A. Eceiza, G. Kortaberria, Generation of nanocomposites based on (PMMA-b-PCL)-grafted Fe2O3 nanoparticles and PS-b-PCL block copolymer, Eur. Polym. J. 58 (2014) 226–232. https://doi.org/https://doi.org/10.1016/j.eurpolymj.2014.06.022.

[32]   J.S. Lee, J. Feijen, Polymersomes for drug delivery: Design, formation and characterization, J. Control. Release. 161 (2012) 473–483. https://doi.org/https://doi.org/10.1016/j.jconrel.2011.10.005.

[33]   T. Anajafi, S. Mallik, Polymersome-based drug-delivery strategies for cancer therapeutics, Ther. Deliv. 6 (2015) 521–534. https://doi.org/10.4155/tde.14.125.

[34]   K. Strebhardt, A. Ullrich, Paul Ehrlich's magic bullet concept: 100 years of progress, Nat. Rev. Cancer. 8 (2008) 473–480. https://doi.org/10.1038/nrc2394.

[35]   D.E. Discher, F. Ahmed, POLYMERSOMES, Annu. Rev. Biomed. Eng. 8 (2006) 323–341. https://doi.org/10.1146/annurev.bioeng.8.061505.095838.

[36]   M.C. García, 13 - Stimuli-responsive polymersomes for drug delivery applications, in: A.S.H. Makhlouf, N.Y. Abu-Thabit (Eds.), Stimuli Responsive Polym. Nanocarriers Drug Deliv. Appl., Woodhead Publishing, 2019: pp. 345–392. https://doi.org/https://doi.org/10.1016/B978-0-08-101995-5.00019-2.

[37]   S.P. Nunes, A.R. Behzad, B. Hooghan, R. Sougrat, M. Karunakaran, N. Pradeep, U. Vainio, K.-V. Peinemann, Switchable pH-Responsive Polymeric Membranes Prepared via Block Copolymer Micelle Assembly, ACS Nano. 5 (2011) 3516–3522. https://doi.org/10.1021/nn200484v.

[38]   J. Cheng, Y. Zhang, P. Gopalakrishnakone, N. Chen, Use of the Upside-Down Method to Prepare Porous Polymer Films with Tunable Surface Pore Sizes, Langmuir. 25 (2009) 51–54. https://doi.org/10.1021/la8035264.

[39]   Polystyrene Microsphere 0.10 μm, (n.d.). https://www.histoline.com/en/00876-15.

Advanced Functional Membranes   Materials Research Forum LLC
Materials Research Foundations **120** (2022) 151-183   https://doi.org/10.21741/9781644901816-5

[40]   H. Yabu, Fabrication of honeycomb films by the breath figure technique and their applications, Sci. Technol. Adv. Mater. 19 (2018) 802–822. https://doi.org/10.1080/14686996.2018.1528478.

[41]   Y. Dou, M. Jin, G. Zhou, L. Shui, Breath Figure Method for Construction of Honeycomb Films, Membranes (Basel). 5 (2015) 399–424. https://doi.org/10.3390/membranes5030399.

[42]   J.J. Smith, I. Zharov, Ion Transport in Sulfonated Nanoporous Colloidal Films, Langmuir. 24 (2008) 2650–2654. https://doi.org/10.1021/la7013072.

[43]   H. Pingle, P.-Y. Wang, H. Thissen, S. Mcarthur, P. Kingshott, Colloidal crystal based plasma polymer patterning to control Pseudomonas aeruginosa attachment to surfaces, Biointerphases. 10 (2015) 04A309. https://doi.org/10.1116/1.4936071.

[44]   X. Peng, J. Jin, Y. Nakamura, T. Ohno, I. Ichinose, Ultrafast permeation of water through protein-based membranes, Nat. Nanotechnol. 4 (2009) 353–357. https://doi.org/10.1038/nnano.2009.90.

[45]   A. Gugliuzza, M.C. Aceto, F. Macedonio, E. Drioli, Water Droplets as Template for Next-Generation Self-Assembled Poly-(etheretherketone) with Cardo Membranes, J. Phys. Chem. B. 112 (2008) 10483–10496. https://doi.org/10.1021/jp802130u.

[46]   J. Russell, Y. Lin, A. Böker, L. Su, P. Carl, H. Zettl, J. He, K. Sill, R. Tangirala, T. Emrick, K. Littrell, P. Thiyagarajan, D. Cookson, A. Fery, Q. Wang, T. Russell, Self-Assembly and Cross-Linking of Bionanoparticles at Liquid-Liquid Interfaces, Angew. Chemie Int. Ed. 44 (2005) 2420–2426. https://doi.org/10.1002/anie.200462653.

[47]   P. Arumugam, D. Patra, B. Samanta, S.S. Agasti, C. Subramani, V.M. Rotello, Self-Assembly and Cross-linking of FePt Nanoparticles at Planar and Colloidal Liquid–Liquid Interfaces, J. Am. Chem. Soc. 130 (2008) 10046–10047. https://doi.org/10.1021/ja802178s.

[48]   J. He, P. Kanjanaboos, N.L. Frazer, A. Weis, X.-M. Lin, H.M. Jaeger, Fabrication and Mechanical Properties of Large-Scale Freestanding Nanoparticle Membranes, Small. 6 (2010) 1449–1456. https://doi.org/https://doi.org/10.1002/smll.201000114.

[49]   J. He, X.-M. Lin, H. Chan, L. Vuković, P. Král, H.M. Jaeger, Diffusion and Filtration Properties of Self-Assembled Gold Nanocrystal Membranes, Nano Lett. 11 (2011) 2430–2435. https://doi.org/10.1021/nl200841a.

[50]   S.K.M. Nalluri, B.J. Ravoo, Light-Responsive Molecular Recognition and Adhesion of Vesicles, Angew. Chemie Int. Ed. 49 (2010) 5371–5374. https://doi.org/https://doi.org/10.1002/anie.201001442.

[51]   E. Krieg, H. Weissman, E. Shirman, E. Shimoni, B. Rybtchinski, A recyclable supramolecular membrane for size-selective separation of nanoparticles, Nat. Nanotechnol. 6 (2011) 141–146. https://doi.org/10.1038/nnano.2010.274.

[52]   P. van Rijn, M. Tutus, C. Kathrein, L. Zhu, M. Wessling, U. Schwaneberg, A. Böker, Challenges and advances in the field of self-assembled membranes, Chem. Soc. Rev. 42 (2013) 6578–6592. https://doi.org/10.1039/C3CS60125K.

[53]   D. Carvajal, R. Bitton, J.R. Mantei, Y.S. Velichko, S.I. Stupp, K.R. Shull, Physical properties of hierarchically ordered self-assembled planar and spherical membranes, Soft Matter. 6 (2010) 1816–1823. https://doi.org/10.1039/B923903K.

[54]   Z. Zheng, C.T. Nottbohm, A. Turchanin, H. Muzik, A. Beyer, M. Heilemann, M. Sauer, A. Gölzhäuser, Janus Nanomembranes: A Generic Platform for Chemistry in Two Dimensions, Angew. Chemie Int. Ed. 49 (2010) 8493–8497. https://doi.org/https://doi.org/10.1002/anie.201004053.

[55]   A. Onoda, K. Fukumoto, M. Arlt, M. Bocola, U. Schwaneberg, T. Hayashi, A rhodium complex-linked β-barrel protein as a hybrid biocatalyst for phenylacetylene polymerization, Chem. Commun. 48 (2012) 9756–9758. https://doi.org/10.1039/C2CC35165J.

[56]   O. Onaca, M. Nallani, S. Ihle, A. Schenk, U. Schwaneberg, Functionalized nanocompartments (Synthosomes): Limitations and prospective applications in industrial biotechnology, Biotechnol. J. 1 (2006) 795–805. https://doi.org/https://doi.org/10.1002/biot.200600050.

[57]   A. de la Escosura-Muñiz, A. Merkoçi, Nanochannels Preparation and Application in Biosensing, ACS Nano. 6 (2012) 7556–7583. https://doi.org/10.1021/nn301368z.

[58]   N. Muhammad, T. Dworeck, M. Fioroni, U. Schwaneberg, Engineering of the E. coli Outer Membrane Protein FhuA to overcome the Hydrophobic Mismatch in Thick Polymeric Membranes, J. Nanobiotechnology. 9 (2011) 8. https://doi.org/10.1186/1477-3155-9-8.

[59]   A. Güven, T. Dworeck, M. Fioroni, U. Schwaneberg, Residue K556-A Light Triggerable Gatekeeper to Sterically Control Translocation in FhuA, Adv. Eng. Mater. 13 (2011) B324–B329. https://doi.org/https://doi.org/10.1002/adem.201080127.

[60]   O. Onaca, P. Sarkar, D. Roccatano, T. Friedrich, B. Hauer, M. Grzelakowski, A. Güven, M. Fioroni, U. Schwaneberg, Functionalized Nanocompartments (Synthosomes) with a Reduction-Triggered Release System, Angew. Chemie Int. Ed. 47 (2008) 7029–7031. https://doi.org/https://doi.org/10.1002/anie.200801076.

[61]   M.S. Kaucher, M. Peterca, A.E. Dulcey, A.J. Kim, S.A. Vinogradov, D.A. Hammer, P.A. Heiney, V. Percec, Selective Transport of Water Mediated by Porous Dendritic Dipeptides, J. Am. Chem. Soc. 129 (2007) 11698–11699. https://doi.org/10.1021/ja076066c.

[62]   N.C. Mougin, P. van Rijn, H. Park, A.H.E. Müller, A. Böker, Hybrid Capsules via Self-Assembly of Thermoresponsive and Interfacially Active Bionanoparticle–Polymer Conjugates, Adv. Funct. Mater. 21 (2011) 2470–2476. https://doi.org/https://doi.org/10.1002/adfm.201002315.

[63]   Langmuir Films, Nanosci. Instruments. (n.d.). https://www.nanoscience.com/techniques/langmuir-films/.

[64]   P. van Rijn, N.C. Mougin, D. Franke, H. Park, A. Böker, Pickering emulsion templated soft capsules by self-assembling cross-linkable ferritin–polymer conjugates, Chem. Commun. 47 (2011) 8376–8378. https://doi.org/10.1039/C1CC12005K.

[65]   Y. Yang, Z. Fang, X. Chen, W. Zhang, Y. Xie, Y. Chen, Z. Liu, W. Yuan, An Overview of Pickering Emulsions: Solid-Particle Materials, Classification, Morphology, and Applications, Front. Pharmacol. 8 (2017) 287. https://doi.org/10.3389/fphar.2017.00287.

[66]   E.A. Jackson, M.A. Hillmyer, Nanoporous Membranes Derived from Block Copolymers: From Drug Delivery to Water Filtration, ACS Nano. 4 (2010) 3548–3553. https://doi.org/10.1021/nn1014006.

[67]   V. Chimisso, V. Maffeis, D. Hürlimann, C.G. Palivan, W. Meier, Self-Assembled Polymeric Membranes and Nanoassemblies on Surfaces: Preparation, Characterization, and Current Applications, Macromol. Biosci. 20 (2020) 1900257. https://doi.org/https://doi.org/10.1002/mabi.201900257.

[68]   M. Pantouvaki, L. Zhao, C. Huffman, K. Vanstreels, I. Ciofi, G. Vereecke, T. Conard, Y. Ono, M. Nakajima, K. Nakatani, G.P. Beyer, M.R. Baklanov, Ultra Low-k Materials Based on Self-Assembled Organic Polymers, MRS Online Proc. Libr. 1335 (2011) 102. https://doi.org/10.1557/opl.2011.1200.

[69]   M. Asadian, K.V. Chan, M. Norouzi, S. Grande, P. Cools, R. Morent, N. De Geyter, Fabrication and Plasma Modification of Nanofibrous Tissue Engineering Scaffolds, Nanomaterials. 10 (2020). https://doi.org/10.3390/nano10010119.

[70]   O. Karaman, C. Celik, A. Sendemir, Self-Assembled Biomimetic Scaffolds for Bone Tissue Engineering, in: Biomed. Eng. Concepts, Methodol. Tools, Appl., 2017: pp. 476–504. https://doi.org/10.4018/978-1-5225-3158-6.ch021.

Advanced Functional Membranes

Materials Research Foundations **120** (2022) 184-213

Materials Research Forum LLC

https://doi.org/10.21741/9781644901816-6

Chapter 6

Porous Membrane and their Applications

S.K. Swain[2], A. Sahoo[1], N. Sarkar[1]*

[1]Centurion University of Technology and management, Odisha, India

[2]National Institute of Science Technology, Berhampur, Odisha, India

*niladri.sarkar@cutm.ac.in

Abstract

Porous membranes have several uses in separation technology. This chapter basically explored various fabrication routes of different types of porous membranes and their applications in industrial and biomedical fields. The urgent need of a low cost perfect membrane needs to be optimised with good mechanical strength, chemical resistance, and thermal stability. This chapter focuses on producing porous polymeric/ceramic membranes with nano/submicron holes using block copolymer self-assembly, track etching, 3D printing, nanoimprinting, lithography, and electrospinning, etc. Porous polymeric/ceramic membranes are used in separation techniques in various industries, drug administration, bioseparation, and biosensing. The effect of membrane material, pore shape and size towards the membrane performance for various technological challenges are thoroughly discussed.

Keywords

Porous Membranes, Polymeric Membrane, Ceramic Membrane, Separation Techniques, Industrial Application, Biomedical Applications

Contents

1.   Introduction

Research progressed on porous membranes is one of the critical importance in the development of science and innovation. Porous membranes have promised tremendous potential in industrial and biomedical applications, because of their primary adaptability, flexible surface science, and bio-compatibility. Recently porous polymers have drawn tremendous research interest due to its various applications, including gas adsorption/stockpiling, detachment, catalysis, ecological remediation, energy, optoelectronics, and wellbeing. Ongoing years have seen enormous exploration leap forwards in these fields on account of the remarkable pore structures and flexible skeletons of permeable polymers. Although there are various research articles are available, explaining characteristics, fabrication and its potential applications in biological and biomedical fields. But most of them have put their interest in a specific technique and its application based on that technique. In this particular chapter, we try to cover all the synthetic procedures available for the preparation of porous membranes out of polymers as well as ceramic materials.

Porous materials consist of a strong framework with characterized openings or pores which have pores of different size from 2 nm to 20 mm [1]. The partition of solutes by permeable membrane (PM) is principally a component of sub-atomic size and layer pore size dispersion [2]. These layers are utilized to segregate colloidal particles or huge subatomic solutes out of solvent. High separation may be acquired when the size of the dissolved or dispersed material (molecules or small particles) are moderately bigger than the pore diameter of the membrane. Utilizing the definition of pore size as embraced by the IUPAC, the PMs are classified based on their pore size. Pore size having 50 nm or more is named macroporous, pore widths somewhere in the range of 2-50 nm are grouped as mesoporous materials. Membrane with normal pore widths somewhere in the range of 0.2-2.0 nm is named microporous. If the pore size is less than 0.2 nm, layers are categorized as nonporous membranes. Based on the membrane architecture, membranes can be categorized as symmetric as the pore diameter does not fluctuate from any cross sectional view. A wide scope of inorganic materials, for example, metals, glasses, ceramics, or natural including various classes of polymers can be utilized to create porous membranes. The material choice also, pore size of the membrane rely upon the application for which it would be utilized.

Since the last two decades, polymer membrane fabrication technique has improved drastically. The first portion of this survey gives a concise summary of various fabrication techniques for polymeric based membranes. These techniques incorporate various methodologies, for example nanoimprinting and track etching based on top-down approaches and block-copolymer self-assembly and electrospinning on bottom-up approaches and the second portion of this survey covers numerous applications.

2.   Different fabrication techniques

This section depicts the most commonly used methods for making porous polymeric membranes, including track-drawing, nanoprint lithography (NL), 3D printing, block-copolymer self-assembly and electrospinning.

2.1   Nanoimprint lithography(NL)

Nanoimprint lithography (NL) is the most unique well-known sensitive lithography-based methodology for producing permeable polymeric membranes. This lithography technique became well-known for its tremendous potential for growth because of simplicity, as well as the time effectiveness and low cost. It's a technique where a template is pressed against the opposing layer which is basically distorted on a substrate, i.e. a metallic surface or a silicon wafer. Mould fabrication is the best fundamental factor, since the precision and composition are essential in defining the structure, i.e. engraved polymers. Two of the most widely used nanoimprint lithography techniques are thermal nanoimprinting (TNL) and photo nanoimprinting (PNL) [3]. To create nano and microstructural patterns in a regulated way, photolithography, also known as electron beam lithography (EBL), at nanoscale is often used to make the master mould. EBL, on the other hand, offers permeable anodic aluminium oxide (pAAO) membranes with uniform straight nanochannels that may be used as a format to build nanoscale masters since pAAO films exhibit uniform nanochannels are very easy to manufacture. Nickel (Ni), silicon (Si), and thermoset polymers are used to make master moulds. Because of its flexibility and strength, Ni is the most widely used of these materials. A tiny nanoscale master is created for soft lithography by mixing hard polydimethylsiloxane (hPDMS) with a PDMS mould, since hPDMS has a considerably better resolution than PDMS [4]. Choi and co-workers used a limited adjustable SU-8 stencil membrane to construct a polymeric double scale nanoimprinting mould [5]. Photolithography, from one perspective, included spinning a layer of photoresist (PR) on a slice of device material and then PR exposed to UV light through a mask. Thermal nanoimprinting, on the other hand, included making microscale openings on the equal proportion, mixing a layer of PR and SU-8 layer which is a epoxy based negative layer. Thermal nanoimprinting was used to create a PMMA (polymethyl methacrylate) substrate with nanoscale cavities for the next layer. Ultimately, to get a two-layer micro and nanostructure layer, a SU-8 film was applied to the PMMA substrate. After UV exposure, PUA511RM, an UV curing resin was apportioned against the collected substrate and took off utilising an adaptable PC membrane as a supporting backplane, after an extremely PDMS thin layer was deposited following that siliconization of the hybrid structure to work with the arrival of the last master shape. Lee and coworkers demonstrated the utilisation of delicate lithography to adjust the pattern thickness by warming thermoplastic substrate [6].

Advanced Functional Membranes   Materials Research Forum LLC
Materials Research Foundations **120** (2022) 184-213   https://doi.org/10.21741/9781644901816-6

2.2   Photo nanoimprint lithography (PLN)

The photo nanoimprinting approach, in contrast to the thermal nanoimprinting method, entails emblazoning the perfect instances on a layer of fluid photoresist and relaxing it by UV irradiation, that causes the oppose to solidify by causing polymer cross-linking. Over the TNL same, the PLN method has a few distinct advantages. It may lead at room temperature at first, which avoids problems caused by thermal development differences between the resist, form and substrate. Furthermore, the PR is often less adhesive than different kinds of polymers, such as PDMS, allowing for a lower imprint pressure to be used [7]. Lower viscosity resistance makes it simpler for the liquid to fill the mould cavity without the need of suction. Zhou et al. used UV nanoimprinting to produce a polymeric membrane with nanoscale holes, using high-thickness PDMS nanopillars [8]. The high thickness PDMS nanopillars were supplied by projecting hPDMS from a nanoporous pAAO layer. Yanagishita and colleagues have demonstrated a technique for image nanoimprinting that produced an extremely uniform pore clusters polymeric pattern of about size equal to100 nm by using a Ni nanopillar form, which is made of pAAO membrane (Figure 1). Then, the Ni form was loaded with PAK-01 (photopolymer monomer) and put on a glass substrate to initiate the polymerisation process in presence of UV radiation [9].

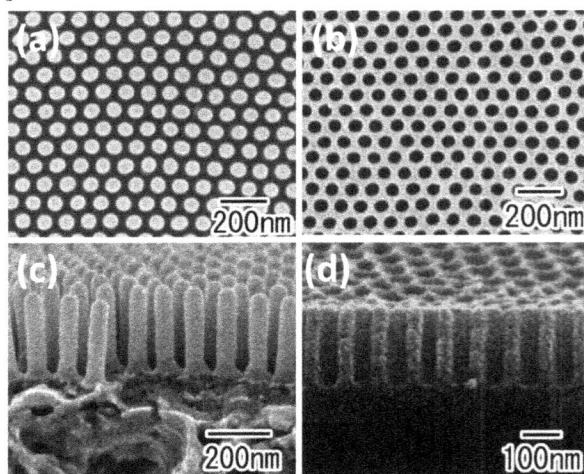

Figure 1 (a) SEM images represent the top view of the Ni nanopillars mould, (b) cross section, and (c) top view and (d) side view of porous Nickel mould with high porosity, reproduced with the permission from IOP Science [9].

Advanced Functional Membranes   Materials Research Forum LLC
Materials Research Foundations **120** (2022) 184-213   https://doi.org/10.21741/9781644901816-6

The permeable polymeric membrane was carefully removed from the Ni mould, which had been pre-treated with a release agent. Despite the previously stated nanoimprinting advantages, the manufacturing of free-standing porous membranes is still a challenge since an extremely thin layer is often present at the bottom portion of holes. And membrane detachment from the substrate may also cause harm to the membrane. To resolve these problems, a sacrificial layer is placed in between the polymer and the substrate. Choi and colleagues used the UV nanoimprinting technique to create an underlying membrane detachment of 10-13 nm holes [10]. A sacrificial LOR layer of about 100 nm thickness was spin coated below the resist to make an SU-8 film. Conical holes were created and penetrated the sacrificial LOR layer via the film layer (SU-8) Following that, a polymer reflow procedure reduced the diameter of a pore to less than 10 nanometers.

2.3   Thermal nanoimprint lithography (TNL)

It is a kind of lithography that uses heat to create a thermoplastic polymeric thin layer which is maintained onto a strong substrate above its glass transition temperature (Tg) to allow decoration of a specified master mould with optimum pressing factor and time and also to mollify the layer [11]. At temperatures below Tg, the thermoplastic polymer's protection against flexible distortion and the esteem pressure where thermoplastic polymer deformation occurs both decrease substantially. Overheating, on the other hand, may damage the film [12]. After that, the engraving temperature is decreased below Tg which cause solidification of the thermoplastic layer. One of the most noticeable advantage of TNL versus PNL is the availability of a broad range of thermoplastics for the former. It's also essential to note how easy it is to regulate the thickness and uniformity of these thermal supports by spin coating them on a substrate [4]. In thermal nanoimprinting, a hard Ni structure is often used as a master shape, specifically for nanopore creation, since elastomeric forms, such as those constructed of PDMS, fail to meet expectations.

2.4   Tracks etching technology (TET)

Analysts can provide layers with perfectly regulated and consistent pore size, shape, area, and thickness using micro and nanofabrication processes based on track etching. The procedure's most notable advantage is that TET provides a broad range of pore sizes with high densities, starting from nano to microscale, that can be regulated autonomously.

TET, first described in the 1960s, works by lighting polymers with powerful X-beam irradiation, heavy ions, UV light and electrons causing straight damaged tracks to form through the exposed polymeric layer. To transfer the damaged tracks into pores, either an applied electric field or a wet chemical etching conditions are generally followed [13]. The use of heavy ions allows for a wider range of direct energy transfer of track-framing particles, pore channel point circulation, and pore length [14]. A specific single light setup includes a magnetic defocusing instrument, a trigger, a sample stack, and an electrostatic deflector a detector as shown in Figure 2 [15].

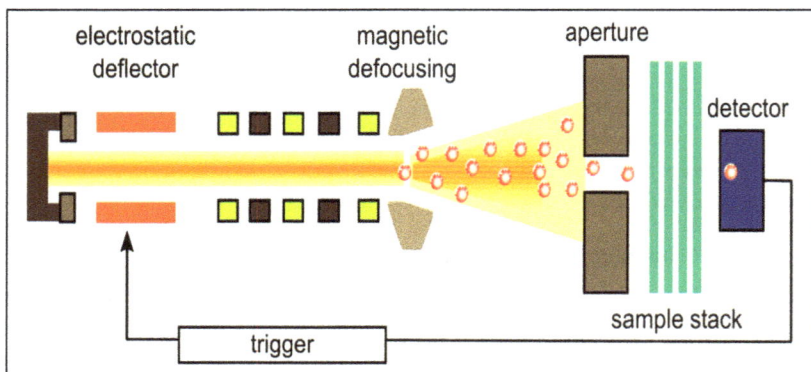

Figure 2 Track etching setup using a single irradiation. reproduced with the permission from Beilstein Institute for the Advancement of Chemical Sciences [15].

To begin, the ion magnetically defocuses and alters such that a single beam of a particular frequency of a single beam passes through the aperture. The ion illuminates the stack of foils a short time later. The particles are identified by a solid state particle detector installed behind the test, and when a single ion is detected, the whole beam of ions are deflected by the electrostatic force of the chopper [16].

2.5   Other soft lithography methods

There are a few other types of imprinting techniques outside the two main nanoimprinting processes. Le et al. described a unique method for fabricating a porous PDMS membrane using two photoresist sacrificial layers (Figure 3).

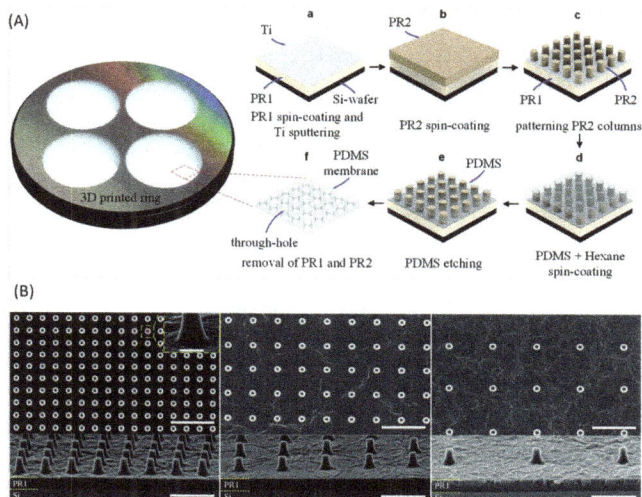

Figure 3 Synthetic procedure of free standing PDMS layers and their top and side view in optical and SEM images, reproduced with the permission from Royal Society of Chemistry [17].

This research includes spinning of a positive photoresist layer as a sacrificial layer onto a silicon wafer. A very thin Titanium layer was sputtered on the photoresist layer. This results in prevention of the intermixing of the first and second photoresist layer [17]. Standard photolithography was used to create perfect instances of nanopillars on the following photoresist layer. The exterior of the nanopillar displays were then spin coated with a PDMS arrangement in hexane. The supported membrane PDMS was etched in a mixture of oxygen and sulfur hexafluoride to reduce the membrane thickness and to create the pores throughout the layer. Despite the fact, for soft lithography the material like PDMS has likely been the most often used because of its flexible surface chemistry, biocompatibility, compound dormancy, low cost and biocompatibility [17,18]. Research has been going on using different varieties of polymers. Yanagishita et. al. submerged a Ni nano array produced out of pAAO layer in monomers of benzoyl peroxide and methyl methacrylate arrangement of about 2 wt. percent, then polymerized it for 12 hours at 50°C to create a PMMA membrane [19]. After that, acid dissolution with Ni mould results in a free-standing membrane. This created membrane exhibited a width of 4.5 m a pitch distance of 500 nm, and pore width of 320 nm that were identical to the master mould membrane of pAAO, which had a pitch distance of 500 nm, a thickness of 4.5 m and pore width of 300 nm. This small difference in measurement of pore between the two layers e.g. (pAAO, PMMA) was believed to be due to shrinking of the PMMA volume during polymerization. This technique has the benefit of being able to create the nanoscale holes having high aspect ratios and that is hard to do with nanoimprinting which are pressure-based. This projection technique may also avoid producing a thin layer which is present at the bottom of the pores. With poly(-caprolactone) (PCL), Bernards et al. used the projection method to produce a nanoporous material which can be biodegraded. A ZnO layer was combined with ZnO nanorods as a layout, and the PCL arrangement was cast. A porous film having nanoscale pore size was created by etching the zinc oxide nanorods (having approximate length of 500 nm and width of 237 nm) with H_2SO_4. The pore size and thickness of the produced PCL film were 21 nm and 5×10^9 pore cm^{-1}, correspondingly. This methodology creates layers with a broad range of pore diameters and thicknesses, that has been used in a variety of applications, such as tumour cell separation [20], nanoparticle exhibit fabrication [21], and biosensing [22].

2.6   Self-assembly of polymers

Polymer self-assembly creates transparent polymeric films by causing polymeric precursors to spontaneously associate in response to explicit non-covalent interactions [23]. A phase inversion technique may often trigger self-assembly. Thermally induced phase separation (TIPS), vapour induced phase separation (VIPS), and solvent induced phase separation (SNIPS), are all noteworthy phase inversion processes. By manipulating the self-assembly routes, the macroscopic shape of the polymers may be successfully tuned, ranging from spherical colloidal particles to colloidal gels with hierarchical porosity. The resultant materials exhibit unique microporous characteristics due to the interior chamber of SNIPS. This synthetic method may enable the creation of soft, flexible materials that retain their porosity indefinitely.

2.7   Self-assembly of block copolymers (BCP)

For many decades, block copolymer (BCP) self-assembly has attracted considerable attention due to its ability to generate ordered structures with a wide variety of morphologies, having spherical, vesicular, cylindrical, lamellar, bicontinuous arrangements, and a variety of hierarchical or complex architectures. These self assembled polymers have prospective or practical uses in a wide variety of areas. Nonetheless, this technique is very beneficial for delivering permeable inorganic compounds i.g silica and carbons [23]. To begin with, block copolymers should be amphiphilic for the most part, with one portion which can interact with the prepolymer in a non-covalently manner. Secondly, this prepolymer should have the ability to cross-connect further forming a stable

Advanced Functional Membranes   Materials Research Forum LLC
Materials Research Foundations **120** (2022) 184-213   https://doi.org/10.21741/9781644901816-6

system with no effect on the BCPs. Furthermore, easy BCP removal is essential without affecting the system structure [24]. The most important prepolymers in the development of the permeable frame are phenolic tars [25,26]. The fabrication of a 2D hexagonal mesoporous polymer sheet was disclosed by Schuster et al. The triblock copolymer (Pluronic P123) template was used to create an oligomeric precursor, followed by removal of BCPs. On a Si wafer, an ethanolic solution of precursor with BCP was coated. Circularly moulded pores of larger diameter = 6.8 nm and shorter diameter = 3 nm were found in the produced polymeric film. Liang and colleagues have developed a comparison approach based on PS-P4VP (polystyrene-block-poly(4-vinylpyridine)) self-assembly [27]. This method's distinctive feature is the incorporating monomers of resorcinol to the PS-P4VP BCP to give a base permeable design prior to resorcinol polymerization and carbonisation. This helps the BCP-resorcinol complex formation, resulting in excessive cross-connected resorcinol-formaldehyde tar that is sometimes hard to dissolve. The volume of the PS entity in the BCP may control the pore width, and the thickness of the pitch region can be controlled by the amount of resin used.

Block copolymer can also be used as a porous membrane framework where to construct the permeable film system by removing just one portion of BCP and, at times, some additional portions. One of the challenges in framing highly organised permeable polymeric membranes is to control the orientation of the pores perpendicular to the substrate all over the membrane. It is possible to tune the pore size of a membrane with systematic consideration of the proper experimental conditions such as parameters of kinetics of phase separation and film formation to achieve the targeted morphology [28,29].

2.8   3D printing technique

Added material manufacturing, often known as 3D printing, is one of the unique techniques which promises a vast production of membranes used in biomedical applications. The capability to create complicated forms or components in three dimensional methods using varieties of materials, such as graphene-based materials, ceramics, metals, conventional thermoplastic of different scales [30], is the greatest advantage of this technique. This technique is classified into four categories: (1) photo polymerisation, which involves restoring photoreactive polymers using a laser or UV light [31]. (2) powder-based printing, where an ideal material's powder is placed on a substrate which follows a binder material spurting and selective sintering to get a 3D material [32], (3) material extrusion, which softens and expels a polymeric material fibre to form an optimal structure when heating the thermoplastic over its T_g [33] and (4) a desired 3d shape is achieved by lamination of the thin layer of the materials [34]. Stereolithography (SLA) is a polymerisation technique based on the laser light based lithography that has got wide applications for production of polymer based membranes currently. In this process an UV laser follows the optimum design and attaches the photoresponsive polymer on the bottom/ top part of the stage in motion. When the perfect decoration is created on the stage it is vertically dropped on the surface of the resin followed by curing by the UV-laser to form a consistent 3D item (Figure 4) [35].

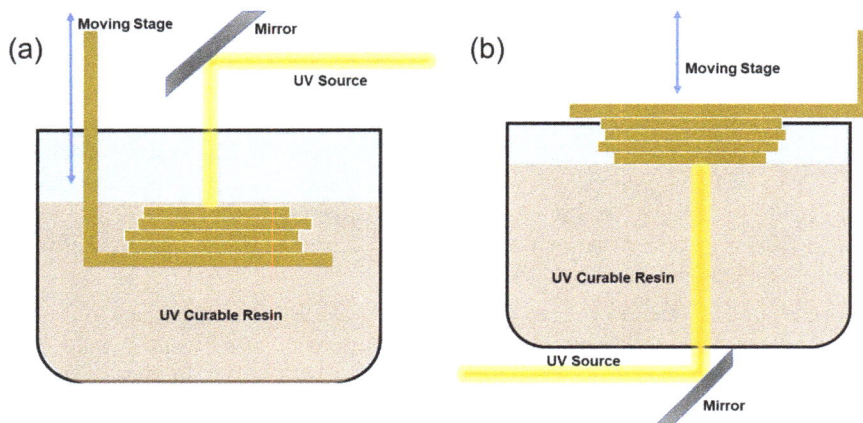

Figure 4 3d Printing by stereolithography showing in normal as well as inverted mode, reproduced with the permission from Wiley-VCH [35].

However, 3D printing has enormous promise in a variety of areas, including design, assembly, manufacturing and medicine design, it is still with major drawbacks such as printing resolution, microscale pore formation [30,35]. Producing membranes with pores in nano range is still a great challenge. Although most three dimensional printing methods have a microscale target limit, the late advancement in the use of the TPP(two-photon polymerisation) technology for three dimensional printing has shown the potential to reach as far as feasible to a 100 nm scale. While this technology has progressed rapidly over the past decade, it still faces a few challenges: (1) maximum production costs, (2) lower stability of final material without proper treatment, and (3) types of material used. According to literature, there have been few reports of 3D printing being utilised to produce permeable polymeric sheets with pore size in nano range. But we believe this technology will pave the way for the potential production of permeable films in future.

Advanced Functional Membranes
Materials Research Foundations **120** (2022) 184-213

Materials Research Forum LLC
https://doi.org/10.21741/9781644901816-6

2.9. Electrospinning (ES)

Porous polymeric structures when synthesized by electrostatically operated spinning of polymeric fibres is called electrospinning. This method may produce permeable polymeric films with pore diameter ranging from micro to nanoscale. High pore interconnectivity, adjustable thickness, high gravimetric porosity, interstitial space in micro range, high surface to volume ratio and a low thickness with good mechanical property in terms of strength are all advantages of such layers [36].

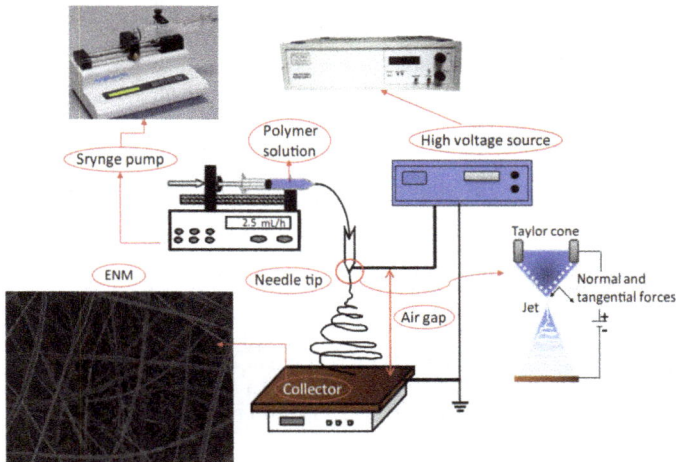

Figure 5 Schematic procedure of electrospinning technique, reproduced with the permission from Wiley-VCH [37]

Figure 5 shows an average electrospinning measurement. The pre-arranged polymer arrangement is first softened or disintegrated in solvent and then positioned into a capillary tube in this method. Then, an electric field strength is employed in between the collector and tip of the capillary. The developed charge is in opposition to the surface strain of polymer fluid held at the tip of the capillary. The hemispherical condition of polymer liquid at the tip transforms into a cone-shaped structure termed Taylor cone as the electric force is increased. The arrangement is launched from the tip of the Taylor cone when a critical value is induced by the applied electrical field that exceeds the surface pressure of the polymer fluid. Electrically controlling the path taken by the charged jet of polymer solution allows uniform polymeric fibres to be collected to produce permeable membranes [36]. In general, there are two types of electrospinning: (1) solution electrospinning (SES) and (2) melt electrospinning (MES).

2.9.1 Solution electrospinning (SES)

SES is a mostly used and well-established technique that can be easily handled at 25°C. Ambient system parameters, solution, and process are the three main entities that govern the electrospinning interaction. The electrospinning process is influenced by environmental factors such as temperature and humidity. Ray et al. discovered that high humidity inhibits polymeric solution elongation, resulting in an increase in nanofiber diameter [38]. Where as according to Mit-uppatham et al. high temperature results in the synthesis of thin fibres [39].

Table 1 Various preparation procedures and materials used.

Preparation Technique	Materials Used	References
Nanoimprint Lithography	hard polydimethylsiloxane (hPDMS)	[4]
Photo nanoimprint Lithography (PLN)	porous anodic aluminium oxide (pAAO) + Ni metal	[9]
Thermal nanoimprint lithography (TNL)	Thermoplastic polymers, Ni substrate	[11,12]
Tracks Etching Technology (TET)	powerful X-beam irradiation/ heavy ions/ UV light on polymeric membrane	[13,14]
Self-assembly of polymers	PS-P4VP (polystyrene-block-poly(4-vinylpyridine))	[24, 27]
3D printing Technique	Organometallic polymers, Resin, Ceramic, Metals, Conventional Thermoplastic Polymers	[32-34]
Electrospinning (ES)	Poly ethelene glycol, Synthetic/Natural polymers, Chitosan	[36,37]

2.9.2 Melt electrospinning (MES)

MES uses a simple cleaning method to remove the solvent after fabrication, and there is no need to choose a solvent. It's best for polymers that don't have a suitable solvent to dissolve at 25°C. Do and co-workers used dissolved coaxial electrospinning to make polyethylene glycol shell nanofibers [40]. This technique allows for changes to the film geometry, such as tube form, as well as the precursor selection from a broad variety of polymers, such as natural and synthetic polymers having biocompatibility [41]. Tissue engineering, drug delivery, water purification, and lithium ion battery technology are only a few of the applications of the membranes created by this process. The significant surface roughness of these films encourages surface fouling, making them unfit specifically for biosensing applications. The Syhtnetic Techmiques and the materials used in various process are summarised in Table 1.

3.   Applications of porous membranes

3.1   Pressure-driven membrane processes

Depending on the desired separation, membranes can be used to (1) remove pollutants from waste streams so that the water can be recycled or discharged, or (2) concentrate or recover necessary components while the remainder is allowed to flow through the membrane. Both activities are crucial because they may help to accomplish two objectives in many industrial processes: (a) water reuse and (b)raw material or (by)product recovery. To remove particles in the 0.05–10 mm range, MF employs porous membranes with pore sizes in the range of 0.05–10 mm are used. Particles are kept depending on their dimensions in this filtering process (shape and size dependent). This membrane technique most closely resembles conventional coarse filtration (Figure 6). [42]

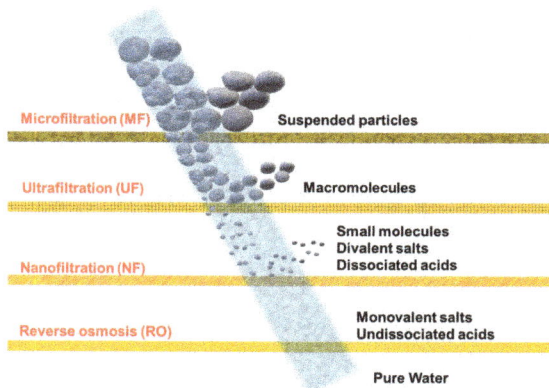

Figure 6 Pressure driven membrane processes.

Advanced Functional Membranes   Materials Research Forum LLC
Materials Research Foundations **120** (2022) 184-213   https://doi.org/10.21741/9781644901816-6

Although polymeric materials make up the bulk of MF membranes, inorganic materials such as aluminium (Al) oxide, zirconium (Zr) oxide, titanium (Ti) oxide, carbon, silicon carbide, and other metals are also frequently used (steel, tin, nickel, etc.). MF may be used in the purification (e.g., of fruit juices, wine, and beer), sterilisation (e.g., of drinks and medicines), or concentration phases (e.g., in the cold sterilisation of beverages and pharmaceuticals) to retain suspended particles (i.e., in cell harvesting). In the production of drinkable water from ground water using NF and RO, MF is also used as a pretreatment step [43,44]. Ultrapure water production from the semiconductor industry, metals recovery as colloidal oxides or hydroxides, wastewater and surface water treatment, and oil–water emulsion separation are some of the other technological applications. Large molecules, having dimensions in the range of 2–100 nm (proteins, polymers, colloids, and emulsions) are kept in UF. Small molecules, such as salts and organic solvents, as well as low-molecular-weight solutes, may pass easily across the membrane. The separation is based on a sieve process, similar to MF. UF membranes may be made of polymers or inorganic materials, much as MF membranes (mainly Al and Ti oxides). UF is used as a concentration/purification step in the dairy (milk and whey processing, and cheese manufacturing) and food industries (i.e., in the production of potato starch, and protein recovery and hydrolysis). It is also used in the metal and surface industries (for oil–water emulsion treatment and electropaint recovery), the textile and automobile industries (for indigo recovery), and wastewater treatment. NF is a membrane method that retains low molecular weight organic molecules (200–1000 Da) as well as multivalent ions like sulphates in a size range between UF and RO. In contrast to RO, the pressure utilised is modest (in the range of 3–20 bars), yet substantial fluxes are obtained. To separate particles in NF membranes, particle size, particle charge, or a combination of both may be employed. The majority of NF membranes are now polymeric composites; nevertheless, a growing number of NF ceramic membranes with unique separation characteristics are becoming available [45]. In water softening (calcium ion removal), solvent recovery, and wastewater treatment, small compounds including proteins, enzymes, antibiotics, dyes, herbicides, pesticides, and insecticides are frequently removed utilising NF methods. Low-molecular-weight compounds, like salts, may be retained using the RO method. Organic solvents, on the other hand, have a low retention rate because they break down quickly in the membrane. cellulose triacetate, aromatic polyamides, and polybenzimidazoles are the most prevalent RO membrane materials utilised. Ceramic RO membranes are currently not available in the market. Desalination of saltwater; ultrapure water manufacturing (electronic industry); concentration of food juice, sugars, and milk; and wastewater treatment are the primary uses of RO [46].

Membranes are used in a variety of industrial applications that need system stability, high accessibility, low requirements for preventive maintenance, and minimal care. For such demanding applications as those found in the chemical and pharmaceutical industries, as well as those found in water and wastewater treatment, ceramic membranes have emerged as the preferred choice due to their extremely high chemical and physical (thermal and mechanical) stability, exceptional separation features, and extended operational period. When compared to other membrane types utilised in pressure driven membrane methods, the following are the most often mentioned benefits of ceramic membranes: The operating life is lengthy and consistent. They show high temperature resistance (up to 300 °C, depending on the material) throughout the pH range, and pressures up to 30 bars (generally they withstand up to 90 bars). The stability in various chemical environments is excellent (i.e., oxidants, hydrocarbons, organic solvents, and including acids like hydrofluoric/ phosphoric acids when pure titania is used), bacteria resistant and, in many instances, bioinert compatibility with highly viscous liquids resistance to corrosion and abrasion. In addition better sterilization and cleaning properties under chemical or steam with back flushing are also shown. Because of these properties, they may be used in a number of applications where polymeric and other inorganic membranes are ineffective. Membrane pore diameters in the MF-UF-NF range may now be found in a variety of sizes (from 5 mm down to a few hundred daltons). Cleaning under chemicals at high temperatures using chlorine, ozone, hydrogen peroxide, and robust inorganic acids, caustic, as well as steam sterilisation, are also possible with ceramic membranes. Owing to their exclusive properties, ceramic membranes (CMs) may be cleansed simply by backpulsing. In this process the flow over the membrane is reversed. As a consequence, the filter cake is driven aside from the surface of the membrane, extending filtering efficiency and reducing cleaning downtimes. Most ceramic membranes have much greater permeability rates than presently available polymeric membranes. In most applications, good organic retention and a low fouling propensity have been verified. Salt rejection, on the other hand, reduces dramatically when electrolyte concentration increases, as it does during the treatment of pickling bath solutions. As a result, using ceramic NF membranes in such a medium isn't always a good idea. Ceramic membrane rejection, on the other hand, is less temperature sensitive than polymeric membrane rejection (Considering both the average cut-off and the size distribution), which may be a benefit in certain applications. The hefty weight and high production costs of ceramic components are the two biggest disadvantages. Nevertheless, in situations where polymeric membranes are not possible, they usually have a lengthy service life. Furthermore, they may be recoated in the case of membrane loss, lowering element (membrane) replacement costs [46]. CMs were first created for uranium enrichment, but they were later utilised in wastewater treatment as well. Various industries like food, chemical, petrochemical, biotechnology, pharmaceuticals, dairy, and metal polishing as well as power generation are all using ceramic membranes. Each industry has its own set of needs and opportunities [47]. Aside from industry-specific applications, a few common techniques are utilised across a variety of sectors. Oil/water separation and cleaning chemical recovery are two examples [48-50]. There are many examples in the food industry, metal fabrication and allied industries, automotive sector, chemical manufacturing sector, and oil refinery, among others, of petroleum oils from container washers, spills, drilling, and other processing operations. CMs are used in wastewater treatment for a variety of purposes, including oil removal from wastewater or purifying washed water from metal or textile cleaning, pre - treatment of landfill effluent. The benefits of CM may also be seen in recycling technologies including enamel, coating, and paint recycling, petrochemicals recycling, and degreasing cleaning baths. Emulsified oils in the wastewater are difficult to separate using traditional treatment methods like coalescers and oil skimmers. Furthermore, many polymeric membranes are inappropriate in hostile chemical conditions, such as heavily polluted oily wastewater, owing to their low stability (e.g., petroleum fractions, lube oils etc). On commercial scale systems fitted with back pulsing devices, long-term flux stability has been established (300 liter/m² h was reported for the handling of generated water comprising 10–100 ppm suspended particles and oils using 0.8 mm Membralox membranes over many months) [51]. Chemical recovery varies by location, but on average, 95 percent of the active chemical solution is recovered, with Kjeldahl nitrogen rejections of up to 75 percent, 97 percent oil and grease rejections, and 100 percent suspended particles rejections. Many factors influence the payback, including the characteristics of the discarded cleaning solution, the cost of new chemicals, the percentage of recovery, the number of times the permeate may be reused, as well as utility and waste disposal costs. However, it is typical to have a payback period of less than a year or two [52]. Ceramic MF membranes are also often used in the processing of various water sources prior to RO. Ceramic membranes provide consistent performance under substantially changing feed circumstances in this application, and as a consequence, RO membrane lifespan may be doubled or tripled [52]. Recovery of catalysts, recycling of brine, colored and pigmented dyes and pigments may be recovered. Desalination of products is a process that involves the removal of salt from items (i.e., optical brighteners, surfactants, and dyestuffs). Organic solvents are cleaned, recovered, and recycled (solvent NF) by using porous membranes.

3.1.1 Textile/pulp and paper manufacturing

Processing of fuel oil and ethanol, fractionation of linosulfonates, COD reduction of bleach plant effluent, recovering whitewater from paper mills, recovering the coating on paper, rinse water from main and secondary wool scouring effluent, are some of the applications of porous membranes. The loss of colour, high-quality water for recycling, recovering and reusing dye baths, finishing the spin bath water recovery is aided by the use of a water jet loom are done by utilizing porous membranes.

3.1.2 Biotechnology, pharmaceuticals, and cosmetics

Pharmaceutical active agents (antibiotics, enzymes, proteins, amino acids, and vitamins), as well as plant extracts, are concentrated, fractionated, isolated, and sterilised, biomass and algae separation, concentration, and dewatering, filtration of blood plasma, fat emulsions disposal, Clarification of fermentation broth, yeast separation, and bacterial suspension concentration, getting rid of endotoxins, recovery of aqueous stream byproducts, production of vaccines, nutraceuticals, are performed by ceramic membranes.

3.1.3 Surface engineering/automotive/mechanical/mineral/metal industry

Degreasing and rinsing of baths, treatment of emulsions of oil and water, heavy metals' resurgence, wastewater from grinding operations, glass/glass fibre manufacturing, wastewater treatment and, removal of oily wastewater are some of the examples where porous membranes are widely used.

3.1.4 Foodstuffs, dairy products, and beverages

Wine, beer, cider, soy sauce, and vinegar purification, juice concentration, microorganisms are removed from milk and whey. Milk and whey components are separated and fractionated. Whey desalination, soybean protein separation, processing of vegetable oil, drinking water purification, processing of sugars and starches, product recovery from beer and cider tank bottoms, water recovery from the waste stream are some of the application of porous membranes in above mentioned industries.

3.1.5 Water, recycling, and the environment

Treatment of wastewater and potable water, membrane bioreactors reduce COD, BOD, and suspended solids, chemical recycling and washing wastewater (i.e., laundry), waste oil treatment, emulsified oil removal, pesticides and drugs removal, getting rid of germs, heavy metals removal, treatment of grey water, sewage plant /sewage drain purification, water recovery from methane digesters, vacuum residual oil upgrade, processing of landfill leachate, treatment of radioactive wastewater are some of the application of porous membranes.

3.2 Chemical and petrochemical industries

The chemical industry is very diversified, using all main process technologies. There is a requirement to (1) recover and reuse chemicals in various chemical process applications, and/or (2) treat highly variable wastewaters including detergents, polymers, and organic solvents. However, there is a growing need for better product quality, faster payback periods, and lower operational costs. Membrane technology is becoming more popular as product separation gets more complicated. This is especially true in the fine chemical sector, where membranes are employed in a wide range of applications, including the concentration and desalting of dyestuffs and optical brighteners, for example. Ceramic membranes, which can function in more severe circumstances such as high temperature (over 200 C), high/low pH, and solvent resistance, may provide substantial benefits over more traditional technologies in such a situation [53]. The following are some of the most common ceramic membrane uses in the chemical industry today: Recovery of pigment, hydroxide, and catalyst. The recovered product is recycled back into the process stream, and the resultant wastewater's chemical oxygen demand (COD) is substantially decreased. Desalting is the process of removing inorganic particles from powder and liquid dyestuffs during production. The membrane rejects the dye while allowing salts and water to flow through, concentrating and desalting the dye at the same time. In the manufacture of powder dyes, NF may be used to improve plant capacity prior to drying. In the production of liquid dyes, the dye's solubility is improved by nearly full desalting. Concentration of silica sol and latex membrane technology offers a cost-effective way to concentrate latex white water with up to 15% solids. After that, the latex may be recovered, and the cleaned water can be reused. While aqueous membrane applications continue to dominate the market for pressure-driven membrane processes, emerging solvent-resistant NF membranes open up new possibilities in the chemical sector [54]. Separation of oil and water, most governments have certain standards for the quality of wastewater when disposing of mineral oil in surface water and sewage systems. A significant quantity of oil may be extracted from wastewater using a water-oil separator. Chemically stabilised oil/water solutions, on the other hand, must be handled carefully. Membrane filtration may be used to purify these solutions, resulting in dischargeable water. Oil/grease 136 ppm to 1 ppm, Pb 0.79 to 0.09, Ni 0.15 to 0.01, Cu 1.49 to 0.36, and Zn 5.9 to 0.37 are typical findings from waste stream to permeate [55]. Ceramic membranes are often used as an intermediate filter or prefilter for more costly RO membranes, as well as for the reduction of oily bilge wastes, process wastes, and wash water wastes. Pretreatment with RO. Backpulsing allows ceramic membranes to be cleaned continuously in situ (CIP), reducing the requirement for off-line cleaning. As a result, ceramic microfilters lower the cost of replacing ROs. The amalgam process produces sodium hydroxide (50 percent by weight), which is loaded with mercury downstream of the amalgam decomposer in chlorine-alkali facilities. Filtration using ceramic membranes may remove this mercury, as well as other solid contaminants like graphite.

Advanced Functional Membranes   Materials Research Forum LLC
Materials Research Foundations **120** (2022) 184-213   https://doi.org/10.21741/9781644901816-6

3.3   Porous membrane in biomedical applications

Nanoporous materials offer several potential biological uses that are presently being investigated. The membrane in implanted devices will behave as a semipermeable section which would keep the material or components in place and allows for the controlled release of necessary molecules. Nanoporous membranes (NPMs) may also be used for biomolecular analysis in a number of ways. In vitro uses include protein separation and diagnostics. Significant research has been conducted over the past decade to automate biological testing and reduce the amount of time and money spent on sample collection and analysis. These efforts have resulted in a large number of microfabricated devices that perform mixing, detection, reaction, separation or preconcentration functions. It has been shown that the alteration in selective transport and diffusivity across the membrane may be utilised to extract analytes from complex mixtures. Lab-on-a-chip concepts have garnered a great deal of interest in recent years [56]. The current section explains an overview of NPMs and their use in biological applications. While the applications are divided into categories depending on how the membrane functions, there is considerable overlap between the categories.

3.3.1   Biomolecule separation and sorting

In various fields, such as the pharmaceutical industry, the food industry, and biotechnology, separation and sorting are vital because they are used to separate and purify molecules from a variety of biological feed streams. Separation and sorting are also important in the food industry. Separation science makes use of a number of techniques, such as size exclusion chromatography and biopolymer gel electrophoresis [56,57] to separate molecules. In recent years, researchers have been looking into the separation of biomolecules in more organised holes for a range of applications. During kidney application, a synthetic NPM acts as a supporting entity for renal cells as well as a filter for blood, keeping important molecules like serum proteins and releasing smaller waste particles to flow through it [58]. The regulation of the flowing materials through the nanopores may be done from outside. For instance, Martin and colleagues have developed a membrane composed of an assembly of cylindrical gold nanotubules having a pore diameter of less than 1 nm that may be used as a sensor [59]. They demonstrated how, simply changing the charge in the nanotubles (positive to negative), enabled them to refuse ions with opposite charges, permitting for voltage gating of molecules with opposite charges.

3.3.2 Biosensing

Sensory systems detect substances and enable cells to respond to stimuli via a variety of protein pores that are linked with the membrane. It is possible to utilise this kind of biosensing in several technical fields, comprising the medical diagnostics, pharmaceutical industry, and the detection of potentially harmful biomolecules, among others. A biological component is used in conjunction with a physico-chemical recognition unit in the majority of these applications to distinguish analytes in biological systems, and the biosensing device is used in these applications. Electrochemical sensors, such as amperometric sensors, are devices that utilise an immobilised enzyme to convert targeted analytes in a biological feed stream into products which can be detected electrochemically. The most common method that is being used for assessing blood sugar levels is the fasting glucose test. Researchers have shown that glucose oxidase immobilised on a porous nanocrystalline TiO_2 screen is capable of detecting glucose in the blood [60]. A similar approach has been used to develop cholesterol biosensors based on immobilised enzymes (cholesterol oxidase) inside ZnO nanoporous thin films [61], which detect cholesterol levels in the blood. It is also important to mention the amperometric-type enzymeless nanoporous systems that have been developed. In recent years, researchers have demonstrated a system for sensing glucose that uses an electrode of nanoporous platinum. This miniaturised electrochemical cell was implanted in a microfluidic chip consisting of a network of microfluidic transport channels [62]. Bohn and his coworkers proposed a nanocapillary membrane based PTAS devices which are electrically actuable, to perform analytical operations and sensing. It may reduce the number of constituent parts in the sensing device [63]. Another category of NPM based biosensor, which analyses samples at a single molecular level at a time. The next section offers a well description of these important materials.

3.5.3 Analysis on a single molecular level

Nanoporous supports have become more popular for single-molecule analysis because they allow researchers to probe biomolecules (proteins, RNA, and DNA) one at a time. For example, it has become well established in recent years that determining the size and frequency of blockages in an electrolyte's ion current when the biologically active molecules are brought through nanopores integrated in insulating membranes can provide information about the structure, concentration, size, and sequence of single-stranded RNA or DNA, among other things. This method has been utilised by manufacturing specially designed receptors on membranes. For example a-hemolysin (cc-HL) protein integrated in a lipid membrane to detect biomolecules [64]. In the early days of analysis of single molecules, lipid membranes embedded in microporous film of polymers like teflon were used in many of the experiments. In addition, it was the fact that the microporous membrane support needed to be further improved for practical applications, as the lipid membrane was unstable and ruptured after a shorter time span, which was one of the major drawbacks. It was suggested that the protein pores can be supported using nanoporous membranes [63]. The design of the protein pores restricts the number of sensing applications that can be implemented using lipid-bound nanopores. Synthetic nanopores have recently been explored as a means of improving the functionality of single molecule detectors, according to the researchers. Artificial nanopores constructed from a variety of materials, including glass, polymers, and solid-state membranes, have been used to detect DNA of both type (i.e. double-stranded and single-stranded), ions, proteins and macromolecules. Analyte-surface interactions may be controlled by nanoscale manipulation of synthetic nanopores, allowing for the development of sensors with desired functionality. Arrays of these sensor platforms with a range of analyte sensitivities, which can be combined into microfluidic devices [65], may be used to create lab-on-a-chip technology.

3.3.4 Immunoisolation

It is the technique of preventing an immune response from harming implanted cells or drug release systems that is known as immunoisolation. It is the most popular technique to encapsulate materials with the use of a nanoporous semipermeable membrane. Using nanoporous membranes, it is possible to protect transplanted cells from being attacked by the immune system of the body. These membranes have pore size suitable for permitting the minute molecules like oxygen, insulin, glucose. At the same time also small enough to block the bigger chemicals present in our immune system such as immunoglobin, which are present in huge quantities. Such immunoisolating devices act as a semipermeable membrane in artificial pancreas. According to researchers these devices act as an interface between the body fluid and insulin releasing pancreatic cells.

According to Desai et al., the application of microfabricated nanoporous silicon interfaces in implantable artificial pancreas for the treatment of diabetes was explored [66]. The desirable properties of these nanoporous membranes are biocompatibility and resistant to biofouling in order to be used in a variety of in vivo applications.

3.3.5 Drug distribution

The invention of in vivo delivery techniques to enable for the controlled distribution of drugs locally where they are needed is now being explored by the pharmaceutical industries. In order to provide more effective therapy while also reducing the danger of wrong dosage, controlled delivery systems are desired. When the pore size, porosity, and thickness of nanoporous membranes are all regulated to a high degree, they offer an attractive approach for creating capsules that may be used to deliver pharmacologic compounds with controlled release [67]. As an example, nanoporous materials are being investigated for their potential use in drug-eluting stents for the treatment of coronary artery disease [68]. The use of nanoporous inorganic membranes for the continuous administration of ophthalmic chugs to treat eye disorders has lately been investigated [69]. When linked to biosensors, smart food delivery systems that respond to physiological changes are possible.

Conclusions

Porous polymeric membranes are considered a hot field of research in recent separation based technological applications. Materials and manufacturing techniques of these porous polymeric membranes were briefly explored in this chapter. Polymers with high mechanical strength and cheap cost were observed to gain high importance. The favored method for producing porous polymeric membranes with controlled shape and performance was also discussed. It covered the system, solution, processing, and the surrounding environment. Despite this, certain issues need to be addressed in future research. Investigation on more efficient and less expensive manufacturing methods are the present need of modern industrial applications. The use of various types of porous membranes may have the potential to replace energy-intensive distillation techniques for separation of organic solvent as these are fast and cost-effective. On the other hand, nano porous materials are more favored for biomedical applications. Controlled pore size distribution, high porosity and low thickness, mechanical and chemical resistance are the essential requirements for selective membranes. True nanoscale devices need to maintain precise pore distribution and molecular sorting. Implantable device membranes must be biocompatible and biofouling resistant. Currently, inorganic, or organic compounds are coated onto nano porous membranes to impart chemical and mechanical stability. The responsiveness of nano porous membranes is currently being investigated in labs for biological applications. Smart membranes may be programmed to interact with various physiological environments like pH, temperature, and ion concentration. Making multi-responsive nano-porous composite membranes is the recent challenge in membrane technology which needs to be solved to an early date.

References

[1] F. Bazzarelli, L. Giorno, E. Piacentini, & E. Drioli, Porous membranes. Encyclopaedia of Membranes; Drioli, E., Giorno, L., Eds, (2015) 1-3. https://doi.org/10.1007/978-3-642-40872-4_2226-1

Advanced Functional Membranes

Materials Research Foundations **120** (2022) 184-213

Materials Research Forum LLC

https://doi.org/10.21741/9781644901816-6

[2] A. Shiohara, B. Prieto-Simon, & N. H. Voelcker, Porous polymeric membranes: fabrication techniques and biomedical applications. J. Mater. Chem. B, 9 (2021) 2129-2154. https://doi.org/10.1039/D0TB01727B

[3] J. Lee, S. Park, K. Choi, & G. Kim, Nano-scale patterning using the roll typed UV-nanoimprint lithography tool. Microelectron. Eng., 85 (2008) 861-865. https://doi.org/10.1016/j.mee.2007.12.059

[4] A. Pandey, S. Tzadka, D. Yehuda, & M. Schvartzman, Soft thermal nanoimprint with a 10 nm feature size, Soft Matter, 15 (2019) 2897-2904. https://doi.org/10.1039/C8SM02590H

[5] J. Choi, Z. Jia, & S. Park, Fabrication of polymeric dual-scale nanoimprint molds using a polymer stencil membrane, Microelectro. Eng., 199(2018) 101-105. https://doi.org/10.1016/j.mee.2018.07.009

[6] M. H. Lee, M. D. Huntington, W. Zhou, J. C. Yang, & T. W. Odom, Programmable soft lithography: solvent-assisted nanoscale embossing, Nano lett., 11(2011), 311-315. https://doi.org/10.1021/nl102206x

[7] M. Vogler, S. Wiedenberg, M. Mühlberger, I. Bergmair, T. Glinsner, H. Schmidt, & G. Grützner, Development of a novel, low-viscosity UV-curable polymer system for UV-nanoimprint lithography. Microelectron. Eng., 84 (2007) 984-988. https://doi.org/10.1016/j.mee.2007.01.184

[8] T. Mäkelä, M. Kainlauri, P. Willberg-Keyriläinen, T. Tammelin, & U. Forsström, Fabrication of micropillars on nanocellulose films using a roll-to-roll nanoimprinting method, Microelectron. Eng., 163 (2016) 1-6. https://doi.org/10.1016/j.mee.2016.05.023

[9] T. Yanagishita , K. Nishio, & H. Masuda , Nano imprinting using Ni molds prepared from highly ordered anodic porous alumina templates, Jpn. J. Appl. Phys., 45 (2006) L804. https://doi.org/10.1143/JJAP.45.L804

[10] M. Tibbe, J. Loessberg-Zahl, M. P. Do Carmo, M. van der Helm, , J. Bomer, A. Van Den Berg,... & J. Eijkel, Large-scale fabrication of free-standing and sub-μm PDMS through-hole membranes, Nanoscale, 10 (2018) 7711-7718. https://doi.org/10.1039/C7NR09658E

[11] N. Kooy, K. Mohamed, L. T. Pin, & O. S. Guan, A review of roll-to-roll nanoimprint lithography. Nanoscale Res. Lett., 9 (2014) 1-13. https://doi.org/10.1186/1556-276X-9-320

[12] K. J. Sohn, J. H. Park, D. E. Lee, H. I. Jang, & W. I. Lee, Effects of the process temperature and rolling speed on the thermal roll-to-roll imprint lithography of flexible polycarbonate film, J. Micromech. Microeng, 23 (2013) 035024. https://doi.org/10.1088/0960-1317/23/3/035024

[13] X. Tan, & D. Rodrigue, A review on porous polymeric membrane preparation. Part II: Production techniques with polyethylene, polydimethylsiloxane, polypropylene, polyimide, and polytetrafluoroethylene, Polymers, 11(2019) 1310. https://doi.org/10.3390/polym11081310

[14] P. Apel, Track etching technique in membrane technology, Radiat. Meas., 34 (2001) 559-566. https://doi.org/10.1016/S1350-4487(01)00228-1

[15] M. E. Toimil-Molares, Characterization and properties of micro-and nanowires of controlled size, composition, and geometry fabricated by electrodeposition and ion-track technology, Beilstein J. Nanotechnol., 3 (2012) 860-883. https://doi.org/10.3762/bjnano.3.97

[16] B. S. Lalia, V. Kochkodan, R. Hashaikeh, & N. Hilal, A review on membrane fabrication: Structure, properties and performance relationship, Desalination, 326 (2013) 77-95. https://doi.org/10.1016/j.desal.2013.06.016

[17] M. Tibbe, J. Loessberg-Zahl, M. P. Do Carmo, M. van der Helm, J. Bomer, A. Van Den Berg, & J.Eijkel, Large-scale fabrication of free-standing and sub-μm PDMS through-hole membranes, Nanoscale, 10 (2018) 7711-7718. https://doi.org/10.1039/C7NR09658E

[18] H. Kavand, , H. van Lintel, S. BakhshiSichani, S. Bonakdar, H.Kavand, J. Koohsorkhi, & P. Renaud, Cell-imprint surface modification by contact photolithography-based approaches: Direct-cell photolithography and optical soft lithography using PDMS cell imprints, ACS Appl. Mater. Interfaces, 11 (2019) 10559-10566. https://doi.org/10.1021/acsami.9b00523

[19] T. Yanagishita, K. Nishio, & H. Masuda, Polymer through-hole membranes with high aspect ratios from anodic porous alumina templates, Jpn. J. Appl. Phys., 45 (2016) L1133. https://doi.org/10.1143/JJAP.45.L1133

[20] F.Tang, Z. Shao, M. Ni, Y. Cui, C. Yuan, & H. Ge, Fabrication of perforated polyethylene microfiltration membranes for circulating tumor cells separation by thermal nanoimprint method, Appl. Phys. A, 125 (2019) 1-7. https://doi.org/10.1007/s00339-018-2343-5

[21] J. J. Kim, K. W. Bong, E. Reátegui, D. Irimia, & P. S. Doyle, Porous microwells for geometry-selective, large-scale microparticle arrays, Nat. Mater., 16 (2017) 139-146. https://doi.org/10.1038/nmat4747

[22] H. Im, J. N. Sutherland, J. A. Maynard, & S. H. Oh, Nanohole-based surface plasmon resonance instruments with improved spectral resolution quantify a broad range of antibody-ligand binding kinetics. Anal. Chem., 84 (2012) 1941-1947. https://doi.org/10.1021/ac300070t

[23] P. R. Sundararajan, Physical aspects of polymer self-assembly. John Wiley & Sons.(2016). https://doi.org/10.1002/9781118994405

[24] A. Stein, & R. C. Schroden, Colloidal crystal templating of three-dimensionally ordered macroporous solids: materials for photonics and beyond, Curr. Opin. Solid State Mater. Sci., 5 (2001) 553-564. https://doi.org/10.1016/S1359-0286(01)00022-5

[25] Y. Meng, D. Gu, F. Zhang, Y. Shi, H. Yang, Z. Li, & D. Zhao, Ordered mesoporous polymers and homologous carbon frameworks: amphiphilic surfactant templating and direct transformation, Angew. Chem. Int. Ed., 44 (2005) 7053-7059. https://doi.org/10.1002/anie.200501561

[26] Y. Meng, D.Gu, F. Zhang, Y. Shi, L. Cheng, D. Feng, & D. Zhao, A family of highly ordered mesoporous polymer resin and carbon structures from organic– organic self-assembly. Chem. of Mater., 18 (2006) 4447-4464. https://doi.org/10.1021/cm060921u

[27] C. Wang, T. M. Wang, & Q. H. Wang, Ultralow-dielectric, nanoporous poly (methyl silsesquioxanes) films templated by a self-assembled block copolymer upon solvent annealing, J. Polym. Res., 26(2019) 1-10. https://doi.org/10.1007/s10965-018-1650-z

[28] W. A. Phillip, B. O'Neill, M. Rodwogin, M. A. Hillmyer, & E. L. Cussler, Self-assembled block copolymer thin films as water filtration membranes. ACS Appl. Mater. Interfaces, 2 (2010) 847-853. https://doi.org/10.1021/am900882t

[29] K. V. Peinemann, V.Abetz, & P. F. Simon, Asymmetric superstructure formed in a block copolymer via phase separation. Nat. Mater., 6(2007) 992-996. https://doi.org/10.1038/nmat2038

[30] Z. X. Low, Y. T. Chua, B. M. Ray, D. Mattia, I. S. Metcalfe, & D. A. Patterson, Perspective on 3D printing of separation membranes and comparison to related unconventional fabrication techniques, J. Membr. Sci., 523(2017) 596-613. https://doi.org/10.1016/j.memsci.2016.10.006

[31] N. A. Chartrain, C. B. Williams, & A. R. Whittington, A review on fabricating tissue scaffolds using vat photopolymerization. Actabiomaterialia, 74(2018) 90-111. https://doi.org/10.1016/j.actbio.2018.05.010

[32] J. A. Lewis, & G. M. Gratson, Direct writing in three dimensions. Mater. Today., 7 (2004) 32-39. https://doi.org/10.1016/S1369-7021(04)00344-X

[33] M. N. Jahangir, K. M. M. Billah, Y.Lin, D. A. Roberson, R. B. Wicker, & D. Espalin, Reinforcement of material extrusion 3D printed polycarbonate using continuous carbon fiber, Addit. Manuf., 28 (2019) 354-364. https://doi.org/10.1016/j.addma.2019.05.019

[34] D. X. Luong, A. K. Subramanian, G. A. L. Silva, J. Yoon, S. Cofer, K. Yang, & J. M. Tour, Laminated object manufacturing of 3D-printed laser-induced graphene foams. Adv. Mater., 30(2018) 1707416. https://doi.org/10.1002/adma.201707416

[35] R. D. Farahani, M. Dubé, & D. Therriault, Three-dimensional printing of multifunctional nanocomposites: manufacturing techniques and applications. Adv. Mater., 28 (2016) 5794-5821. https://doi.org/10.1002/adma.201506215

[36] S. Remanan, , M. Sharma, S. Bose, & N. C. Das, Recent advances in preparation of porous polymeric membranes by unique techniques and mitigation of fouling through surface modification. Chem. Select., 3(2018) 609-633. https://doi.org/10.1002/slct.201702503

[37] P. Arribas, , M. Khayet, M. C. García-Payo, & L. Gil, Self-sustained electro-spun polysulfone nano-fibrous membranes and their surface modification by interfacial polymerization for micro-and ultra-filtration., Sep. Purif. Technol., 138(2014) 118-129. https://doi.org/10.1016/j.seppur.2014.10.010

[38] S. S. Ray, S. S. Chen, N. C. Nguyen, & H. T. Nguyen, Electrospinning: A versatile fabrication technique for nanofibrous membranes for use in desalination. In Nanoscale Materials in Water Purification, Elsevier. (2019) 247-273. https://doi.org/10.1016/B978-0-12-813926-4.00014-8

[39] C. Mituppatham, , M. Nithitanakul, & P. Supaphol, Ultrafine electrospun polyamide6 fibers: effect of solution conditions on morphology and average fiber diameter. Macromol. Chem. Phy., 205 (2004) 2327-2338. https://doi.org/10.1002/macp.200400225

[40] C.Van Do, T. T. T. Nguyen, & J. S. Park, Fabrication of polyethylene glycol/polyvinylidene fluoride core/shell nanofibers via melt electrospinning and their

characteristics, Sol. Energy Mater. Sol. Cells, 104 (2012) 131-139. https://doi.org/10.1016/j.solmat.2012.04.029

[41] T. A. M. Valente, D. M. Silva, P. S. Gomes, M. H. Fernandes, J. D. Santos, & V. Sencadas, Effect of sterilization methods on electrospun poly (lactic acid)(PLA) fiber alignment for biomedical applications. ACS Appl. Mater. Interfaces, 8 (2016) 3241-3249. https://doi.org/10.1021/acsami.5b10869

[42] S. Luque, D. Gómez, and J. R. Álvarez. "Industrial applications of porous ceramic membranes (pressure-driven processes)." Membr. Sci. Technol., 13 (2008) 177-216. https://doi.org/10.1016/S0927-5193(07)13006-0

[43] P. B. Belibi, M. M. G. Nguemtchouin, M. Rivallin, J. N. Nsami, J. Sieliechi, S. Cerneaux, M. B. Ngassoum, and M. Cretin. "Microfiltration ceramic membranes from local Cameroonian clay applicable to water treatment." Ceram. Int. 41 (2015) 2752-2759. https://doi.org/10.1016/j.ceramint.2014.10.090

[44] B. Das, B. Chakrabarty, and P.Barkakati, Preparation and characterization of novel ceramic membranes for micro-filtration applications. Ceram. Int., 13 (2016) 14326-14333. https://doi.org/10.1016/j.ceramint.2016.06.125

[45] H. Guo, S. Zhao, X. Wu, and H. Qi., Fabrication and characterization of TiO_2/ZrO_2 ceramic membranes for nanofiltration, Microporous and Mesoporous Mater., 260 (2018) 125-131. https://doi.org/10.1016/j.micromeso.2016.03.011

[46] Zi. Yang, Y. Zhou, Z. Feng, X. Rui, T. Zhang, and Z. Zhang. "A review on reverse osmosis and nanofiltration membranes for water purification." Polymers., 8 (2019) 1252. https://doi.org/10.3390/polym11081252

[47] M. W. Hakami, A. Alkhudhiri, S. Al-Batty, M. P. Zacharof, J. Maddy, and N. Hilal. "Ceramic microfiltration membranes in wastewater treatment: Filtration behavior, fouling and prevention." Membranes., 9 (2020) 248. https://doi.org/10.3390/membranes10090248

[48] M. Cheryan, & N. Rajagopalan, Membrane processing of oily streams. Wastewater treatment and waste reduction, J.Membr. Sci., 151(1998) 13-28. https://doi.org/10.1016/S0376-7388(98)00190-2

[49] R. Bhave, Inorganic Membranes Synthesis, Characteristics and Applications: Synthesis, characteristics, and applications. Springer Science & Business Media, 2012.

[50] N. A. Ahmad, P. S. Goh, Z. A. Karim, & A. F. Ismail, Thin film composite membrane for oily waste water treatment: Recent advances and challenges, Membranes, 8 (2018) 86. https://doi.org/10.3390/membranes8040086

Materials Research Forum LLC
https://doi.org/10.21741/9781644901816-6

[51] R. Sondhi, and R. Bhave, Role of backpulsing in fouling minimization in crossflow filtration with ceramic membranes. J. Membr. Sci 186 (2001) 41–52. https://doi.org/10.1016/S0376-7388(00)00663-3

[52] J. Finley, Ceramic membranes: A robust filtration alternative. Filtr. Sep. 42(2005) 34–37. https://doi.org/10.1016/S0015-1882(05)70695-9

[53] R. Sondhi, R. Bhave, and L. Jung, Applications and benefits of ceramic membranes. Membr. Technol. 11(2003) 5–8. https://doi.org/10.1016/S0958-2118(03)11016-6

[54] M. B. Hägg, & L. Deng, Membranes in gas separation. Handbook of Membrane Separations: Chemical, Pharmaceutical, Food, and Biotechnological Applications, (2015), 143-180

[55] Hilliard Hilco Division Technical Bulletin: Ceramic Membrane Crossflow Liquid Filtration System CMS-3. (2005). THC-500-10/05

[56] M. Ventra, E. Stephane, and R. H. James, eds. Introduction to nanoscale science and technology. Springer Science & Business Media, 2006

[57] R. Ghosh, Protein Bioseparation Using Ultrafiltration: Theory, Applications and New Developments. Imperial College Press; London: 2002. https://doi.org/10.1142/p257

[58] W. H. Fissell, H.D. Humesa, A.J. Fleischmanb, S. Roy, Dialysis and Nanotechnology: Now, 10 Years, or Never? Blood Purifi., 25 (2007) 1. https://doi.org/10.1159/000096391

[59] M. Nishizawa, V.P. Menon, C.R. Martin, Metal Nanotubule Membranes with Electrochemically Switchable Ion-Transport Selectivity. Science., 268 (1995), 700-702. https://doi.org/10.1126/science.268.5211.700

[60] Q. Li, G. Luo, J. Feng, Q. Zhou, L. Zhang, Y. Zu, Amperometric Detection of Glucose with Glucose Oxidase Absorbed on Porous Nanocrystalline TiO_2 Film. Electroanalysis., 13(2001), 413 416. https://doi.org/10.1002/1521-4109(200104)13:5<413::AID-ELAN413>3.0.CO;2-I

[61] S.P. Singh, S.K. Arya, P. Pandey, B.D. Malhotra, S. Saha, K. Sreenivas, V. Gupta, Cholesterol Biosensor Based on RF Sputtered Zinc Oxide Nanoporous Thin Film. App. Phys. Lett. 91 (2007) 1-3. https://doi.org/10.1063/1.2768302

[62] S. Joo, S. Park, T.D. Chung, H.C. Kim, Integration of a Nanoporous Platinum Thin Film into a Microfluidic System for Non-enzymatic Electrochemical Glucose, Sensing. Anal., 23 (2007) 277-281. https://doi.org/10.2116/analsci.23.277

Advanced Functional Membranes

Materials Research Forum LLC

Materials Research Foundations **120** (2022) 184-213

https://doi.org/10.21741/9781644901816-6

[63] H. Bayley, B.S. Cremer, Stochastic Sensors Inspired by Biology. Nature. 413 (2001) 226-230. https://doi.org/10.1038/35093038

[64] J.J. Kasianowicz, E. Brandin, D. Branton, D.W. Deamer, Characterization of Individual Polynucleotide Molecules Using a Membrane Channel. Proc. Natl. Acad. Sci., 1996. https://doi.org/10.1073/pnas.93.24.13770

[65] M.J. Kim, M. Wanunu, D.C. Bell, A. Meller, Rapid Fabrication of Uniformly Sized Nanopores and Nanopore Arrays for Parallel DNA Analysis. Adv. Mater. Dec. 2006. https://doi.org/10.1002/adma.200601191

[66] I. Tsujino, J. Ako, Y. Honda, P. J. Fitzgera, Drug Delivery Via Nano- Micro and Macroporous Coronary Stent Surfaces. Expert Opin. Drug Deliv., 4 (2007) 287-295. https://doi.org/10.1517/17425247.4.3.287

[67] T. A. Desai, S. Sadhana, J. W. Robbie, B. Anthony, C. Michael, J. Shapiro, T. West, "Nanoporous implants for controlled drug delivery." In BioMEMS and Biomedical Nanotechnology, pp. 263-286. Springer, Boston, MA, 2006. https://doi.org/10.1007/978-0-387-25844-7_15

[68] T. A. Desai, W. H.Chu, J. K. Tu, G. M. Beattie, A. Hayek, M. Ferrari, Microfabricated immunoisolating biocapsules, Biotechnol. Bioeng. 57 (1998) 118-120. https://doi.org/10.1002/(SICI)1097-0290(19980105)57:1<118::AID-BIT14>3.0.CO;2-G

[69] R. E. Orosz, S. Gupta, M. Hassink, M. Abdel-Ralman, L. Moldovan, N. I. Moldovan, Delivery of Antiangiogenic and Antioxidant Drugs of Ophthalmic Interest through a Nanoporous Inorganic Filter. Mol. Vis. 10 (2004) 555-565.

Materials Research Forum LLC

https://doi.org/10.21741/9781644901816-7

Chapter 7

Advanced Functional Membranes for Water Purification

E. Kavitha, S. Vishali[*]

Department of Chemical Engineering, College of Engineering and Technology, SRM Institute of Technology, Kattankulathur 603 203, Tamil Nadu, India

[*] meet.vishali@gmail.com

Abstract

Worldwide, the scarcity of water due to the contamination of water and water resources from human activities has become a challenge. To overcome this, we need adequate water treatment technologies. For the past few decades, research has becoming wider to innovate the advanced technologies for water purification. Several conventional methods have been adopted for water treatment. Extensive investigations have been done, and some are still in development to find an eco-friendly and feasible process for water purification. The advent of membrane separation processes has changed the scenario in the field of desalination and water treatment. The proper selection of membrane material and configuration could be an excellent approach towards water purification. This chapter deals with the various advanced functional membranes, the functionalization of membranes employed in UF, NF and membrane distillation. This chapter also deals with their applications in wastewater treatment.

Keywords

Advanced Functional Membranes, UF, NF, Membrane Distillation, Water Purification

Contents

1.   Introduction

The swift growth in population and industrialization lead to the extensive demand for quality water. Due to the increase in industrialization and change of lifestyle, a vast volume of wastewater has been generated. Globally, the major crises for the human community are the inadequate availability of potable water, which forces the public to drink unhealthy water. The consumption of contaminated water is a health concern for humans and the entire ecosystem. The familiar wastewater sources are the industrial sector, commercial zone, residential area, irrigation release, etc. The major pollutants present in wastewater are organic, inorganic and biological contaminants. Some of the pollutants like heavy metals, radioactive materials are more toxic and hazardous. Also, some of the pollutants are non-biodegradable, which severely affects the healthy life of the ecosystem. However, proper treatment, reuse and recycle of water could reduce the water scarcity problem significantly [1].

Various conventional treatment technologies include chemical precipitation, adsorption, ion exchange, membrane separation processes, electrochemical treatment, etc. All these techniques have their own merits and also limitations. In the past few years, emerging pollutants in wastewater have become a severe threat. The existing traditional

Advanced Functional Membranes   Materials Research Forum LLC
Materials Research Foundations **120** (2022) 214-236   https://doi.org/10.21741/9781644901816-7

methodologies are not compatible with the treatment of emerging pollutants. So, we require advanced technologies or advancements in the existing technologies to overcome this issue. In the past few decades, membrane separation processes are gaining importance because of their ambient condition operation, no phase change involved in the process, ease of installation and operation, less footprint area, compatibility to combine with other conventional separation processes and the possibility of both separation and recovery of the value-added components from the effluent. These merits triggered the attention of researchers to work towards advanced membrane technologies and developments in the fabrication of membranes [2].

The membrane separation processes are pressure-driven, concentration-driven, thermal-driven, electric potential driven processes. The separation mechanism can be sieving or diffusion. The proper selection of membrane material and modules could be an excellent solution for treating various pollutants. In recent years, plenty of research work is going on to improve the surface characteristic of the membrane so that it could be adopted for the selective separation of various contaminants in wastewater [3].

2. Developments in the synthesis of nano-based membranes

In recent years, nanomaterial science has grasped the attention in wastewater treatment by membrane separation processes. By coating, the suitable nanomaterial over the surface of the membrane could widen the applications of membrane separation processes in wastewater treatment. Recently, there are many rapid developments in the synthesis of nanomaterials and their incorporation in the polymeric matrix to produce the mixed matrix membranes for the application of water treatment [1].

2.1 Nanocomposite ultrafiltration membrane (NUF)

The ultrafiltration (UF) membrane has been increasing for the past few years due to the merits such as low-pressure operation, low energy consumption, ease of separation mechanism based on size exclusion and backwashing of the membrane is possible, which leads to reduced membrane fouling. Polymeric materials such as polysulfone (PSF), polyethersulfone (PES), polyvinylidene fluoride (PVDF), polypropylene (PP), polycarbonate, cellulose acetate, cellulose nitrate, carboxymethylcellulose, polyvinyl alcohol, polyacrylic acid, etc., have been utilized in the fabrication of UF membranes. In recent years, UF membrane finds extensive application in wastewater treatment.

2.1.1 Polysulfone nanocomposite UF membrane

The developments and growth in the petrochemical and petroleum refinery, metallurgical, food process, paint, and fine chemicals industries have increased oily effluent generation

Materials Research Forum LLC
https://doi.org/10.21741/9781644901816-7

[2]. Membrane-based separation technologies have proven to be superior in the treatment of oily effluent compared to other convention techniques [3–5]. Y. Zhang et al. (2009) synthesized the advanced organic-inorganic composite membrane, using sulfated, Y -Y-doped zirconia as an additive to PSF. N, N-Dimethylacetamide is heated to 40-50°C and polysulfone is dissolved by proper mixing. Then, polyethylene glycol 400 is added to enhance the yield of pores. The Zirconia nanoparticles are added to this solution and mixed thoroughly to get the homogenous solution. The membrane is cast on a glass plate to form a thin film and immersed in a water bath. Then the membrane is leached and stored in formaldehyde to avoid biofouling [6].

The characterization studies on this novel membrane showed tensile strength twice that of PSF, and contact angle also got reduced almost 50%. This membrane has been employed to study the treatment of oily wastewater. The schematic diagram of the treatment process is shown in Fig. 1. The investigations have shown significant retention of oil, which is nearly 99% and it could be a feasible membrane for the separation of oil from the effluent stream. From the literature, the incorporation of inorganic nanoparticles on the surface of the membrane could increase the permeability of the membrane and enhance the surface characteristics of the membrane [7–9].

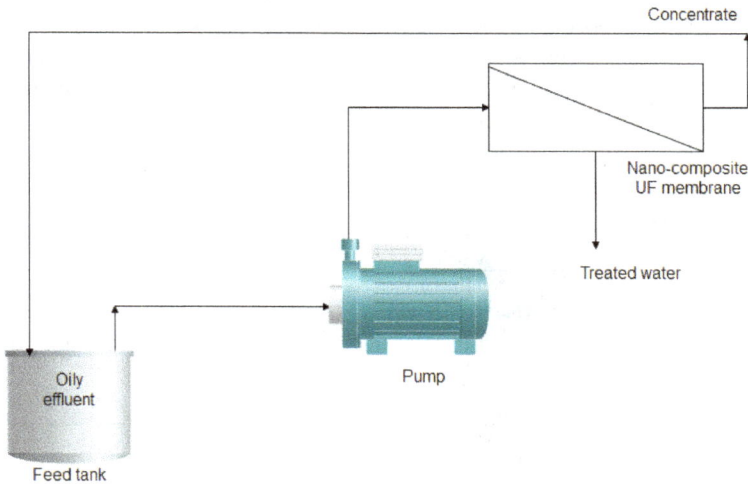

Figure 1. The schematic representation of the experimental process for the treatment of oily effluent

2.1.2 PVDF UF membrane

Biopolymers are the natural polymers produced from living organisms and consist of monomers forming macromolecules by a covalent bond. Cellulose is an abundantly available biopolymer. It is crystalline, fibrous, hydrophilic and has a high molecular weight [13]. It is an eco-friendly biopolymer and also sustainable and biodegradable material. The Nano-crystalline cellulose combines both cellulose components such as its hydrophilicity, recoverability, biodegradability, and the benefits of nanoparticles such as high specific surface area, higher activity, mechanical and tensile strength [14]. The synthesis of NCC can be done by either chemical modification or mechanical modification methods. The chemical modification involves strong acid hydrolysis [15] and mechanical modification involves high-pressure refinement or homogenization [16,17].

Haolong Bai et al. (2012) synthesized NCC using chemical and mechanical modification techniques [18]. The synthesis process involves acid hydrolysis and mechanical homogenization. The cellulose pulp is reacted with a diluted sulphuric acid solution at 85°C. Then the pH of the solution is adjusted to neutral. After drying the solution, the solid content is suspended in N, N-dimethylacetamide and then it is subjected to high-pressure homogenization. The fabrication of PVDF/NCC composite membrane follows Loeb – Sourirajan Phase inversion technique. The experimental findings of Haolong Bai et al. (2012) are as follows: (i) the permeability and water flux could be improved significantly by composite membrane compared to PVDF membrane, and (ii) the pore size of the membrane and porosity were found to be more than PVDF membrane [18].

Table 1 Advanced functional Nano composite membranes

Nano-composite membrane	Contribution of the research work	Type of effluent	Target ion/ contaminant	References
Polysulfone composite membrane with sulfated Y-doped zirconia/	Synthesis, characterization of membrane and treatment of oily wastewater	Oily wastewater	Oil	[6]
Polyvinylidene fluoride nano composite membrane coated with alumina nano particles	Synthesis, characterization of membrane and treatment of oily wastewater	Oily wastewater from an oil field	Oil	[12]

Materials Research Forum LLC
https://doi.org/10.21741/9781644901816-7

Polyvinylidene fluoride composite membrane with cellulose nano-crystals	Synthesis and characterization of membrane	-	-	[18]
Polyvinylidene fluoride nano composite membrane coated with TiO$_2$ nano particles	Synthesis and characterization of membrane	-	-	[19]
Polyethersulfone coated with functionalized graphene oxide mixed matrix membrane	Synthesis, characterization of membrane and treatment of wastewater	Raw wastewater from Membrane bioreactor water treatment plant	Fe(II), Zn(II), Cd(II), Cr(VI),	[20]
ZIF-8 Nano composite membrane	Synthesis, characterization of membrane and treatment of dye effluent	Dye effluent	Dye	[21]
Hydroxyapatite nano composite membrane coated with silver nano particles	Synthesis, characterization of membrane and treatment of palm oil mill effluent	Palm oil mill effluent	Oil	[22]
Polyethersulfone nanocomposite membrane with sulfonated graphene oxide	Synthesis, characterization of membrane. Treatment of wastewater and electricity generation	Wastewater	COD removal	[23]
Polysulfone nano-composite membrane with graphene oxide	Synthesis, characterization of membrane and treatment of wastewater containing Bisphenol A	Wastewater	Bisphenol A	[24]
Polyvinylidene fluoride-polyaniline membrane with graphene oxide	Synthesis, characterization of membrane and treatment of dye effluent	Dye effluent	Allura red, Methyl orange	[25]

Advanced Functional Membranes   Materials Research Forum LLC
Materials Research Foundations **120** (2022) 214-236   https://doi.org/10.21741/9781644901816-7

3.   Advanced functional nano filtration membranes

Organic, inorganic and biological contaminants are present in wastewater. The inorganic contaminants, one of the major pollutants present in wastewater, are released from process industries such as petroleum refineries, fertilizers, textile and dye, paint, pharmaceutical industries, etc. [26].   The chief inorganic contaminants in wastewater are heavy metals, trace elements, mineral acids, salts, metal compounds, etc. [26]. Most of these pollutants are toxic, hazardous and non-biodegradable. These contaminants harm humans, animals and also aquatic life. However, some of these pollutants are micropollutants present in only trace quantity such as in mg/L or ng/L. Even in trace concentration, these contaminants are not desirable for the healthy life of the ecosystem. There are no adequate technologies for the treatment of micro-pollutants either from municipal wastewater or industrial effluent.

However, pressure-driven membrane technologies such as RO and NF have proven promising technology for separating these micro-pollutants since they possess higher rejection, compactness and ease of operation, and feasibility to combine with other conventional separation processes [27]. NF membranes find applications in desalination, treatment of fine chemicals, food process and biotechnology industrial effluents. In the past few decades, various developments in the surface modification and fabrication of NF membranes by technologies such as interfacial polymerization by applying nanomaterials and other additives, grafting, electron beam radiation, layer-by-layer modification, plasma surface modification, etc. [28]. All these technologies have been developed to enhance the characteristics and functionalization of NF membranes to improve the rejection rate, selectivity and to overcome fouling tendency.

3.1   Interfacial polymerization

In recent years, the fabrication of thin-film composite membranes by interfacial polymerization has gained attention due to the improvement in the selectivity and anti-fouling tendency of the membrane. The ease of operation and self-inhibiting reaction made this technique more feasible. Various monomers include bisphenol A, diethylenetriamine, triethylenetetramine,   tetraethylenepentamine,   m-phenylenediamine,   piperazidine, polyhexamethylene guanidine hydrochloride, etc., have been employed in the interfacial polymerization for the fabrication of thin-film composite membranes [29,30].

In recent years, the incorporation of nanoparticles in interfacial polymerization has attracted many researchers. Since they possess specific unique characteristics such as antimicrobial and catalytic tendencies, they gained attention. Also, the composite membranes incorporated with nanoparticles exhibit excellent selectivity, rejection rate, permeability, mechanical strength, and hydrophilicity. The widely used nanoparticles are silica, and zinc oxide, titanium dioxide. In one of the recent studies, aluminum oxide hydroxide particle known as Boehmite nanoparticles is used to synthesize polyethersulfone membrane to improve the separation efficiency of the membrane [31].

This advanced functionalized mixed matrix NF membrane possesses a highly hydrated surface and is hydrophilic. Recently, a combination of nanoparticles has been studied to improve the characteristics of the NF membrane. For example, polyether sulphone membrane incorporated with polyaniline/Fe3O4 nanoparticles aids in the removal of heavy metal ions from wastewater [32]. The combination of nanoparticles utilized in the fabrication of NF membrane is carbon nanotubes, halloysite nanotube, nanofiber, etc. [33–35].

3.2   Grafting polymerization

Grafting polymerization can be done either by UV-assisted/initiated grafting polymerization or by photo-grafting polymerization. Qiu et al. (2007) synthesized the NF membrane from the cardo-polyether ketone UF membrane via photo-grafting co-polymerization [36]. Deng et al. (2011) fabricated an NF membrane with a positive charge on its surface from polysulfone UF membrane using UV grafting polymerization [37].

3.3   Electron beam radiation

The electron beam radiation process involves the creation of active sites on the polymer without any additives. This leads to the formation of NF membrane with improved porosity, pore size, permeability, and rejection rate. Few examples for NF synthesis by electron beam radiation are as follows: (i) nylon 66 polymer cross-linked by electron beam irradiation [38], (ii) polysulfone layer grafted with 2-acrylamido-2-methylpropanesulfonic acid.

3.4   Plasma surface modification

In this process, to increase the hydrophilicity and decrease the membrane fouling, nitrogen functional groups are incorporated with the membrane by applying nitrogen-containing plasma [39]. The surface characteristics of the NF membrane can be permanently modified by plasma-induced graft polymerization.

3.5   Layer-by-layer surface modification

This process involves the addition of amine coupling and silane coupling in a layer -by-layer surface modification of NF membrane. One of the examples layer-by-layer surface modification of NF membrane is the fabrication of polybenzimidazole/polyethersulfone NF membrane to separate heavy metal ions from wastewater.

3.6   Applications of advanced functional NF membranes in wastewater treatment

Polyelectrolyte multilayer-based NF membrane fabricated by layer-by-layer surface modification has been employed to separate micro-pollutants from wastewater [40]. Polydopamine coated polyether sulfone incorporated with amino-functionalized multi-wall carbon nanotubes has been employed to treat oily wastewater [41]. The cross-flow filtration setup is shown in Figure 2.

NF hollow fiber membrane prepared by mixing sulfonated polysulfone with polyethersulfone followed with the formation of hollow fiber membrane using polyethyleneimine in the bore fluid has been investigated by Gao et al. (2020) for the separation of heavy metal ions from wastewater [42].

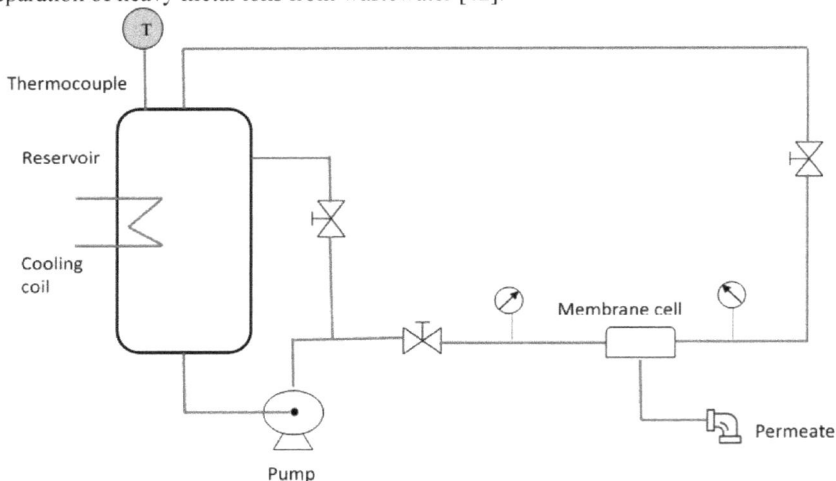

Figure 2   Cross-flow filtration setup [41]

4.   Membrane distillation

Membrane distillation is an emerging technology in separation processes and gaining importance in recent years. It is the thermally driven membrane separation process. This process involves the separation at lower pressure and temperature compared to the traditional distillation processes and hydrophobic porous membranes are employed for this process. The separation mechanism is based on the difference in vapor pressure of volatile species across the hydrophobic membrane surface. The pore size of the membrane is relatively larger than the pores of the RO and NF membrane. The feed side of the membrane is at a slightly higher temperature compared to the permeate side and the

Advanced Functional Membranes

Materials Research Forum LLC

Materials Research Foundations **120** (2022) 214-236

https://doi.org/10.21741/9781644901816-7

permeate side is at lower temperature or ambient temperature. This temperature difference across the membrane surface ensures the vapor pressure difference. Due to the hydrophobic nature of the membrane and vapor pressure difference across the membrane surface, water vapor alone can permeate through the membrane. The water vapor exiting the permeate side of the membrane is condensed and collected. The schematic representation of the separation mechanism of membrane distillation is shown in Fig. 3. The rate-controlling mass transfer mechanisms are the Knudsen diffusion model, molecular diffusion model and viscous flow model [43].

This process has been employed for seawater desalination, wastewater treatment, the concentration of fruit juices, etc. Recently, this process has been adopted for the treatment of radioactive wastes [44, 45]. In the past few years, the application of this process has been widening due to its unique merits, such as relatively low-pressure operation compared to other membrane separation processes like RO, low-temperature operation compared to other conventional distillation processes; also, there is no need to heat the feed solution till its boiling point for achieving the separation, cost-effective process, could be easily coupled with other membrane separation processes like UF and RO, alternate energy sources like solar energy could be utilized [45–49].

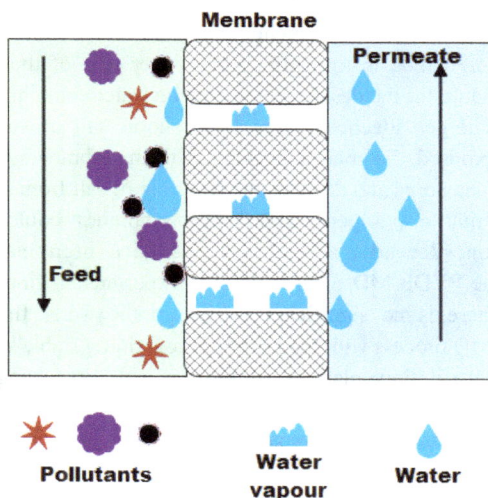

Figure 3 Schematic representation of separation mechanism by membrane distillation [43]

4.1   Configurations and modules of membrane distillation

There are mainly four different configurations in the membrane distillation process. They are direct contact membrane distillation (DCMD), air gap membrane distillation (AGMD), sweep gas membrane distillation (SGMD), and vacuum membrane distillation (VMD) [50].

Polymeric compounds such as polyvinylidene fluoride, Polypropylene, and polytetrafluoroethylene are widely used to fabricate membranes employed in MD units [51–53]. The application of hydrophobic membranes in the MD process suffers from certain limitations, such as membrane fouling and wetting of pores of the membrane. Both these characteristics are not desirable for the effective separation process. So, recently, much research is carried out to investigate the feasibility and functional modification of commercially available membranes to overcome these limitations. Various studies have been conducted to assess the characteristics of superhydrophobic and omniphobic membranes for their successful implementation in the membrane distillation process for desalination and wastewater treatment.

4.2   Super hydrophobic membranes in membrane distillation

The main drawback of the hydrophobic membrane in the membrane distillation process for wastewater treatment is the wetting of membrane surface and fouling. The foulant blocks the membrane's pores, which in turn affects the water flux of the membrane [54]. In applying the MD module for the desalination of saline water containing surfactants, pores of the membrane will get attached to the hydrophobic tail of surfactants. Since the hydrophilic head is exposed, that makes the pores of the membrane hydrophilic. This leads to wetting of membrane pores and declines the retention of salt from saline water [55]. To overcome these limitations, superhydrophobic membranes could be employed for membrane distillation. Recently, several attempts have been made to enhance the hydrophobicity of the PVDF MD membrane. The experimental findings of Wang et al. (2017) show that there is no significant change in the water flux and retention of contaminants in the MD process with feed solution containing stable surfactants [56]. The preparation of dope solution plays a significant role in changing the structural characteristics of the membrane. For the past few years, researchers have been investigating the modification in the preparation of dope solutions to enhance membrane characteristics in the MD module. M. Khayet et al. (2002) investigated the addition of 1,2 -ethanediol with a dope solution and found an increase in the roughness of the surface of the membrane [57].

The hydrophobicity of the membrane could be enhanced by using hydrophobic nanoparticles since they are superhydrophobic [58]. The addition of hydrophobic

Advanced Functional Membranes   Materials Research Forum LLC
Materials Research Foundations **120** (2022) 214-236   https://doi.org/10.21741/9781644901816-7

nanoparticles, such as silica nanoparticles, calcium nanoparticles, into the dope solution could improve the roughness of the membrane surface, pore size, and hydrophobicity [58,59]. Hence, it has been proved from several studies that by using the appropriate additives in the dope solution, roughness, porosity, and hydrophobicity of the membrane could be improved significantly. Adopting the proper additive could induce crystallization, leading to the formation of the nodule to improve the roughness of the membrane.

Several methodologies have been attempted in the fabrication of superhydrophobic membranes for the MD process. Various preparation techniques for the fabrication of PVDF superhydrophobic membrane include phase inversion, non-solvent induced separation, vapor induced phase separation, etc. The following methodology has been adopted from W. Zhang et al. (2013). This methodology follows a modified phase inversion technique for the fabrication of a superhydrophobic membrane. Ammonia water, an inert solvent additive, has been mixed with the dope solution. Incorporating additive in the dope solution creates the PVDF clusters by incipient precipitation due to the localized micro phase separation. Hence, the formation of a superhydrophobic membrane is obtained as the clusters of PVDF grow to form a spherical structure with increased contact angle (up to 158°) [43].

Hou et al. (2014) studied the preparation of PVDF / calcium carbonate nanoparticles flat sheet composite membrane and employed for the desalination process by DCMD process [58]. Wu et al. (2017) fabricated the superhydrophobic membrane PVDF membrane by silica particle-assisted non-solvent induced phase separation process and applied it in the MD process. The preparation of dope solution involves the proper mixing of propylene glycol and dimethyl acetamide followed by silica particles. Then the solution is kept for ultrasonication for about half an hour. With this solution, PVDF is mixed by stirring the solution at a speed of 1000 rpm at a temperature of 80°C for 4 h to yield a homogeneous solution. Then the dope solution is cast on a glass plate and kept at ambient air for the 30s. The solidified sheet is removed and immersed in deionized water at atmospheric temperature for a day. Then the membrane sheet is removed from the water and kept for atmospheric drying, followed by vacuum drying [62]. Wu et al. (2017) studied the treatment of saline water by fabricating PVDF superhydrophobic membrane by the process mentioned earlier and employed in the MD module to treat saline water. It was observed from the experimental findings that the salt retention is more than 99% and the water flux of the membrane was observed 2.7 times more than that of pristine PVDF [62]. The schematic diagram of the experimental setup for the direct contact membrane distillation is shown in Figure 4.

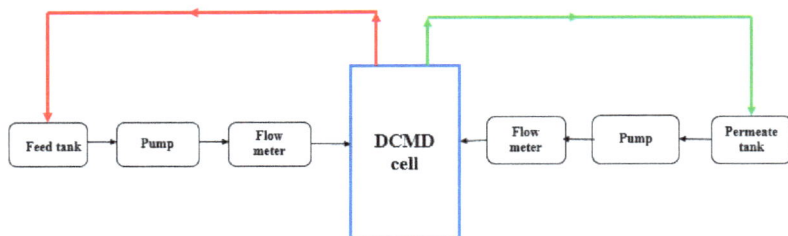

Figure 4 Schematic representation of experimental setup of direct contact membrane distillation setup [62]

Zhang et al. (2020) studied the fabrication of superhydrophobic PVDF membrane by non solvent induced phase inversion technique by utilizing NaCl-H2O as a green additive, with improved water flux and more than 99% salt rejection [63]. A recent study by Li et al. (2020) involves the fabrication of 1H, 1H, 2H, 2H-perfluorodecyltriethoxysilane (PDTS) - Zinc oxide – PVDF and its application in DCMD. The dope solution is prepared by mixing PVDF, zinc oxide nanoparticles in the solvent dimethyl acetamide and acetone in a proper ratio. Then the solution is stirred for the complete blending of the solution. The thoroughly blended solution undergoes electro spinning and then membrane casting. The fabricated membrane is adopted for the DCMD process to treat saline water, saline water with low surface tension sodium dodecyl sulphate and calcium sulphate [64]. A Novel super hydrophobic composite hollow membrane was synthesized quite recently by Ding et al. (2021) and employed in a vacuum membrane distillation module for the desalination process. The synthesis involves the filter coating and phase inversion technique. A thin composite layer is prepared from poly-dimethyl silane and PVDF on a substrate by suction coating. The substrate is immersed in a bath with silicon dioxide, which leads to the formation of a super hydrophobic membrane [65]. Wen et al. (2021) performed the preparation of super hydrophobic membrane by using graphene oxide membrane with fluorinated silica nanoparticles. The membrane performance was investigated by employing the membrane in an air gap membrane distillation process.

4.3 Omniphobic membranes

Superhydrophobic membranes hold excellent water repellent tendencies. However, they are susceptible to wetting by organic contaminants in wastewater. The contaminants possessing low surface tension can lead to the wetness of the pores of the membrane. The wettability of membrane pores can deteriorate the function of the membrane in effectively separating the liquid-vapor interface and which in turn can restrict its application in the MD process [66]. Since the omniphobic membranes can exhibit repellent both to water and organic compounds, the studies on the synthesis of omniphobic membranes and fabrication of MD modules for wastewater treatment and desalination have significantly gained importance in recent years. The re-entrant geometries and functional modification of the membrane are the key factors in the synthesis of omniphobic membranes to enhance their repellent tendency towards water and species with low surface tension [67]. Also, it is equally important to impart the characteristics such as chemical, thermal and mechanical strength to withstand the MD process. In the functional modification process to impart omniphobicity, a thin gas film is created adjacent to the surface of the membrane, which restricts the contact area of liquid on the surface of the membrane. However, exhibiting the omniphobic behaviour towards highly saline water with oil and surfactants is still uncertain. Boo et al. (2016) studied the performance of the membrane and found the steady performance of the membrane with excellent wetting resistance towards the feed containing surfactants [68].

Several studies have been demonstrated to impart the surface omniphobicity by creating the re-entrant geometries on various substrates [69], especially by the deposition of nanoparticles [70]. Zinc oxide is a well-known compound that exhibits rich growth morphology [71] and produces by chemical bath deposition. Studies have been conducted to synthesis omniphobic surfaces by chemical bath deposition of zinc oxide nanoparticles on the silicon substrates [72, 73]. Chen et al. (2018) studied membrane fabrication by depositing zinc oxide nanoparticles on the glass fiber to create the re-entrant geometry to impart omniphobicity [70]. The fabrication of omniphobic membrane for the MD process is adopted from Chen et al. (2018). The fabrication involves the deposition of zinc oxide nanoparticles on the glass fibre membrane using a chemical bath deposition method. After the deposition, the process follows fluorination of the surface and then a coating of polymer.

The schematic representation of the process flow diagram is shown in Fig. 5. The saline water as a feed solution is taken in the feed tank. The feed solution is heated and passed through the direct contact membrane distillation (DCMD) module. In the DCMD module, the feed solution is in direct contact with the hot side of the membrane. Due to the vapor pressure difference, water vapor alone permeates through the membrane. Then it flows through the chiller, where it is condensed and the treated water is collected in the permeate collection tank.

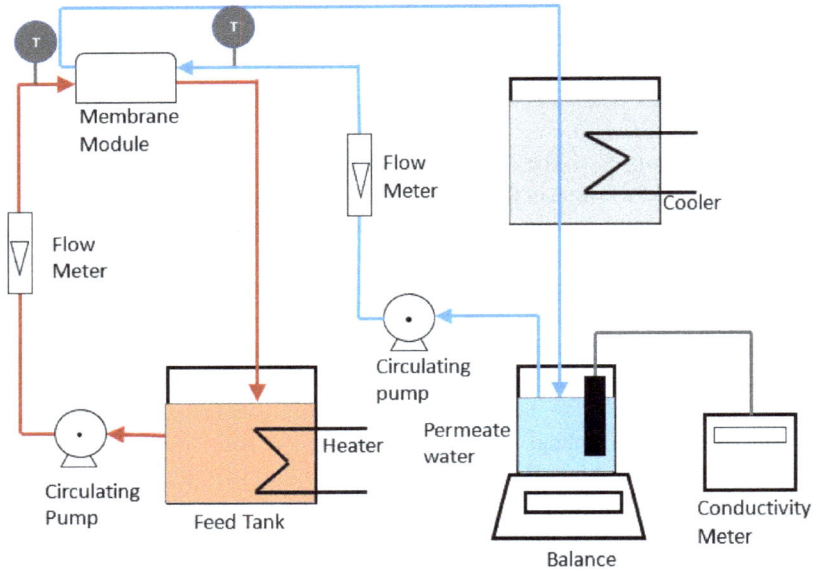

Figure 5 Process flow diagram of DCMD for the treatment of saline water [70]

Conclusion

In the past few decades, there has been an immense growth in the development of membrane separation processes and the fabrication of advanced functional membranes for the application in water purification. However, some of the technologies are either under research or in pilot scale and need to be enhanced for their specific applications. In recent years, the advent of nano-composite based UF and NF membranes have been gaining significant attention in the field of water purification and also produced substantial outcome. The advancements in the characteristics of membranes for the application in membrane distillation process has becoming the emerging technology. The implementation of advanced functional membranes in the industrial scale by overcoming the challenges such as membrane fouling, cost of operation, etc., could be an excellent choice for addressing the water scarcity issues. The outcome of all the advanced functional membranes in water purification has evidenced the success of technology. In the future, the advanced membranes with more economical and technical viability, could find wide applications in water purification.

Advanced Functional Membranes
Materials Research Foundations **120** (2022) 214-236

Materials Research Forum LLC
https://doi.org/10.21741/9781644901816-7

References

[1] J. Dechnik, J. Gascon, C.J. Doonan, C. Janiak, C.J. Sumby, Mixed-matrix membranes, Angew. Chem. Int. Ed. 56 (2017) 9292-9310. https://doi.org/10.1002/anie.201701109

[2] M. Padaki, R. Surya Murali, M.S. Abdullah, N. Misdan, A. Moslehyani, M.A. Kassim, N. Hilal, A.F.Ismail, Membrane technology enhancement in oil-water separation. A review, Desalination 357 (2015) 197-207. https://doi.org/10.1016/j.desal.2014.11.023

[3] W.J. Lee, P.S. Goh, W.J. Lau, C.S. Ong, A.F. Ismail, Antifouling zwitterion embedded forward osmosis thin film composite membrane for highly concentrated oily wastewater treatment, Sep. Purif. Technol. 214 (2019) 40-50. https://doi.org/10.1016/j.seppur.2018.07.009

[4] S. Gao, Y. Zhu, J. Wang, F. Zhang, J. Li, J. Jin, Layer-by-layer construction of Cu^{2+}/Alginate multilayer modified ultrafiltration membrane with bio inspired super wetting property for high-efficient crude-oil-in-water emulsion separation, Adv. Funct. Mater. 28 (2018) 1801944. https://doi.org/10.1002/adfm.201801944

[5] S. Meng, Y. Ye, J. Mansouri, V. Chen, Fouling and crystallisation behaviour of superhydrophobic nano-composite PVDF membranes in direct contact membrane distillation, J. Memb. Sci. 463 (2014) 102-112. https://doi.org/10.1016/j.memsci.2014.03.027

[6] Y. Zhang, P. Cui, T. Du, L. Shan, Y. Wang, Development of a sulfated Y-doped nonstoichiometric zirconia/polysulfone composite membrane for treatment of wastewater containing oil, Sep. Purif. Technol. 70 (2009) 153-159. https://doi.org/10.1016/j.seppur.2009.09.010

[7] A. Bottino, G. Capannelli, V. Asti, P. Piaggio, Preparation and properties of novel organic–inorganic porous membranes, Sep. Purif. Technol. 22 (2001) 269-275. https://doi.org/10.1016/S1383-5866(00)00127-1

[8] A. Bottino, G. Capannelli, A. Comite, Preparation and characterization of novel porous PVDF-ZrO_2 composite membranes, Desalination 146 (2002) 35-40. https://doi.org/10.1016/S0011-9164(02)00469-1

[9] D.J. Lin, C.L. Chang, F.M. Huang, L.P. Cheng, Effect of salt additive on the formation of microporous poly (vinylidene fluoride) membranes by phase inversion from LiClO4/Water/DMF/PVDF system, Polymer 44 (2003) 413-422. https://doi.org/10.1016/S0032-3861(02)00731-0

[10] S.S. Chin, K. Chiang, A.G. Fane, The stability of polymeric membranes in a TiO_2 photocatalysis process, J. Memb. Sci., 275 (2006) 202-211. https://doi.org/10.1016/j.memsci.2005.09.033

[11] G. Wu, S. Gan, L. Cui, Y. Xu, Preparation and characterization of PES/TiO_2 composite membranes, Appl. Surf. Sci. 254 (2008) 7080-7086. https://doi.org/10.1016/j.apsusc.2008.05.221

[12] Y.S. Li, L. Yan, C.B. Xiang, L.J. Hong, Treatment of oily wastewater by organic-inorganic composite tubular ultrafiltration (UF) membranes, Desalination 196 (2006) 76-83. https://doi.org/10.1016/j.desal.2005.11.021

[13] M.S. Peresin, Y. Habibi, J.O. Zoppe, J.J. Pawlak, O.J. Rojas, Nanofiber composites of polyvinyl alcohol and cellulose nanocrystals: manufacture and characterization, Biomacromolecules 11 (2010) 674-681. https://doi.org/10.1021/bm901254n

[14] S. Li, Y. Gao, H. Bai, L. Zhang, P. Qu, L. Bai, Preparation and characteristics of polysulfone dialysis membranes modified with nanocrystalline cellulose, BioRes. 6 (2011) 1670-1680.

[15] C.S. Beck, M. Roman, D.G. Gray, Effect of reaction conditions on the properties and behavior of wood cellulose nanocrystal suspensions, Biomacromolecules 6 (2005) 1048-105. https://doi.org/10.1021/bm049300p

[16] A. Chakraborty, M. Sain, M. Kortschot, Cellulose microfibrils: A novel method of preparation using high shear refining and cryo crushing, Holzforchung, 59 (2005) 102-107. https://doi.org/10.1515/HF.2005.016

[17] K. Abe, S. Iwamoto, H. Yano, Obtaining cellulose nanofibers with a uniform width of 15 nm from wood, Biomacromolecules 8 (2007) 3276-3278. https://doi.org/10.1021/bm700624p

[18] H. Bai, X. Wang, Y. Zhou, L. Zhang, Preparation and characterization of poly (vinylidene fluoride) composite membranes blended with nano-crystalline cellulose, Prog. Nat. Sci. Mater. Int. 22 (2012) 250-257. https://doi.org/10.1016/j.pnsc.2012.04.011

[19] J.P. Mericq, J. Mendret, S. Brosillon, C. Four, Composite polymeric membrane with entrapped TiO_2 nano-sized particles for water treatment: Optimized elaboration through a structural and functional characterization, Procedia Eng. 44 (2012) 1272-1274. https://doi.org/10.1016/j.proeng.2012.08.752

Materials Research Forum LLC

https://doi.org/10.21741/9781644901816-7

[20]   A. Giwa, S.W. Hasan, Novel polyethersulfone-functionalized graphene oxide (PES-fGO) mixed matrix membranes for wastewater treatment, Sep. Purif. Technol. 241 (2020) 116735. https://doi.org/10.1016/j.seppur.2020.116735

[21]   K. Ma, N. Wang, C. Wang, Q.F. An, Freezing assisted in situ growth of nano-confined ZIF-8 composite membrane for dye removal from water, J. Memb. Sci. 632 (2021) 119352. https://doi.org/10.1016/j.memsci.2021.119352

[22]   F. Anwar, G. Arthanareeswaran, Silver nano-particle coated hydroxyapatite nano-composite membrane for the treatment of palm oil mill effluent, J. Water Process Eng. 31 (2019) 100844. https://doi.org/10.1016/j.jwpe.2019.100844

[23] A.K.M. Ali, M.E.A. Ali, A.A. Younes, M.M. Abo El fadl, A.B. Farag, Proton exchange membrane based on graphene oxide/polysulfone hybrid nano-composite for simultaneous generation of electricity and wastewater treatment, J. Hazard. Mater. 419 (2021) 126420. https://doi.org/10.1016/j.jhazmat.2021.126420

[24] S. Nasseri, S. Ebrahimi, M. Abtahi, R. Saeedi, Synthesis and characterization of polysulfone/graphene oxide nano-composite membranes for removal of bisphenol A from water, J. Environ. Manage. 205 (2018) 174-182. https://doi.org/10.1016/j.jenvman.2017.09.074

[25] H. Nawaz, M. Umar, A. Ullah, H. Razzaq, K.M. Zia, X. Liu, Polyvinylidene fluoride nanocomposite super hydrophilic membrane integrated with Polyaniline-Graphene oxide nano fillers for treatment of textile effluents, J. Hazard. Mater. 403 (2021) 123587. https://doi.org/10.1016/j.jhazmat.2020.123587

[26] K. L. Wasewar, S. Singh, S.K. Kansal, Chapter 13 - Process intensification of treatment of inorganic water pollutants, in: P. Devi, P. Singh, S.K. Kansal (Eds.), Elsevier, 2020, pp. 245-271. https://doi.org/10.1016/B978-0-12-818965-8.00013-5

[27] J.H. Kim, P.K. Park, C.H. Lee, H.H. Kwon, S. Lee, A novel hybrid system for the removal of endocrine disrupting chemicals: Nanofiltration and homogeneous catalytic oxidation, J. Memb. Sci. 312 (2008) 66-75. https://doi.org/10.1016/j.memsci.2007.12.039

[28] A.W. Mohammad, Y.H. Teow, W.L. Ang, Y.T. Chung, R.D.L. Oatley, N. Hilal, Nanofiltration membranes review: Recent advances and future prospects, Desalination 356 (2005) 226-254. https://doi.org/10.1016/j.desal.2014.10.043

[29] X. Li, Y. Cao, H. Yu, G. Kang, X. Jie, Z. Liu, Q. Yuan, A novel composite nanofiltration membrane prepared with PHGH and TMC by interfacial polymerization, J. Memb. Sci. 466 (2014) 82-91. https://doi.org/10.1016/j.memsci.2014.04.034

[30] Y. Li, Y. Su, Y. Dong, X. Zhao, Z. Jiang, R. Zhang, J. Zhao, Separation performance of thin-film composite nanofiltration membrane through interfacial polymerization using different amine monomers, Desalination 333 (2014) 59-65. https://doi.org/10.1016/j.desal.2013.11.035

[31] V. Vatanpour, S.S. Madaeni, L. Rajabi, S. Zinadini, A.A. Derakhshan, Boehmite nanoparticles as a new nanofiller for preparation of antifouling mixed matrix membranes, J. Memb. Sci. 401 (2012) 132-143. https://doi.org/10.1016/j.memsci.2012.01.040

[32] P. Daraei, S.S. Madaeni, N. Ghaemi, E. Salehi, M.A. Khadivi, R. Moradian, B. Astinchap, Novel polyethersulfone nanocomposite membrane prepared by $PANI/Fe_3O_4$ nanoparticles with enhanced performance for Cu(II) removal from water, J. Memb. Sci. 415 (2012) 250-259. https://doi.org/10.1016/j.memsci.2012.05.007

[33] V. Vatanpour, M. Esmaeili, M.H.D.A Farahani, Fouling reduction and retention increment of polyethersulfone nanofiltration membranes embedded by amine-functionalized multi-walled carbon nanotubes, J. Memb. Sci. 466 (2014) 70-81. https://doi.org/10.1016/j.memsci.2014.04.031

[34] J. Zhu, N. Guo, Y. Zhang, L. Yu, J. Liu, Preparation and characterization of negatively charged PES nano filtration membrane by blending with halloysite nanotubes grafted with poly (sodium 4-styrenesulfonate) via surface-initiated ATRP, J. Memb. Sci. 465 (2014) 91-99. https://doi.org/10.1016/j.memsci.2014.04.016

[35] N.N. Bui, M.L Lind, E.M.V. Hoek, J.R. McCutcheon, Electrospun nanofiber supported thin film composite membranes for engineered osmosis, J. Memb. Sci. 385 (2011) 10-19. https://doi.org/10.1016/j.memsci.2011.08.002

[36] C. Qiu, Q.T. Nguyen, Z. Ping, Surface modification of cardo polyetherketone ultrafiltration membrane by photo-grafted copolymers to obtain nanofiltration membranes, J. Memb. Sci. 295 (2007) 88-94. https://doi.org/10.1016/j.memsci.2007.02.040

[37] H. Deng, Y., Xu, Q. Chen, X. Wei, B. Zhu, High flux positively charged nanofiltration membranes prepared by UV-initiated graft polymerization of methacrylatoethyl trimethyl ammonium chloride (DMC) onto polysulfone membranes, J. Memb. Sci. 366 (2011) 363-372. https://doi.org/10.1016/j.memsci.2010.10.029

[38] A. Linggawati, A.W. Mohammad, C.P. Leo, Effects of APTEOS content and electron beam irradiation on physical and separation properties of hybrid nylon-66 membranes, Mater. Chem. Phys. 133 (2012) 110-117. https://doi.org/10.1016/j.matchemphys.2011.12.071

[39] E.S. Kim, Q. Yu, B. Deng, Plasma surface modification of nanofiltration (NF) thin-film composite (TFC) membranes to improve anti organic fouling, Appl. Surf. Sci. 257 (2011) 9863-9871. https://doi.org/10.1016/j.apsusc.2011.06.059

[40] S.M. Abtahi, L. Marbelia, A.Y. Gebreyohannes, P. Ahmadiannamini, C.C. Joannis, C. Albasi, W.M., De, V.I.F.J. Vankelecom, Micropollutant rejection of annealed polyelectrolyte multilayer based nanofiltration membranes for treatment of conventionally-treated municipal wastewater, Sep. Purif. Technol. 209 (2019) 470-481. https://doi.org/10.1016/j.seppur.2018.07.071

[41] S. Zarghami, T. Mohammadi, M. Sadrzadeh, B. Vander, Bio-inspired anchoring of amino-functionalized multi-wall carbon nanotubes (N-MWCNTs) onto PES membrane using polydopamine for oily wastewater treatment, Sci. Total Environ. 711 (2020) 134951. https://doi.org/10.1016/j.scitotenv.2019.134951

[42] J. Gao, K.Y. Wang, T.S. Chung, Design of nanofiltration (NF) hollow fiber membranes made from functionalized bore fluids containing polyethyleneimine (PEI) for heavy metal removal, J. Memb. Sci. 603 (2020) 118022. https://doi.org/10.1016/j.memsci.2020.118022

[43] Z.L.E. Zhang, W. Zhang, Membranes for Environmental Applications, Springer Nature, Switzerland, 2020. https://doi.org/10.1007/978-3-030-33978-4

[44] H. Liu, J. Wang, Treatment of radioactive wastewater using direct contact membrane distillation, J. Hazard. Mater. 261 (2013) 307-315. https://doi.org/10.1016/j.jhazmat.2013.07.045

[45] A. Alkhudhiri, N. Darwish, N. Hilal, Membrane distillation: A comprehensive review, Desalination 287 (2012) 2-18. https://doi.org/10.1016/j.desal.2011.08.027

[46] M. Gryta, K. Karakulski, A.W. Morawski, Purification of oily wastewater by hybrid UF/MD, Water Res. 35 (2001) 3665-3669. https://doi.org/10.1016/S0043-1354(01)00083-5

[47] A. Criscuoli, E. Drioli, Energetic and exergetic analysis of an integrated membrane desalination system, Desalination 124 (1999) 243-249. https://doi.org/10.1016/S0011-9164(99)00109-5

[48] H. Kurokawa, T. Sawa, Heat recovery characteristics of membrane distillation, Heat Transfer-Japanese Research: Co-sponsored by the Society of Chemical Engineers of Japan and the Heat Transfer Division of ASME 25 (1996) 135-150. https://doi.org/10.1002/(SICI)1520-6556(1996)25:3<135::AID-HTJ1>3.0.CO;2-Y

[49] G.J. Blanco, R.L. Garcia, M.I. Martin, Seawater desalination by an innovative solar-powered membrane distillation system: the MEDESOL project, Desalination 246 (2009) 567-576. https://doi.org/10.1016/j.desal.2008.12.005

[50] A. Alkhudhiri, N. Hilal, Membrane distillation-Principles, applications, configurations, design, and implementation, Elsevier Inc., 2018. https://doi.org/10.1016/B978-0-12-815818-0.00003-5

[51] J. Kim, H. Kwon, S. Lee, S. Lee, Hong, Membrane distillation (MD) integrated with crystallization (MDC) for shale gas produced water (SGPW) treatment Desalination, 403 (2017) 172-178. https://doi.org/10.1016/j.desal.2016.07.045

[52] J. Kim, J. Kim, S. Hong, Recovery of water and minerals from shale gas produced water by membrane distillation crystallization Water Res., 129 (2018) 447-459. https://doi.org/10.1016/j.watres.2017.11.017

[53] S. Adnan, M. Hoang, H. Wang, Z. Xie, Commercial PTFE membranes for membrane distillation application: Effect of microstructure and support material, Desalination 284 (2012) 297-308. https://doi.org/10.1016/j.desal.2011.09.015

[54] Z. Wang, D. Hou, S. Lin, Composite Membrane with Underwater-Oleophobic Surface for Anti-Oil-Fouling Membrane Distillation, Environ. Sci. Technol. 5 (2016) 3866-3874. https://doi.org/10.1021/acs.est.5b05976

[55] J. Lee, C. Boo, W.H. Ryu, A.D. Taylor, M. Elimelech, Development of Omniphobic Desalination Membranes Using a Charged Electrospun Nanofiber Scaffold, ACS Appl. Mater. Interfaces. 8 (2016) 11154-11161. https://doi.org/10.1021/acsami.6b02419

[56] Z. Wang, S. Lin, Membrane fouling and wetting in membrane distillation and their mitigation by novel membranes with special wettability, Water Res. 112 (2017) 38-47. https://doi.org/10.1016/j.watres.2017.01.022

[57] M. Khayet, C.Y. Feng, K.C. Khulbe, T. Matsuura, Study on the effect of a non-solvent additive on the morphology and performance of ultrafiltration hollow-fiber membranes, Desalination 148 (2002) 321-327. https://doi.org/10.1016/S0011-9164(02)00724-5

[58] D. Hou, G. Dai, H. Fan, J. Wang, C. Zhao, H. Huang, Effects of calcium carbonate nano-particles on the properties of PVDF/nonwoven fabric flat-sheet composite membranes for direct contact membrane distillation, Desalination 347 (2014) 25-33. https://doi.org/10.1016/j.desal.2014.05.028

[59] J.E. Efome, M. Baghbanzadeh, D. Rana, T. Matsuura, C.Q. Lan, Effects of superhydrophobic SiO_2 nanoparticles on the performance of PVDF flat sheet membranes for vacuum membrane distillation, Desalination 373 (2015) 47-57. https://doi.org/10.1016/j.desal.2015.07.002

[60] W. Zhang, Z. Shi, F. Zhang, X. Liu, J. Jin, L. Jiang, Superhydrophobic and Superoleophilic PVDF Membranes for Effective Separation of Water-in-Oil Emulsions with High Flux, Adv. Mater., 25 (2013) 2071-2076. https://doi.org/10.1002/adma.201204520

[61] K.G. Zakrzewska, Nuclear Waste Processing: Pressure-Driven Membrane Processes, in: E. Drioli, L. Giorno (Eds.), Encycl. Membr., Springer, Berlin, 2015, pp.1-3.

[62] C. Wu, W. Tang, J. Zhang, S. Liu, Z. Wang, X. Wang, X. Lu, Preparation of super-hydrophobic PVDF membrane for MD purpose via hydroxyl induced crystallization-phase inversion, J. Memb. Sci. 543 (2017) 288-300. https://doi.org/10.1016/j.memsci.2017.08.066

[63] R. Zhang, W. Tang, H. Gao, C. Wu, S. Gray, X. Lu, In-situ construction of superhydrophobic PVDF membrane via $NaCl$-H_2O induced polymer incipient gelation for membrane distillation, Sep. Purif. Technol. 274 (2021) 117762. https://doi.org/10.1016/j.seppur.2020.117762

[64] J. Li, L.F. Ren, H.S. Zhou, J. Yang, J. Shao, Y. He, Fabrication of super hydrophobic PDTS-ZnO-PVDF membrane and its anti-wetting analysis in direct contact membrane distillation (DCMD) applications, J. Memb. Sci. 620 (2021) 118924. https://doi.org/10.1016/j.memsci.2020.118924

[65] Z. Ding, Z. Liu, C. Xiao, Excellent performance of novel superhydrophobic composite hollow membrane in the vacuum membrane distillation, Sep. Purif. Technol. 268 (2021) 118603. https://doi.org/10.1016/j.seppur.2021.118603

[66] S. S. Mosadegh. D. Rodrigue, J. Brisson, M.C. Iliuta, Wetting phenomenon in membrane contactors – Causes and prevention, J. Memb. Sci. 452 (2014) 332-353. https://doi.org/10.1016/j.memsci.2013.09.055

[67] A. Tuteja, W. Choi, M. Ma, J.M. Mabry, S.A. Mazzella, G.C. Rutledge, G.H. McKinley, R.E. Cohen, R.E., Designing Superoleophobic Surfaces, Science 318 (2007) 1618-1622. https://doi.org/10.1126/science.1148326

[68] C. Boo, J. Lee, M. Elimelech, Omniphobic Polyvinylidene Fluoride (PVDF) Membrane for Desalination of Shale Gas Produced Water by Membrane Distillation,

Environ. Sci. Technol. 50 (2016) 12275-12282.
https://doi.org/10.1021/acs.est.6b03882

[69] P.S. Brown, B. Bhushan, Durable, superoleophobic polymer–nanoparticle composite surfaces with re-entrant geometry via solvent-induced phase transformation, Sci. Rep. 6 (2016) 21048. https://doi.org/10.1038/srep21048

[70] L.H. Chen, A. Huang, Y.R. Chen, C.H. Chen, C.C. Hsu, F.Y. Tsai, K.L. Tung, Omniphobic membranes for direct contact membrane distillation: Effective deposition of zinc oxide nanoparticles, Desalination 428 (2018) 255-263.
https://doi.org/10.1016/j.desal.2017.11.029

[71] I. Daou, O. Zegaoui, A. Elghazouani, Physicochemical and photocatalytic properties of the ZnO particles synthesized by two different methods using three different precursors, Comptes Rendus Chim. 20 (2017) 47-54.
https://doi.org/10.1016/j.crci.2016.04.003

[72] G. Perry, Y. Coffinier, V. Thomy, R. Boukherroub, Sliding Droplets on Superomniphobic Zinc Oxide Nanostructures, Langmuir 28 (2017) 389-395.
https://doi.org/10.1021/la2035032

[73] R. Dufour, G. Perry, M. Harnois, Y. Coffinier, V. Thomy, V. Senez, R. Boukherroub, From micro to nano reentrant structures: hysteresis on superomniphobic surfaces, Colloid Polym. Sci. 291 (2013) 409-415. https://doi.org/10.1007/s00396-012-2750-7

Advanced Functional Membranes   Materials Research Forum LLC
Materials Research Foundations **120** (2022) 237-266   https://doi.org/10.21741/9781644901816-8

Chapter 8

Advanced Functional Membranes for Energy Applications

E. Kavitha[1], S. Kiruthika[1], S. Vishali[1*]

[1]Department of Chemical Engineering, College of Engineering and Technology, SRM Institute of Technology, Kattankulathur 603 203, Tamil Nadu, India.

* meet.vishali@gmail.com

Abstract

Global warming has become a serious threat to the environment as well as human life. The application of renewable and green energy sources has been emphasized in recent years to overcome the energy demand and save the environment. As a competent alternative to renewable energy sources, membrane-based energy generation has attracted the attention of researchers. In the past few decades, the application of advanced functional membranes in green energy production has gained importance. The application of polyelectrolyte membranes in fuel cells has become an emerging technology due to high proton conductivity, excellent thermal, chemical stability, and mechanical strength. Pressure retarded osmosis is also one of the membrane-based energy generation techniques which have been upgraded with significant developments. The various polymeric membranes, both inorganic and organic, have been employed in the energy production processes. In the past few years, the application of biopolymeric membranes made up of chitosan blends has shown excellent progress. The storage of energy also plays an equivalent role in energy production. The application of membranes has a vital role in energy storage batteries. This chapter deals with all the advanced functional membranes for energy production and storage.

Keywords

Osmosis Membrane, Hybrid Membrane, Fuel Cell, Batteries

Contents

1. Introduction

With swift growth in population, industrialization, and change of living style, the global energy crisis has become a serious concern. Worldwide, global warming has become a serious threat that needs more attention. To overcome this issue, now a day there has been a concern to adopt renewable energy resources. Also, there is a concern to reduce the consumption of the prime energy source fossil fuels. It involves the combustion of fuel to get energy and the emission of CO_2 from burning fossil fuels causing environmental pollution. In recent years, there is a thirst for green energy sources like fuel cells, biofuels, geothermal and solar energy, etc. [1]. Even these clean energy sources suffer from certain limitations. In the H_2 fuel cell, it produces by-product gases in the energy generation process, and there is a problem of catalyst poisoning due to the presence of contaminants [2]. Generally, biofuels are produced from the fermentation of biomass, followed by azeotropic products that involve a higher cost of production. It is essential to develop the technologies to produce, store, and efficiently consume the energy in green energy production by overcoming these issues.

Renewable energy resources such as wind energy, solar power, hydropower, etc., have been employed to minimize the consumption of fossil fuels for energy generation. The highly advanced membrane processes for energy generation have received significant attention in the past few decades [3]. Pressure retarded osmosis is one among those advanced membrane technologies but not yet commercialized. Statkraft Company, located in Norway, is the first unit to demonstrate it at the pilot plant level.

The global community is concerned about energy shortages and environmental pollution. The fuel cell mechanization may address these issues since it is capable of generating energy efficiently without emitting pollutants (if hydrogen is used as fuel, the final product is water) [4]. In modern electrochemistry, fuel cells (FCs) are electrochemical devices that convert chemical energy to electricity provided that it is supplied with enough fuel and oxidant [5]. Historically, fuel cells were developed in the 19th century by William Grove. He created a "gaseous voltaic battery" using platinum as an electrode material with oxygen and hydrogen as reactants [5]. The solid oxide fuel cell was developed in 1937 by Baur and Preis using a ZrO_2 and Y_2O_3 blend as electrolytes [5]. The FC was operated at a high temperature (1050 °C). In a fuel cell system, phenol-sulfonic acid and formaldehyde were polymerized to make phenolic polymer membranes. They were also temperamental in mechanical strength and lifespan, and showed a relatively low power density [6]. DuPont's Nafion membrane, developed in the 1970s, doubled the maximum conductivity and operating lifespan to 105 hours.

The Nafion membrane developed by DuPont in the 1970s doubled the maximum protons conductivity and doubled the operational lifetime to 10^5–10^6 hours [7]. Fuel cell systems were rapidly equipped with Nafion membranes, and fuel cells established themselves as a competitive energy source. At present, power plants, compact transformers, and vehicles, notably the Chevrolet Equinox and Ford CEVs, are all being tested with fuel cells [8]. The different fields of fuel cell applications are as follows:

- Enhanced power stability (telecoms, high-tech manufacturing, data analysis, and customer support)
- Reduction of emission (metropolitan areas, industrial establishment, aerodromes, means of transport)
- Access to the power supply is limited in some areas (mobile applications and remote locations)
- Monitoring and management of biowaste gases (effluent treatment)

The Polymer electrolyte membrane fuel cell (PEMFC) represents the most frequent fuel cell used in stationary power applications. Still, molten-carbonate fuel cells, solid oxide fuel cells, phosphoric acid fuel cells, and alkaline fuel cells are also used [9]. Stationary applications could, in principle, rely on both low- and high-temperature fuel cells. The main issue with low-temperature fuel cells is their fast startup time. Still, high-temperature fuel cells (molten-carbonate and solid oxide fuel cells) generate heat without any external reformation, resulting in higher efficiency than low-temperature fuel cells.

Fuel cells may be utilized in various ways: They may replace the grid or provide supplemental power, serve as both primary and secondary power, or provide standby power [10-11]. Mobile power production has been driven primarily by the need for higher quality, higher density, and higher time performance resulting from increased product development (CD players, mini-disk players, notebook computer systems, and cellphones). As technology advances, there is always a race for smaller, cheaper, lighter products with greater functionality [12]. Telecommunications, computers, social networks, and the Internet require ample power sources, which implies a need for fuel cells as portable power systems [13]. In general, fuel cell systems are perfect for mobile power applications since they have a high capacity, reliability, ease of operation, and are relatively inexpensive [14].

Fuel cells can reduce harmful emissions in the transportation industry while maintaining vehicle efficiency without compromising the vehicle's propulsion system. According to recent research, fuel cells have demonstrated efficiencies almost twice that of internal combustion engines. Because fuel cells do not require moving parts, they become a suitable alternative to current combustion engines due to their easy functionality, fuel versatility, ease of installation, and minimal maintenance.

Energy derived from renewable sources is intermittent by nature, and the electricity generated from them is not dispatchable, causing supply and demand to be unpredictable. Due to unbalanced supply and demand, it is, therefore, necessary to store energy. As an added benefit, energy storage can also help utility planners bridge the gap between the drop in electricity generated by photovoltaic panels or a reduced wind speed caused by cloud cover and the ramp-up rate of gas-fired peaker plants, which usually take 15 minutes to ramp up. The chapter describes the different type of membranes used in energy applications.

2.   Pressure retarded osmosis membranes

The increased demand for energy resources, has attracted the attention of researchers towards alternate renewable energy resources [15]. Water is the prime source for hydrothermal energy generation. In recent years, it has also become the energy resource for power generation from wastewater by membrane processes. The concentration gradient between seawater and fresh water has been utilized to capture energy by membrane based processes. Pressure retarded osmosis and reverse electrodialysis are the emerging membrane based technologies for producing energy from saline water [16]. Both are promising and also efficient membrane technologies to generate energy by utilizing the salinity gradient [17-19]. However, there are certain limitations of these technologies such as fouling and also cost of the membrane. The processes are still in pilot scale stage and will be commercialized in near future.

The pressure retarded osmosis involves the separation of saline water and fresh water by a semi permeable membrane. Osmosis is a phenomenon which involves the flow of water from higher chemical potential to lower chemical potential of water. The molar concentration difference of solute acts as the driving force for the transport of water across the membrane and solute molecules are retained by the membrane. Osmosis utilizes the osmotic pressure difference across the membrane for the flow of water form pure water to saline water. The pressure retarded osmosis process phenomenon lies intermediate between osmosis and reverse osmosis. In this process, the hydraulic pressure is applied to the saline water side, like reverse osmosis process, however, the net flow of water is from feed water to the saline water side like osmosis [20].

The source of renewable energy produced by mixing fresh water and saline water could be utilized by pressure retarded osmosis [21]. Due to the salinity gradient, water will flow from fresh water side to saline water. The membrane allows only water to permeate through

Advanced Functional Membranes   Materials Research Forum LLC
Materials Research Foundations **120** (2022) 237-266   https://doi.org/10.21741/9781644901816-8

it. The flow of water increases the volume of water in the saline water side. This in turn increases the pressure which is sufficient to operate the turbine to generate power. The economic and efficient membrane are the key factor in this process. In the case of reverse electrodialysis, the saline and fresh water are allowed to flow through the chamber containing anion exchange and cation exchange membranes arranged alternately. The salinity gradient acts as the driving force for the generation of voltage by the movement of ions.

To achieve the goal of zero liquid discharge and green energy, utilizing the salinity gradient by pressure retarded osmosis, an emerging and efficient process could cater the energy demand. Over the past few years, the technology has been lacking in achieving adequate flowrate. The composite matrix of RO membrane has the bulk supporting layer that significantly affects the water flux due to the internal concentration polarization. Hence, in the recent years, research has been directed towards, developments in the fabrication of membrane with better water flux and also reduced internal concentration polarization. A specially fabricated cellulose acetate forward osmosis membrane and thin film composite polyamide membrane were employed in pressure retarded osmosis process to increase the water flux and decrease the internal concentration polarization [22-24]. However, to avoid fouling and drop in the water flux, the fresh water stream and saline water stream should be pretreated extensively. The schematic diagram of power generation by pressure retarded osmosis process is shown in figure 1.

This process consumes river water as the feed water and sea water as the draw solution. Several studies have been attempted to improve both the technical and economical feasibility of this process. The studies investigated by Bajraktari et al., (2017) and Manzoor et al., (2020) employed highly saline water as the draw solution to extract high energy [26, 27]. Fouling of the membrane is the main drawback of this process. In the composite matrix of the membrane, the porous supporting layer faces the feed side. The feed is most often the treated effluent which may contain organic and inorganic contaminants. These contaminants can permeate through the pores of the membrane and get stuck below the dense layer of the composite matrix. This leads to significant reduction in the water flux [28]. Several attempts have been tried to overcome the fouling by surface modification, synthesis of double-skinned membrane and also the pretreatment of the feed stream [29-32]. Cellulose triacetate composite membrane matrix with active layer facing the feed solution configuration has been investigated to minimize fouling and also to increase the mechanical stability of the membrane [33]. Li et al., (2018) fabricated the integral pressure retarded hollow fiber membrane with active layer facing feed solution configuration to reduce the fouling with increased energy harvesting [34]. The membrane was prepared by phase inversion technique using polyamide-imide substrate, since it possesses excellent mechanical and chemical stability.

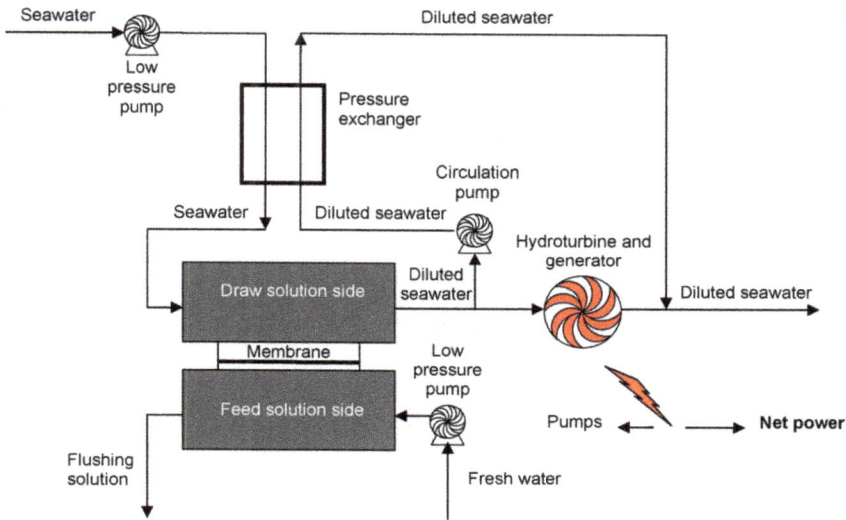

Figure 1 Schematic diagram of pressure retarded osmosis [25]

2.1   Hybrid pressure retarded osmosis processes

In the past few years, the closed loop pressure retarded osmosis process has been gaining attention, since it provides better performance and also it is a eco-friendly process [35]. In the closed loop process, due to the necessary for the additional separation technique to recover the draw solution and also to improve the energy generation, hybrid pressure retarded osmosis has been adopted. Several hybrid systems have been investigated by researchers in the recent years to utilize the benefits of the both the processes of hybrid systems. The hybrid systems such as pressure retarded osmosis-membrane distillation [36, 37], pressure retarded osmosis – thermosiphon [38, 39], pressure retarded osmosis – solar pond [40,41], etc., have been investigated in the recent years.

2.1.1   Pressure retarded osmosis-membrane distillation

The hybrid process forward osmosis-membrane distillation has proven to be a promising technology for desalination, reuse of effluent from industries, reclamation of wastewater [42-44]. Based on this concept, pressure retarded osmosis -membrane distillation system has been attempted to produce energy as well as clean water. The retentate from the reverso osmosis process, which the concentrated brine is fed to the membrane distillation module. From the membrane distillation module, the clean water and concentrated and high temperature draw solution are obtained. The concentrated and high temperature draw solution is passed through the pressure retarded osmosis unit. The low salt concentration wastewater is used as a feed solution in this unit. The salinity gradient act as the driving force for the transport of water to the permeate side followed with the operation of hydroturbine leading to the power generation [45].

2.1.2 Pressure retarded osmosis – thermosiphon

This process involves the natural circulation currents generated due to the difference in the density to produce energy due to thermal gradient. The procedure has been adopted from Arias et al., (2018). The aqueous salt solution is passed through the solar collected where part of water evaporates. The evaporated stream with very less salinity has been taken as the feed stream to the pressure retarded osmosis system. The concentrated solution from the solar collector has been taken as the draw solution. Water transport takes place from feed side to draw solution side, followed with the operation of hydroturbine to generate power. The hot stream that comes out is fed to the heat exchanger to cool down the solution. Then the solution is sent to the solar collector and the cycle of operation has been continued [46]. The future feasibility of the process depends on the optimized temperature to yield maximum output of power in the long term process.

2.1.3 Pressure retarded osmosis – solar pond

This hybrid system has been developed to generate power as well as to reduce the adverse environmental effect of brine solution. The solar pond is the manmade solar collector that can absorb the heat and retain the heat in the bottom layer of the pond due to the salinity gradient. Mohamed and Bicer, 2019, proposed the hybrid system pressure retarded osmosis-solar pond. The solar pond as a solar collector receives the solar heat and retains the heat in the bottom layer of the pond with higher saline concentration. This high temperature stream has been sent to the multiple effect desalination units to heat seawater. The concentrated brine from the desalination units has been fed to the pressure retarded osmosis system as a draw solution and the treated effluent with low salt concentration has been fed to the feed stream. The water transport occurs from feed side to draw solution side. This dilutes the draw solution and increases the volume of it, which is sufficient to drive the hydroturbine to generate power. This hybrid system could enjoy the benefit of improved solar efficiency and minimizes the greenhouse gas emission [47].

Advanced Functional Membranes   Materials Research Forum LLC

Materials Research Foundations **120** (2022) 237-266   https://doi.org/10.21741/9781644901816-8

3. Fuel cell applications

Fuel cells are one of the most powerful tools that must be developed to meet the growing demand for renewable energies. In contrast to combustion-based heat engines, chemical energy in fuel cells is directly transformed into electrical energy without any intermediate steps [48]. In contrast to other energy sources derived electrochemical reactions such as batteries, which temporarily store their reactants in the cells, these sources are continuously supplied with reactants from outside the cells. Fuel cells also do not consume electrodes, and do not participate in reactions like batteries. Fuel cell technology is considered to be a promising technological advancement for the development of renewable energy sources [49].

Fuel cells have developed to the point where they are becoming more advanced. They appear to offer promising markets where electricity can be generated efficiently and environmentally friendly. In addition, there are a number of plants that employ fuel cell technology that has been developed and is operational, ranging from megawatts to milliwatt capacity [50]. Essentially, fuel cells are composed of an electrolyte and a porous electrode, which has catalysts that accelerate electrochemical reactions. There is a negative charge on one electrode (the anode) and a positive charge on the other (the cathode). An electric current is produced by electrochemical reactions at the electrodes.

For the lower to medium scale power generation applications, the best source is the fuel cells with methanol as a fuel. Due to their lower weight and volume as compared to other energy sources such as indirect fuel cells, they can be employed for various transportation applications [51]. In the recent years, direct methanol fuel cells have attracted the attention of many researchers for the generation of green energy. This technology involves the conversion between chemical and electrical energy by utilizing fuel based on methanol. In this process we employ mediator such as polyelectrolyte membrane for the transport of proton and also this prevents the combustion of fuel by acting as a barrier between the reactants. The synthesis of polyelectrolyte membrane with excellent proton conductivity, chemical, mechanical, thermal and electrical stability, and also less permeability of methanol is the crucial part in the fabrication of fuel cells that run on methanol [52, 53].

3.1 Proton exchange membrane

The key component, in fuel cells, is the proton exchange membranes (PEMs). PEMs must meet a number of characteristics, including highly proton-conductive, exceptional strength and durability, chemical and electrochemical stability, minimal fuel crossover, and the ability to fabricate membrane-based electrode assemblies. PEMFCs have become a popular choice of fuel cell systems for portable energy storage because of their portability, long-term durability, fast response time, extensive power output, and low emissions [54], together with their silent operation [55]. An anode and a cathode are connected by a PEMFC proton exchange membrane (PEM). Many of the functional groups in PEM work as proton-conductive groups, allowing the protons to move from one group to another.

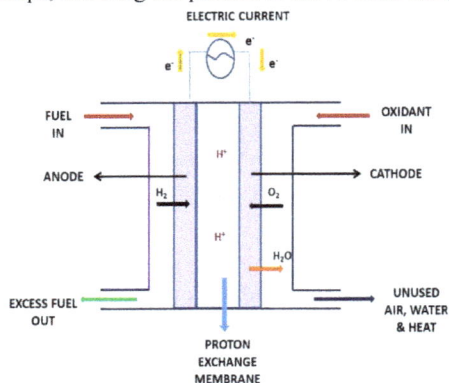

Figure 2 Schematic representation of Proton exchange membrane fuel cell

3.2 Nafion membrane

Nafion membranes were developed by Dupont in the 1970s. In this membrane, the sulfonic acid groups assocaited with the polymers act as proton-conductive groups. Nafion membranes have a strong influence on proton conductivity depending on their water content and temperature. Nafion membranes offer many of the desirable features that make them suitable for FC systems. This membrane has excellent chemical and mechanical stability, high proton conductivity under normal conditions, and low permeability to fuel and oxidant [57]. Nafion™ membranes are at the heart of a new generation of fuel cells offering a clean energy source to power anything that moves. Using Nafion™ membrane in fuel cells for power generation benefits vehicles ranging from automobiles, buses, forklifts, boats, planes, and spacecraft due to:

High efficiency: Without combustion, fuel cells convert chemical energy to electrical energy.

Low emissions: Emissions from fuel cells are negligible, consisting mostly of water.

Quiet operations: Due to its lack of moving parts, a fuel cell produces only the noise produced by its auxiliary equipment.

Minimal intervention: The operation and maintainence of a fuel cell is simple because there are no moving parts, reducing operational costs and maintenance.

The nafion, however, only works at a relatively low temperature (80-100 °C). When the temperature is above 80 °C, the Nafion dehydrates (and their proton conductivity drops) and at 120 °C, the proton conductivity starts to drop significantly [58]. Consequently, Nafion has very poor proton conductivity when relative humidity (RH) is low [59]. Furthermore, Nafion has a relatively high permeability to methanol [60], which significantly reduces the performance of methanol fuel cells.

The Nafion membrane has many drawbacks, but new polymer electrolytes, or PEMs, have been developed that could replace it, especially when operating at high temperatures. In general, recent PEMs can be categorized into three broad groups: (1) polymer membrane, (2) ceramic, and (3) organic-inorganic composite membrane.

3.3   Polymeric membranes

There are several types of membranes that are used in fuel cells, but polymer PEM is the most commonly used. Polymers typically contain proton-conductive groups such as sulfonic acid in their side or main chains. To make the polymers chemically and thermally stable, they are typically fluorinated. The most commonly used PEM in fuel cells is Nafion. There has been extensive research on the possibility of modifying the properties of the Nafion membrane to make it more suitable for conduction of protons at high temperatures and low humidity.Table 1 represents the different modifications of the Nafion membrane with their characteristics.

3.4   Polystyrene-based membranes

A polymer membrane made of polystyrene-sulfonic acid (PSSA) was the first polymer membrane available for commercial use. There is a growing interest in membrane materials fabricated from polystyrene which can function at medium and high temperatures, as well as low water content [64]. Styrene-based polymers are relatively cheap, which makes PSSA-based membranes an attractive option for FC. Membranes based on PSSA have permeability to methanol lower as compared to Nafion [65]. In real-world applications of FC, PSSA membranes are unstable. In addition to the loss of IEC and conductivity, the performance of the membrane is negatively affected by the loss of SSA fragments (up to approximately 0.1% per hour) [66].

3.5   Polyimide-based membranes with sulfonated groups

Polyimides are thermostable polymers that have excellent thermal and mechanical characteristics and are resistant to harsh chemicals and extreme temperatures [67]. Sulfonic groups can be added to these polymers to improve their insulating properties. As a result of this modification, the molecule becomes hydrophilic and proton-conductive. The development of polyimide-based membranes was driven by the expectation that material prices would be significantly lower than Nafion [68]. In humid environments, sulfonated polyamide membranes inflated and needed hydration to allow for proton conductivity (which is comparable to the properties of Nafion). However unlike Nafion, its water absorption, water sorption and drag resistance of the water were not affected significantly within a wide range of temperatures. Due to these features, it is suitable for operations at moderate temperatures [69].

Table 1 Types of Nafion modified membranes

Type of membrane	Operating conditions	Reference
Cs^+ cations doped Tricoli Nafion	➢ Reduces the surface area of hydrated micellar regions, and reduces methanol permeability considerably	[61]
A fluorinated polymer ethylene and propylene EP) or Teflon and fluoroalkyl resin ked wit Nafion (PFA)	➢ Lower the diffusion rate of methanol	[62]
The side chains of Nafion are changed, resulting in Hyflon, a novel perfluorinated membrane.	➢ To increase thermal stability ➢ The membrane exhibits a greater glass transition temperature than Nafion, which allows it to fuction at higher temperatures without degradation. ➢ The proportion of -CF2- groups in Hyflon's backbone affects the hydrogen bonds between water and sulfonic acid groups, presenting more water molecules needed for proton transport to occur.	[63]

3.6   Polyphosphazene membranes

Polyphospheric structures are made up of nitrogen and phosphorus elements that have attached organic, inorganic, or organometallic side groups. Through their modifications across different phosphonate groups, phosphazene offers a variety of beneficial properties. The glass transition temperature of these polymers is low [70]. Each subunit contains phosphorus and nitrogen in their maximum oxidized form, making the subunits non-flammable. Phosphates have the advantage of being exceptionally resistant to the deprotonation reaction, making them a perfect candidate for PEM fuel cell formulations [71]. The combination of these features and their low cost make phosphates a good platform for PEM fuel cells [72].

3.7   Ceramic PEM

Researchers have been focusing more and more on investigating the performance of ceramic materials for use in fuel cell technology in recent decades [73,74]. PEM made of ceramic can be classified into two types as non metallic metal oxide. Non-metal ceramic PEMs, like crystalline silica, have good chemical and mechanical durability, a low cost of construction, and great temperature tolerance. Non-metal ceramic PEM, on the other hand, has a substantially lower protonic conductivity than Nafion. [75].

For metal oxide ceramic PEM oxygen ions are the major transfer ions in metal oxide ceramic membranes, which are primarily used in solid oxide fuel cells. At various humidities, many metal oxides and metal oxyhydroxides, such as TiO_2, Al_2O_3, $BaZrO_3$, and $FeOOH$, have demonstrated the ability to conduct protons [76]. Both TiO_2 and Al_2O_3 have low protonic conductivities, similar to silica glass [77]. The protonic conductivity of $FeOOH$ is substantially higher than that of TiO_2 and Al_2O_3, and much higher than that of Nafion [78]. The ferroxane ceramic membranes, on the other hand, have low elasticity and compression resistance. Because the precursor of those membranes, lepidocrocite, has poor mechanical qualities, they are brittle and easily break into little pieces.

3.8   Inorganic–organic composite PEM

Membranes that contain an organic polymer and an inorganic solid are called inorganic-organic membranes. Fuel cells perform better under high temperature, low relative humidity conditions when an inorganic phase is present because it enhances the interaction between components, limits dimensional change, and enhances performance. As a result, multiple concepts of organic and inorganic fuel cell membranes were developed and implemented into fuel cell stacks. A novel approach uses bimodal/spinodal transformations, nanoporous technologies, or nanoporous synthesis to organize the organic and inorganic components.

There are two basic categories of inorganic–organic composite membranes:

- Membranes made up of proton-conducting polymers and inorganic particles that are less proton-conducting.

- Membranes made up of proton-conducting particles and organic polymers that are less proton-conducting

Nafion is generally combined with solid inorganic acids, silicon dioxides or metal oxides to create composite membranes because it improves the conductivity of protons and maintains chemical and mechanical stability at high temperatures and relatively low humidity [79].

4.   Batteries

The growing usage of variable renewable energy sources necessitates battery storage devices to fulfill high demand during energy consumption peaks. The energy storage will be essential for the future network and a carbon-free world, enabling intermittent renewable energy sources to supply energy to household consumers and businesses and hybrid vehicles. Battery storage is proving to be a promising option for electricity storage, offering flexibility to the system due to its unique ability to support, hold and emit electricity, as well as the benefits of partnering with renewable energy sources. The spatial versatility sets it apart from traditional energy storage technologies, such as pumping storage, allowing the batteries to be installed closer to their customers.

The most fundamental barrier to widespread of marketing strategy is the cost of batteries. Even if stationary applications do not require a long-standing potential, the possibility of periodic discharge requires high output voltages, greater power density, and long service life. Electrochemical energy storage systems can release energy quickly if required. System based on redox flow batteries (RFB) offers the added benefit of versatility, making them one of the most attractive electrical energy storage technologies. In RFBs, different redox pairs such as Sn/Cl, Cr/Ti, V/Sn, Fe/Cr, and V/Fe, were examined [80]. The vanadium redox batteries (Fig. 1) were popular because the other methods get away from the crossing of contaminants as the electrolyte contains various components and, as a result, experiences self-discharge and loss of capacity.

Another method is to find a new membrane technology that is usually followed by structural changes. The optimal membrane includes the following features:

- Substantial ion-exchange specificity
- Higher ionic conductance
- A low rate of water absorption
- The lower proportion of dilatation
- The conductivity of high order
- Chemically and thermally stable
- Cost-effective

Advanced Functional Membranes | Materials Research Forum LLC
Materials Research Foundations **120** (2022) 237-266 | https://doi.org/10.21741/9781644901816-8

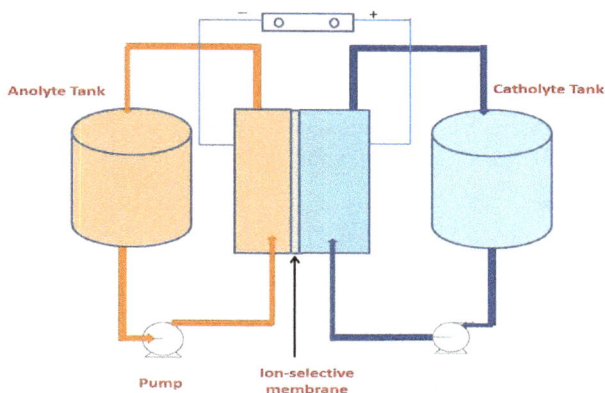

Figure 3 Schematic representation of a VRB system [81]

4.1   Types of exchange membranes

The different membrane types used in redox flow battery (RFB) systems are as follows (Fig. 4.):

- Cationic exchange membranes (CEM)
- Anionic exchange membranes (AEM)
- Amphoteric ion-exchange membranes (AIEM)
- Zeolite membranes

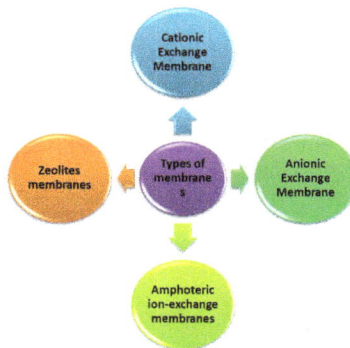

Fig. 4. Classes of Exchange Membranes

4.1.1 Cationic exchange membranes

In the initial phase, sulfonated polyethylene and polystyrene components were used to make the membranes for vanadium redox batteries. Selemion and Nafion membranes exhibit properties that are suitable for VRB applications [81]. The following materials have been widely applied to overcome the limitations of conventional membranes:

- Polyethylene based membranes
- Fluorocarbon based membranes
- Nafion with polymeric composite membranes
- Nafion with inorganic integrated membranes

4.1.1.1 Polyethylene membranes

Polyethylene's affordability and easy production with desirable characteristics made it the subject of extensive studies in the 1990s. Sulfonic acid groups are often introduced into these materials as cation exchange sites, resulting in proton conductivity. Sulfonic groups can be introduced using various reactants, for example, sulfuric acid and sulfuryl chloride [82]. It is possible to modify the pores of the membrane by using different polymers, providing lower-cost supports, and the use of polymers like Nafion to fill the pores. As a result, the materials costs will be reduced. Examples of affordable and readily available membrane supports are shown in Table 2.

Table 2 Polyethylene Membranes [83]

Membrane	Properties
Daramic	Polyethylene with a high molecular weight has high strength. Furthermore, the material is oxidation-resistant.
Daramic incorporated with divinylbenzene mixture of Amberilite and methanol	Enhance the selectivity of the membrane.
Daramic pores infused with Polysodium 4-styrene-sulfate (PSSS)	Increase the cation capability with divinylbenzene cross link.

4.1.1.2 Fluorocarbon membranes

Polymers with fluorinated carbon backbones are used largely in membranes for VRB and fuel cells. These materials can be used as base material to enhance the stability of the VRB battery under oxidative conditions. Dupont Nafion and Nafion 117 are fluorocarbon polymers with a hydrophobic backbone similar to Teflon and hydrophilic side chains adorned with sulfonic acid groups [84]. Polymer names are coded with numbers that indicate the membrane's thickness and equivalent weight.

4.1.1.3 Polymeric composite membranes

Costs can be reduced by substituting cheaper materials for the Nafion. Nafion membranes can be laminated with poly(ether ketone) SPEEK membranes to reduce their cost. SPEEK has excellent conductivity, but it is chemically unstable when exposed to oxidative VO_2^+ environments. In VRB applications, coating the SPEEK material in the Nafion layer prevents substantial performance losses without affecting its stability [85]. A commercial hybrid membrane based on Nafion is now available, such as VANADion [86]. Multiple charges and discharges in a VANADion hybrid membrane have shown similar durability to Nafion 115 [87].

4.1.1.4 Inorganic hybrid membranes

The surface of Nafion has polar molecule clusters in the membrane that enhance V-crossover. Inorganic nanoparticles block these polar nanoclusters when filled with inorganic nanoparticles. The following are the examples of the Nafion based inorganic hybrid membranes:

- A silica sol-gel was used to fill the poles in the Nafion 117 membrane. Despite the composite Nafion/SiO_2 membrane's high protons conductivity, the IEC properties were preserved. The V crossover, on the other hand, was drastically reduced [88].

- The Nafion/ORMOSIL membrane is made up of tetraethyl orthosilicate (TEOS) and diethoxydimethylsilane (DEDMS). Hybrid membranes achieve superior to Nafion membranes in terms of efficiency and Nafion-silica membranes as well as a reduced rate of self-discharge [89].

4.1.1.5 Polymerization of the Nafion surface

A polymer layer can be applied to Nafion or polyelectrolytes can be deposited to reduce the permeability of the membrane, however, these systems may cause membrane inflammation. The effects of inflammation on membrane longevity are detrimental. An interface polymerization of polyelectrolyte polyimide can reduce swelling when placed on the surface of Nafion. On the surface of Nafion, a surface cationic charge is formed by forming a layer that is strongly bonded with Nafion. Consequently, the membrane decreased crossover rate of vanadium ions, but a higher resistance per unit area. The membrane's lower permeability causes a reduction in water transfer and self-discharge [80]. Graphene materials have attracted attention in recent decades due to their excellent conductivity, mechanical properties, and barrier characteristics [85]. Graphene oxide (GO) is blended with nylon to form a well organized membrane that maintains an efficient barrier, preventing the electrons in the direction of the V from cross-over [85].

4.1.1.6 Poly aryl ether membranes

SPEEK, SPES, and SPESK are sulfonated aryl backbone polymers that could be used as an alternative to Nafion. The quaternary ammonium groups, , imidazole, and sulfonic acid can easily be incorporated into these materials to achieve excellent thermomechanical stability and proton permeability [90]. The SPES, SPESK and SPEEK are attractive alternatives for vanadium redox battery usage due to their low cost, excellent electrochemical stability, and potent mechanical behavior. Depending on the amount of sulfonation, these materials exhibit varying proton conductivities; however, they are generally lower than Nafion [91].

Cell discharge and membrane degradation cause lower membrane lifetime due to interactions with cationic exchange membranes and V electrolytes. It is envisaged that commercial membranes would be treated with organic nanoparticles, such as graphene oxide, to produce the most promising outcomes. Current VRB membrane technology uses Nafion 117, which is costly and has a high V permeability. Also, hybrid membranes made from Nafion and SPEEK were researched in the hopes of reducing membrane costs. One other promising method uses relatively inexpensive polypropylene or polyethylene membranes.

4.1.2 Anion membrane

The ionic exchange membrane (AEM) is an alternative membrane class. A positive charge on their functioning groups repels the positively charged species of V from the membrane. It is also known as the Donnan effect [92]. In spite of the AEM's reduced V permeability being of high interest, they are not suitable for VRB use. These AEMs exhibit reduced proton conductivity along with poor chemical stability, making them unsuitable for commercial applications. Polymer backbones and functional groups are the primary factors affecting membrane stability. AEM materials for VRB have recently been investigated for their stability and conductivity. Due to the chemical stability of polybenzimidazole (PBI), it is suitable for use in VRB, which is strongly acidic and oxidative and is therefore more attractive [93].

PBI also reduces water permeability by providing narrow pore sizes [94]. This type of membrane is also known as an acid-doped membrane. Nonfluorinated hydrocarbons undergo highly reactive reactions when exposed to VRB conditions due to their high charge. In addition, quaternary ammonium and imidazole functionalized poly(p-phenylene) based membrane was prepared as an AEM. Results showed that quaternary ammonium groups were best suited to prepare the membrane. In recent years, researchers have focused on improving the stability of these groups by changing the polymer's chemical structure [95].

4.1.3 Amphoteric ion-selective membrane (AISM)

Membranes containing cationic and anionic selective membranes that have the combined features of AEMs and CEMs are known as amphoteric ion selective membranes (AISMs). CEMs exhibit better resistance to chemical degradation and higher conductivity, while AEMs exhibit lower V cross-over rates [96]. Sulphate and vanadium species can be made less permeable by adding both AEM and CEM properties at the same time [97].

4.1.4 Zeolite membranes

Molecular separation is possible in zeolite due to their uniform porosity, which allows them to be used for separating molecules by selecting the shape and size of molecules that penetrate. Zeolites contain many Bronsted Acidic Sites, which contribute to an increased concentration of protons [98]. Within the configuration, these sites are created by the formation of Silicon-Oxygen-Aluminum linkages. VRBs made from zeolite have recently attracted attention for their use as membranes [99]. Proton at a high concentration and a lower thickness led to a high conductivity of the material [100]. The material's higher conductivity is due to the increased proton concentration as well as the reduced width. The polymer also possesses an excellent affinity for membranes because of its small pore sizes

Conclusions

Globally, the change in the climatic condition, especially the CO_2 emission has becoming the impending and serious crisis. The utilization of renewable energy resources could reduce the environmental impact. The viable alternative to renewable energy is the membrane based power generation processes. In the recent years, the membrane based energy production has proven to be a promising technology interms of ease of operation, thermal,chemical and mechanical stability, and also easy scale up of operation. The bio-polymeric membranes have shown significant results in the application of fuel cells compared to the commercially available membranes. Various developments have been implemented in the fabrication of advanced functional membranes to enhance their performance, permeability, technical and economical feasibility. However, there are still lack of experimental results to enhance the scale up of processes. Also, the successful application of the process depends upon the adoptability of efficient and inexpensive membranes. In the past few decades, tremendous improvements have been carried out to increase the feasible applications of membrane based energy generation processes. In the future, it could be implemented successfully in large scale.

Declaration about copyright

All the figures given in this manuscript are self-drawn and not copied from any other articles.

References

[1] L. Dai, K. Huang, Y. Xia, Z. Xu, Two-dimensional material separation membranes for renewable energy purification, storage, and conversion, Green Energy Environ. 6 (2021) 193–211. https://doi.org/10.1016/j.gee.2020.09.015

[2] H. Wang, J. Xu, L. Sheng, X. Liu, Y. Lu, W. Li, A review on bio-hydrogen production technology, Int. J. Energy Res. 42 (2018) 3442–3453. https://doi.org/10.1002/er.4044

[3] S.J. Einarsson, B. Wu, Thermal associated pressure-retarded osmosis processes for energy production: A review, Sci. Total Environ. 757 (2021) 143731. https://doi.org/10.1016/j.scitotenv.2020.143731

[4] J. Zaidi, T. Matsuura (Eds.), Polymer membranes for fuel cells, Springer Sci. Rev. 2008.

[5] G. Hoogers, Fuel Cell Technology Handbook, CRC Press, Boca Raton (FL), 2003. https://doi.org/10.1201/9781420041552

[6] C. Berger, Handbook of Fuel Cell Technology, Prentice-Hall, Englewood Cliffs (NJ), 1968.

[7] B. Smitha, S. Sridhar, A.A. Khan, Solid polymer electrolyte membranes for fuel cell applications – a review, J. Membr. Sci. 259 (2005) 10-26. https://doi.org/10.1016/j.memsci.2005.01.035

[8] N.M. Sammes, Fuel Cell Technology: Reaching Towards Commercialization,

Springer, London, 2006. https://doi.org/10.1007/1-84628-207-1

[9] G.F. McLean, T. Niet, S. P. Richard, N. Djilali, An assessment of alkaline fuel cell technology, Int. J. Hydrog. Energy 27 (2002) 507-526. https://doi.org/10.1016/S0360-3199(01)00181-1

[10] M.W. Ellis, M.R.V. Spakovsky, D.J. Nelson, Fuel cell systems: efficient, flexible energy conversion for the 21st century, Proc. IEEE. 89(12) (2001) 1808-1818. https://doi.org/10.1109/5.975914

[11] F. Barbir, PEM fuel cells, In Fuel Cell Technology, Springer, London, 2006, pp. 27-51. https://doi.org/10.1007/1-84628-207-1_2

[12] S. Giddey, S.P.S. Badwal, A. Kulkarni, C. Munnings, A comprehensive review of direct carbon fuel cell technology, Prog. Energy Combust. Sci. 38(3) (2012) 360-399. https://doi.org/10.1016/j.pecs.2012.01.003

[13] P. Hoffmann, Tomorrow's energy: hydrogen, fuel cells, and the prospects for a cleaner planet, MIT press, 2012. https://doi.org/10.7551/mitpress/8625.001.0001

[14] Y. Song, C. Zhang, C.Y. Ling, M. Han, R.Y. Yong, D. Sun, J. Chen, Review on current research of materials, fabrication and application for bipolar plate in proton exchange membrane fuel cell, Int. J.Hydrog. Energy 45(54) (2020) 29832-29847. https://doi.org/10.1016/j.ijhydene.2019.07.231

[15] B.E. Logan, M. Elimelech, Membrane-based processes for sustainable power generation using water, Nature. 488 (2012) 313-319. https://doi.org/10.1038/nature11477

[16] K.P. Lee, T.C. Arnot, D. Mattia, A review of reverse osmosis membrane materials for desalination—Development to date and future potential, J. Membr. Sci. 370 (2011) 1-22. https://doi.org/10.1016/j.memsci.2010.12.036

[17] S. Chou, R. Wang, L. Shi, Q. She, C. Tang, A. Fane, Thin-film composite hollow fiber membranes for Pressure Retarded Osmosis (PRO) process with high power density, J. Membr. Sci. 389 (2012) 25–33. https://doi.org/10.1016/j.memsci.2011.10.002

[18] J.W. Post, H.V.M. Hamelers, C.J.N. Buisman, Energy Recovery from Controlled Mixing Salt and Fresh Water with a Reverse Electrodialysis System, Environ. Sci. Technol. 42 (2008) 5785-5790. https://doi.org/10.1021/es8004317

[19] A. Achilli, T.Y. Cath, A.E. Childress, Power generation with pressure retarded osmosis: An experimental and theoretical investigation, J. Memb. Sci. 343 (2009) 42-52. https://doi.org/10.1016/j.memsci.2009.07.006

[20] A. Achilli, A.E. Childress, Pressure retarded osmosis: From the vision of Sidney Loeb to the first prototype installation - Review, Desalination 261 (2010) 205–211. https://doi.org/10.1016/j.desal.2010.06.017

[21] K. Nijmeijer, S. Metz, Chapter 5 Salinity Gradient Energy, in: I.C. Escobar, A.I.B.T.-S.S. and E. Schafer (Eds.), Sustain. Water Futur. Water Recycl. versus Desalin., Elsevier, 2010, pp. 95-139. https://doi.org/10.1016/S1871-2711(09)00205-0

[22] S. Loeb, Large-scale power production by pressure-retarded osmosis, using river water and sea water passing through spiral modules, Desalination 143 (2002) 115–122. https://doi.org/10.1016/S0011-9164(02)00233-3

[23] N. Bajraktari, C. Helix-Nielsen, H.T. Madsen, Pressure retarded osmosis from hypersaline sources — A review, Desalination 413 (2017) 65–85. https://doi.org/10.1016/j.desal.2017.02.017

[24] H. Manzoor, M.A. Selam, S. Adham, H.K. Shon, M. Castier, A. Abdel-Wahab, Energy recovery modeling of pressure-retarded osmosis systems with membrane modules compatible with high salinity draw streams, Desalination 493 (2020) 114624. https://doi.org/10.1016/j.desal.2020.114624

[25] Q. She, Y.K.W. Wong, S. Zhao, C.Y. Tang, Organic fouling in pressure retarded osmosis: Experiments, mechanisms and implications, J. Membr. Sci. 428 (2013) 181–189. https://doi.org/10.1016/j.memsci.2012.10.045

[26] X. Li, T. Cai, T.-S. Chung, Anti-Fouling Behavior of Hyperbranched Polyglycerol-Grafted Poly(ether sulfone) Hollow Fiber Membranes for Osmotic Power Generation, Environ. Sci. Technol. 48 (2014) 9898–9907. https://doi.org/10.1021/es5017262

[27] L. Zhang, Q. She, R. Wang, S. Wongchitphimon, Y. Chen, A.G. Fane, Unique roles of aminosilane in developing anti-fouling thin film composite (TFC) membranes for pressure retarded osmosis (PRO), Desalination 389 (2016) 119–128. https://doi.org/10.1016/j.desal.2015.12.024

[28] G. Han, Z.L. Cheng, T.-S. Chung, Thin-film composite (TFC) hollow fiber membrane with double-polyamide active layers for internal concentration polarization and fouling mitigation in osmotic processes, J. Membr. Sci. 523 (2017) 497–504. https://doi.org/10.1016/j.memsci.2016.10.022

[29] E. Abbasi-Garravand, C.N. Mulligan, C.B. Laflamme, G. Clairet, Role of two different pretreatment methods in osmotic power (salinity gradient energy) generation, Renew. Energy. 96 (2016) 98–119. https://doi.org/10.1016/j.renene.2016.04.031

[30] Q. She, X. Jin, C.Y. Tang, Osmotic power production from salinity gradient resource by pressure retarded osmosis: Effects of operating conditions and reverse solute diffusion, J. Membr. Sci. 401–402 (2012) 262–273. https://doi.org/10.1016/j.memsci.2012.02.014

[31] Y. Li, S. Zhao, L. Setiawan, L. Zhang, R. Wang, Integral hollow fiber membrane with chemical cross-linking for pressure retarded osmosis operated in the orientation of active layer facing feed solution, J. Membr. Sci. 550 (2018) 163–172. https://doi.org/10.1016/j.memsci.2017.12.074

[32] I. MS, S. Sultana, S. Adhikary, R. MS, Highly effective organic draw solutions for renewable power generation by closed-loop pressure retarded osmosis, Energy Convers. Manag. 171 (2018) 1226–1236. https://doi.org/10.1016/j.enconman.2018.06.031

[33] Z. Yuan, Y. Yu, L. Wei, X. Sui, Q. She, Y. Chen, Pressure-retarded membrane distillation for simultaneous hypersaline brine desalination and low-grade heat harvesting, J. Membr. Sci. 597 (2020) 117765. https://doi.org/10.1016/j.memsci.2019.117765

[34] S.K. Hubadillah, Z.S. Tai, M.H.D. Othman, Z. Harun, M.R. Jamalludin, M.A. Rahman, J. Jaafar, A.F. Ismail, Hydrophobic ceramic membrane for membrane distillation: A mini review on preparation, characterization, and applications, Sep. Purif. Technol. 217 (2019) 71–84. https://doi.org/10.1016/j.seppur.2019.02.014

[35] F.J. Arias, S. de las Heras, The brinesiphon: A homolog of the thermosiphon driven by induced salinity and downward heat transfer, Sol. Energy. 153 (2017) 454–458. https://doi.org/10.1016/j.solener.2017.05.091

[36] F.J. Arias, A first estimate for a pressure retarded osmosis-driven thermosyphon, Sol. Energy. 159 (2018) 962–965. https://doi.org/10.1016/j.solener.2017.10.064

[37] A.M.O. Mohamed, Y. Bicer, Integration of pressure retarded osmosis in the solar ponds for desalination and photo-assisted chloralkali processes: Energy and exergy analysis, Energy Convers. Manag. 195 (2019) 630–640. https://doi.org/10.1016/j.enconman.2019.05.032

[38] N. Sezer, M. Koç, Development and performance assessment of a new integrated solar, wind, and osmotic power system for multigeneration, based on thermodynamic principles, Energy Convers. Manag. 188 (2019) 94–111. https://doi.org/10.1016/j.enconman.2019.03.051

[39] P. Wang, Y. Cui, Q. Ge, T. Fern Tew, T.S. Chung, Evaluation of hydroacid complex in the forward osmosis–membrane distillation (FO–MD) system for desalination, J. Membr. Sci. 494 (2015) 1–7. https://doi.org/10.1016/j.memsci.2015.07.022

[40] N. Cong Nguyen, H. Cong Duong, S.-S. Chen, H. Thi Nguyen, H. Hao Ngo, W. Guo, H. Quang Le, C. Cong Duong, L. Thuy Trang, A. Hoang Le, X. Thanh Bui, P. Dan Nguyen, Water and nutrient recovery by a novel moving sponge – Anaerobic osmotic membrane bioreactor – Membrane distillation (AnOMBR-MD) closed-loop system, Bioresour. Technol. 312 (2020) 123573. https://doi.org/10.1016/j.biortech.2020.123573

[41] F. Gao, L. Wang, J. Wang, H. Zhang, S. Lin, Nutrient recovery from treated wastewater by a hybrid electrochemical sequence integrating bipolar membrane electrodialysis and membrane capacitive deionization, Environ. Sci. Water Res. Technol. 6 (2020) 383–391. https://doi.org/10.1039/C9EW00981G

[42] S.H. Chae, J. Seo, J. Kim, Y.M. Kim, J.H. Kim, A simulation study with a new performance index for pressure-retarded osmosis processes hybridized with seawater reverse osmosis and membrane distillation, Desalination 444 (2018) 118–128. https://doi.org/10.1016/j.desal.2018.07.019

[43] S. Surampudi, S.R. Narayanan, E. Vamos, H. Frank, G. Halpert, Advances in Direct Methanol Fuel Cells, J.Power sources 47 (1994) 377–385. https://doi.org/10.1016/0378-7753(94)87016-0

[44] M.A. Hickner, H. Ghassemi, Y.S. Kim, B.R. Einsla, J.E. McGrath, Alternative polymer systems for proton exchange membranes (PEMs), Chem. Rev. 104 (2004) 4587–4611. https://doi.org/10.1021/cr020711a

[45] H. Matsuyama, Y. Kitamura, Y. Naramura, Diffusive Permeability of Ionic Solutes in Charged Chitosan Membrane, J. Appl. Polym. Sci. 72 (1999) 397–404. https://doi.org/10.1002/(SICI)1097-4628(19990418)72:3<397::AID-APP9>3.0.CO;2-C

[46]   R. Xu, Y. Wu, X. Wang, J. Zhang, X. Yang, B. Zhu, Enhanced ionic conductivity of yttria-stabilized ZrO2 with natural CuFe-oxide mineral heterogeneous composite for low temperature solid oxide fuel cells, Int. J. Hydrogen Energy 42 (2017) 17495–17503. https://doi.org/10.1016/j.ijhydene.2017.05.218

[47]   L. Li, Q. Shi, L. Huang, C. Yan, Y. Wu, Green synthesis of faujasite-La0.6Sr0.4Co0.2Fe0.8O3-δ mineral nanocomposite membrane for low temperature advanced fuel cells, Int. J. Hydrogen Energy 46 (2021) 9826–9834. https://doi.org/10.1016/j.ijhydene.2020.05.275

[48]   U. Lucia, Overview on fuel cells, Renew. Sustain. Energy Rev. 30 (2014) 164-169. https://doi.org/10.1016/j.rser.2013.09.025

[49]   O.Z. Sharaf, M.F. Orhan, An overview of fuel cell technology: Fundamentals and applications, Renew. Sustain. Energy Rev. 32 (2014) 810-853. https://doi.org/10.1016/j.rser.2014.01.012

[50]   G. Cacciola, V. Antonucci, S. Freni, Technology up date and new strategies on fuel cells, J. Power Sources 100(1-2) (2001) 67-79. https://doi.org/10.1016/S0378-7753(01)00884-9

[51]   V.S. Bagotsky, Fuel cells: problems and solutions, 56, John Wiley & Sons, 2012. https://doi.org/10.1002/9781118191323

[52]   J. A. Flores, Comparative study of different fuel cell technologies, Bol. Soc. Esp. Ceram. Vidr. 52(3) (2013) 105-117. https://doi.org/10.3989/cyv.142013

[53]   L. An, T.S. Zhao, Transport phenomena in alkaline direct ethanol fuel cells for sustainable energy production, J. Power Sources 341 (2017) 199-211. https://doi.org/10.1016/j.jpowsour.2016.11.117

[54]   E.H. Majlan, D. Rohendi, W.R.W. Daud, T. Husaini, M.A. Haque, Electrode for proton exchange membrane fuel cells: A review, Renew. Sustain. Energy Rev. 89 (2018) 117-134. https://doi.org/10.1016/j.rser.2018.03.007

[55]   S. Shahgaldi, A. Ozden, X. Li, F. Hamdullahpur, A novel membrane electrode assembly design for proton exchange membrane fuel cells: Characterization and performance evaluation, Electrochim. Acta. 299 (2019) 809-819. https://doi.org/10.1016/j.electacta.2019.01.064

[56] A. Abaspour, N.T. Parsa, M. Sadeghi, A new feedback Linearization-NSGA-II based control design for PEM fuel cell, Int. J. Comput. Appl. 97(10) (2014) 25-32. https://doi.org/10.5120/17044-7354

[57] L. Zhang, S.R. Chae, Z. Hendren, J.S. Park, M.R. Wiesner, Recent advances in proton exchange membranes for fuel cell applications, Chem. Eng. J. 204 (2012) 87-97. https://doi.org/10.1016/j.cej.2012.07.103

[58] S. Bose, T. Kuila, T.X.H. Nguyen, N.H. Kim, K.T. Lau, J.H. Lee, Polymer membranes for high temperature proton exchange membrane fuel cell: recent advances and challenges, Prog. Polym. Sci. 36(6) (2011) 813-843. https://doi.org/10.1016/j.progpolymsci.2011.01.003

[59] B. Zhang, Y. Cao, S. Jiang, Z. Li, G. He, H. Wu, Enhanced proton conductivity of Nafionnanohybrid membrane incorporated with phosphonic acid functionalized graphene oxide at elevated temperature and low humidity. J. Membr. Sci. 518 (2016) 243-253. https://doi.org/10.1016/j.memsci.2016.07.032

[60] S. Ren, G. Sun, C. Li, Z. Liang, Z. Wu, W. Jin, X. Qin, X. Yang, Organic silica/Nafion® composite membrane for direct methanol fuel cells, J. Membr. Sci. 282(1-2) (2006) 450-455. https://doi.org/10.1016/j.memsci.2006.05.050

[61] V. Tricoli, Proton and methanol transport in poly(perfluorosulfonote) membranes containing Cs^+ and H^+ cations, J. Electrochem. Soc. 145 (1998) 3798-3801. https://doi.org/10.1149/1.1838876

[62] L. Merlo, A. Ghielmi, L. Cirillo, M. Gebert, V. Arcella, Membrane electrode assemblies based on HYFLON® ion for an evolving fuel cell technology, Sep. Sci. Technol. 42(13) (2007) 2891-2908. https://doi.org/10.1080/01496390701558334

[63] S.J. Paddison, J.A. Elliott, Molecular modeling of the short-side-chain perfluorosulfonic acid membrane, J. Phys. Chem. A. 109(33) (2005) 7583-7593. https://doi.org/10.1021/jp0524734

[64] M.A. Hickner, H. Ghassemi, Y.S. Kim, B.R. Einsla, J.E. McGrath, Alternative polymer systems for proton exchange membranes (PEMs), Chem. Rev. 104(10) (2004) 4587-4612. https://doi.org/10.1021/cr020711a

[65] G.S. Prakash, M.C. Smart, Q.J. Wang, A. Atti, V. Pleynet, B. Yang, ... S. Surampudi, High efficiency direct methanol fuel cell based on poly (styrenesulfonic) acid (PSSA)–poly (vinylidene fluoride)(PVDF) composite membranes, J. Fluor. Chem. 125(8) (2004) 1217-1230. https://doi.org/10.1016/j.jfluchem.2004.05.019

Advanced Functional Membranes   Materials Research Forum LLC
Materials Research Foundations **120** (2022) 237-266   https://doi.org/10.21741/9781644901816-8

[66] A. Shukla, P. Dhanasekaran, S. Sasikala, N. Nagaraju, S.D. Bhat, V.K. Pillai, Covalent grafting of polystyrene sulfonic acid on graphene oxide nanoplatelets to form a composite membrane electrolyte with sulfonated poly (ether ether ketone) for direct methanol fuel cells, J. Membr. Sci. 595 (2020) 117484. https://doi.org/10.1016/j.memsci.2019.117484

[67] K.I. Okamoto, Sulfonated polyimides for polymer electrolyte membrane fuel cell, J Photopolym Sci Technol. 16(2) (2003) 247-254. https://doi.org/10.2494/photopolymer.16.247

[68] F. Zhang, N. Li, S. Zhang, S. Li, Ionomers based on multi sulfonated perylenedianhydride: Synthesis and properties of water resistant sulfonated polyimides, J. Power Sources 195(8) (2010) 2159-2165. https://doi.org/10.1016/j.jpowsour.2009.10.026

[69] Y. He, C. Tong, L. Geng, L. Liu, C. Lu, Enhanced performance of the sulfonated polyimide proton exchange membranes by graphene oxide: Size effect of graphene oxide, J. Membr. Sci. 458 (2014) 36-46. https://doi.org/10.1016/j.memsci.2014.01.017

[70] H.R. Allcock, Polyphosphazene elastomers, gels, and other soft materials, Soft Matter. 8(29) (2012) 7521-7532. https://doi.org/10.1039/c2sm26011e

[71] H.R. Allcock, Generation of structural diversity in polyphosphazenes, Appl. Organomet. Chem. 27(11) (2013) 620-629. https://doi.org/10.1002/aoc.2981

[72] H. Tang, P.N. Pintauro, Polyphosphazene membranes. IV. Polymer morphology and proton conductivity in sulfonated poly [bis (3-methylphenoxy) phosphazene] films, J. Appl. Polym. Sci. 79(1) (2001) 49-59. https://doi.org/10.1002/1097-4628(20010103)79:1<49::AID-APP60>3.0.CO;2-J

[73] F.M. Vichi, M.I. T.Tejedor, M.A. Anderson, Effect of pore-wall chemistry on proton conductivity in mesoporous titanium dioxide, Chem. Mater. 12(6) (2000) 1762-1770. https://doi.org/10.1021/cm9907460

[74] S.M. Haile, Fuel cell materials and components, Acta Mater. 51(19) (2003) 5981-6000. https://doi.org/10.1016/j.actamat.2003.08.004

[75] E.M. Tsui, M.R. Wiesner, Fast proton-conducting ceramic membranes derived from ferroxane nanoparticle-precursors as fuel cell electrolytes, J. Membr. Sci. 318(1-2) (2008) 79-83. https://doi.org/10.1016/j.memsci.2008.02.025

[76] S. Shamim, K. Sudhakar, B. Choudhary, J. Anwar, A review on recent advances in proton exchange membrane fuel cells: materials, technology and applications, Adv. Appl. Sci. Res. 6(9) (2015) 89-100.

[77] S.P. Jiang, Functionalized mesoporous structured inorganic materials as high temperature proton exchange membranes for fuel cells, J. Mater. Chem. A. 2(21) (2014) 7637-7655. https://doi.org/10.1039/C4TA00121D

[78] J. Lu, J. Zhang, Mesoporous Structured Materials as New Proton Exchange Membranes for Fuel Cells, In Mesoporous Materials for Advanced Energy Storage and Conversion Technologies, CRC Press, 2017, pp. 97-151. https://doi.org/10.1201/9781315368580-3

[79] X. Sun, S.C. Simonsen, T. Norby, A. Chatzitakis, Composite membranes for high temperature PEM fuel cells and electrolysers: a critical review, Membranes. 9(7), (2019) 83. https://doi.org/10.3390/membranes9070083

[80] C.H.L. Tempelman, J.F. Jacobs, R.M. Balzer, V. Degirmenci, Membranes for all vanadium redox flow batteries, J. Energy Storage. 32 (2020) 101754. https://doi.org/10.1016/j.est.2020.101754

[81] Z. Mai, H. Zhang, X. Li, C. Bi, H. Dai, Sulfonated poly (tetramethydiphenyl ether ether ketone) membranes for vanadium redox flow battery application, J. Power Sources. 196(1) (2011) 482-487. https://doi.org/10.1016/j.jpowsour.2010.07.028

[82] D. Ariono, I.G. Wenten, Surface modification of ion-exchange membranes: Methods, characteristics, and performance, J. Appl. Polym. Sci. 134(48) (2017) 45540. https://doi.org/10.1002/app.45540

[83] J. Zuo, S. Bonyadi, T.S. Chung, Exploring the potential of commercial polyethylene membranes for desalination by membrane distillation, J. Membr. Sci. 497, (2016) 239-247. https://doi.org/10.1016/j.memsci.2015.09.038

[84] J. Liao, Y. Chu, Q. Zhang, K. Wu, J. Tang, M. Lu, J. Wang, Fluoro-methyl sulfonated poly (arylene ether ketone-co-benzimidazole) amphoteric ion-exchange membranes for vanadium redox flow battery, Electrochim. Acta. 258 (2017) 360-370. https://doi.org/10.1016/j.electacta.2017.11.063

[85] A. Eftekhari, Y.M. Shulga, S.A. Baskakov, G.L. Gutsev, Graphene oxide membranes for electrochemical energy storage and conversion, Int. J. Hydrog. Energy. 43(4) (2018) 2307-2326. https://doi.org/10.1016/j.ijhydene.2017.12.012

[86] M. Ulaganathan, V. Aravindan, Q. Yan, S. Madhavi, M. S. Kazacos, T.M. Lim, Recent advancements in all-vanadium redox flow batteries, Adv. Mater. Interfaces 3(1) (2016) 1500309. https://doi.org/10.1002/admi.201500309

[87] P. Yang, J. Long, S. Xuan, Y. Wang, Y. Zhang, J. Li, H. Zhang, Branched sulfonated polyimide membrane with ionic cross-linking for vanadium redox flow battery application, J. Power Sources 438 (2019) 226993. https://doi.org/10.1016/j.jpowsour.2019.226993

[88] J. Xi, Z. Wu, X. Qiu, L. Chen, Nafion/SiO$_2$ hybrid membrane for vanadium redox flow battery, J. Power Sources 166(2) (2007) 531-536. https://doi.org/10.1016/j.jpowsour.2007.01.069

[89] M.A. Aziz, S. Shanmugam, Zirconium oxide nanotube–Nafion composite as high performance membrane for all vanadium redox flow battery, J. Power Sources 337 (2017) 36-44. https://doi.org/10.1016/j.jpowsour.2016.10.113

[90] J. Ran, L. Wu, Y. He, Z. Yang, Y. Wang, C. Jiang ... T. Xu, Ion exchange membranes: New developments and applications, J. Membr. Sci. 522 (2017) 267-291. https://doi.org/10.1016/j.memsci.2016.09.033

[91] H. Prifti, A. Parasuraman, S. Winardi, T.M. Lim, M. Skyllas-Kazacos, Membranes for redox flow battery applications, Membranes 2(2) (2012) 275-306. https://doi.org/10.3390/membranes2020275

[92] B. Shanahan, T. Böhm, B. Britton, S. Holdcroft, R. Zengerle, S. Vierrath, ... M. Breitwieser, 30 μm thin hexamethyl-p-terphenyl poly (benzimidazolium) anion exchange membrane for vanadium redox flow batteries, Electrochem. Commun. 102 (2019) 37-40. https://doi.org/10.1016/j.elecom.2019.03.016

[93] L. Wang, A.T. Pingitore, W. Xie, Z. Yang, M.L. Perry, B.C. Benicewicz, Sulfonated PBI gel membranes for redox flow batteries, J. Electrochem. Soc. 166(8), (2019) A1449. https://doi.org/10.1149/2.0471908jes

[94] J.K. Jang, T.H. Kim, S.J. Yoon, J.Y. Lee, J.C. Lee, Y.T. Hong, Highly proton conductive, dense polybenzimidazole membranes with low permeability to vanadium and enhanced H$_2$SO$_4$ absorption capability for use in vanadium redox flow batteries, J. Mater. Chem. A. 4(37) (2016) 14342-14355. https://doi.org/10.1039/C6TA05080H

[95] D. Chen, X. Chen, L. Ding, X. Li, Advanced acid-base blend ion exchange membranes with high performance for vanadium flow battery application, J. Membr.

Sci. 553 (2018) 25-31. https://doi.org/10.1016/j.memsci.2018.02.039

[96] M. Bhushan, S. Kumar, A.K. Singh, V. K. Shahi, High-performance membrane for vanadium redox flow batteries: Cross-linked poly (ether ether ketone) grafted with sulfonic acid groups via the spacer, J. Membr. Sci. 583 (2019) 1-8. https://doi.org/10.1016/j.memsci.2019.04.028

[97] P.K. Leung, Q. Xu, T.S. Zhao, L. Zeng, C. Zhang, Preparation of silica nanocomposite anion-exchange membranes with low vanadium-ion crossover for vanadium redox flow batteries, Electrochim. Acta. 105 (2013) 584-592. https://doi.org/10.1016/j.electacta.2013.04.155

[98] S.H. Cha, Recent development of nanocomposite membranes for vanadium redox flow batteries, J. Nanomater. (2015) 207525. https://doi.org/10.1155/2015/207525

[99] J. Kim, J.D. Jeon, S.Y. Kwak, Sulfonated poly (ether ether ketone) composite membranes containing microporous layered silicate AMH-3 for improved membrane performance in vanadium redox flow batteries, Electrochim. Acta. 243 (2017) 220-227. https://doi.org/10.1016/j.electacta.2017.05.079

[100] X. Teng, Y. Guo, D. Liu, G. Li, C. Yu, J. Dai, A polydopamine-coated polyamide thin film composite membrane with enhanced selectivity and stability for vanadium redox flow battery, J. Membr. Sci. 601 (2020) 117906. https://doi.org/10.1016/j.memsci.2020.117906

Advanced Functional Membranes
Materials Research Foundations **120** (2022) 267-314

Materials Research Forum LLC
https://doi.org/10.21741/9781644901816-9

Chapter 9

Advanced Functional Membrane for CO₂ Capture

H.J. Bora[1], N. Sultana[1], K.J. Goswami[1], N.S. Sarma[1], A. Kalita[1*]

Physical Sciences Division, Institute of Advanced Study in Science and Technology, Paschim Boragaon, Guwahati-781035, Assam, India

anamik.kalita01@gmail.com

Abstract

The capture of carbon dioxide directly from the air has been shown a growing interest in the mitigation of greenhouse gases but remains controversial among the research community. Due to the high dilution factor of CO_2 in air, simultaneously increases the energy requirement as well as the charge of the respective technology. Membrane/Thin film technology has been conceded as the most investigated as well as most appealing technology to attenuate carbon dioxide from the atmosphere. The membrane and membrane process technique are found to be alluring and eco-friendly to mitigate the carbon due to its cost efficiency, low expenditure of energy as well as comprehensibility in operation. Traditionally, the materials are cast into dense membranes with a standard thickness and after the formation of the membranes, their applications such as carbon capture/separation are evaluated by commutation between permeability and selectivity. In present scenario, efficient separation of CO_2 from other gases has become a worldwide issue. Coal/Natural/Flue gases are evolving as the primary source of CO_2, so the capture of CO_2 from the mentioned sources are extensively contemplated as the next opportunity for the large-scale deployment of gas separation membranes. Although, current researches indicate the advances in material process designs that can crucially enhance the membrane capture systems as well as the separation systems, which make membrane process technique contentious with other technologies present till date for carbon capture. The aforementioned application requires novel polymeric materials which have the ability for efficient carbon capture and possesses high CO_2 separation properties from different mixed gases, along with high mechanical and thermal stability for a longer time. Herein, the present report precisely highlights the recent advancement on the membrane technology based on the functional materials and their applications in the field of CO_2 capture.

Keywords

Functional Materials, Composite, Membrane, Carbon-Dioxide Capture

Contents

1. Introduction

Carbon dioxide (CO_2) is a prime chemical gas present in the atmosphere, which is censorious to continuance of life on earth. It is the main component of photosynthesis with the help of which plants fuel themselves, which assist as the prime origin of food for all living things. Moreover, it yields oxygen which is crucial for human beings for respiration. Researches have proved that a little assemblage of carbon dioxide in the earth's ambience

Advanced Functional Membranes   Materials Research Forum LLC
Materials Research Foundations 120 (2022) 267-314   https://doi.org/10.21741/9781644901816-9

is needful, where glaciation is inhibited, thereby yielding a surrounding where plants and animals life can flourish. However, in this present world, mankind related to the creation of energy generates ampleness of carbon dioxide in the atmosphere which no longer can be stabilized by the earth's natural cycles, which can cause serious environmental problems that confront mankind in the future. Chemical industries are always the dominant sources for creating global warming by increasing the average global temperature by 5.8 °C. According to the International Energy Agency (IEA), CO_2 emission can be controlled by half by 2050 only when the reduction of industrial growth of CO_2 occurs by 21% in 2050 compared to today's CO_2 levels. Global carbon outpouring has been growing expeditiously anticipated by the growing demands of energy output and also by the use of fossil fuel dominantly for energy supply [1]. CO_2 is among the highest produced greenhouse gas, that has been directly implemented in global warming [2,3]. CO_2 is also an element of biogas, natural gas, as well as in landfill gas. It emitted to the atmosphere from flue gas is the main issue contributing to climate change together with global warming. The calorific value of biogas or natural gas is reduced by CO_2, making such gas streams corrosive and acidic, which leads to the removal of CO_2 from such gases. Also, the apprehend of CO_2 from such gases leads to CO_2 separation technologies [4]. There is a vital importance of removing CO_2 for the economic value and an issue of immediate investigation [5,6]. Some reports show natural gas formed 95% CH_4 and 5% of C_2H_6 while biogas contains 45%-65% CH_4 and 30%-40% CO_2. However, energy resources like natural gas, syngas, biogas, shale gas, and have impurity as CO_2 which reduces the heating qualities and values of the gases, demands high energy utilization for the altering as well as transport of corrodes pipelines and other equipment's [7-10]. It is foretelling that if the enormous CO_2 production persists in the future, it could have deleterious consequences which include rising sea levels, melting ice caps, weather changes, ozone layer depletion, poor air quality index, acidification of the ocean, and desertification, which can simultaneously harm the human lives, animals, plants, therefore, mitigation of CO_2 is of utmost importance [11,12]. Till now, there are many methods present for the separation of CO_2 [13-25].

But, due to the intricacy of the gas components and conditions, various technologies invented so far yet suffers from higher cost, high energy consumption, and severe secondary pollution [26-29]. Therefore, developing new technologies to separate such toxic gases and fabrication of new materials and novel processes are necessary. Several methods investigated and evolved in the past few decades to capture CO_2 from various gas mixtures, such as solid absorption, chemical and physical absorption, membrane, cryogenic, and gas hydrates chemical looping (Fig. 1). Among all the methods membrane separation holds many advantages like smaller unit size, lower capital and processing costs, more elementary operation, better energy efficiency, more superficial up-down scaling, and much lower environmental impact [30].

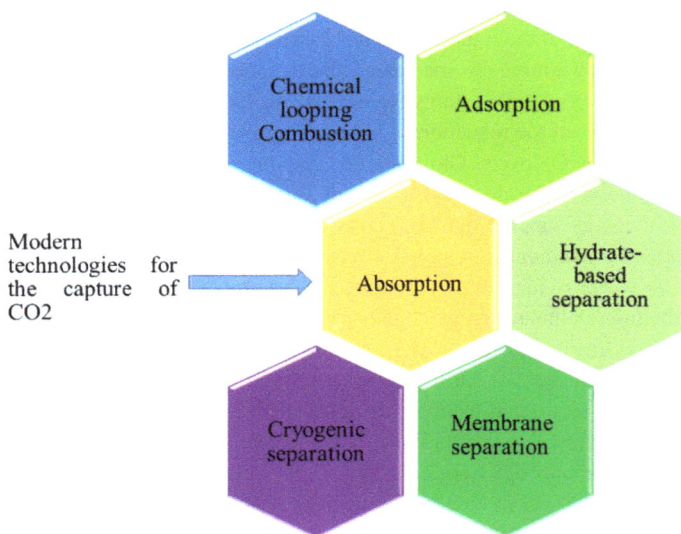

Figure 1 Different modern and conventional techniques for CO_2 capture.

The CO_2 separation from a gas mixture is conducted by numerous techniques such as cryogenic distillation, membrane technology, absorption, and adsorption [31]. Among all the separation methods, adsorption is of strategic importance in the industry. Different materials like carbons, molecular sieves, zeolites, metal-organic framework, and clays gain more attention to separate mixtures of CO_2 and other gases [32-40]. Selectivity of the adsorption process over other separation methods is due to the unique and vital features of the adsorption process. Besides adsorption, membrane and membrane process is contemplated as one of the most appealing mechanization to mitigate CO_2 emission. The separation of CO_2 using membrane technique has accomplished as notably improved, with fast progress in the past decennium. The following figure (Fig. 2) shows several reported membrane/membrane processes placements into various classes.

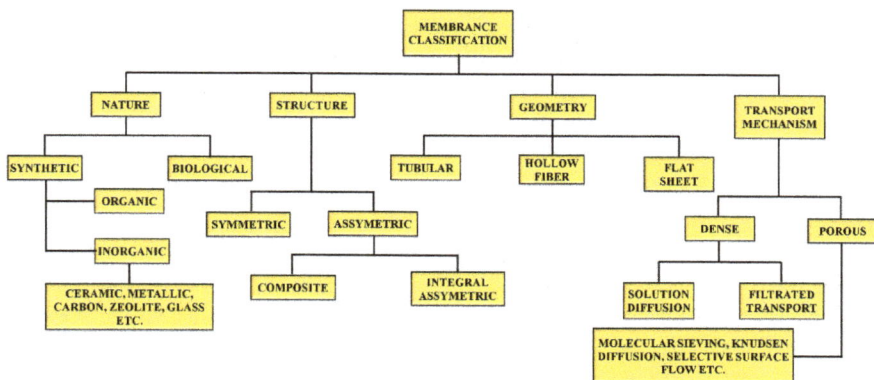

Figure 2 Classification of membranes.

Han *et al.,* have showed the advances of using polymeric membranes and utilized the membranes for the capture of CO_2. They mainly focused on material design and formation for CO_2 capture. They grouped the materials on the basis of gas transport mechanism, i.e., facilitated transport and solution diffusion transport. Study of solution diffusion membrane encloses the current endeavors to shift the upper bound barrier and improved solubility of CO_2 in divergent polymers and introduces the new procedures to fabricate the shape persisting macromolecules with high sieving ability.

2.   Porous polymeric membrane

The separation of CO_2 through membrane exhibits superiority over the other investigated traditional techniques. Although using membranes with high permeability and selectivity is still a significant challenge [41]. Selective utilization of porous membrane is an unconventional concept applied to flue/fuel streams for separating CO_2. There are five mechanisms for separating CO_2 through the permeable membrane: ionic transport, molecular sieve, adsorption/surface diffusion, Knudsen diffusion solution/diffusion separation, and capillary condensation. The membrane performances are still a challenge and are based on two primary characteristics; selectivity and permeability. The former one in this context allows gas particles to pass over the other, and the latter is the flux. Flux can be defined as the transfer of a definite volume of gas per unit time across the membrane. Gas streams can significantly affect the selectivity and permeability of a particular membrane due to their properties (e.g., the molecular weight of the gas component, the kinetic diameter of the gas molecules, velocity), synthetic material, particular method used for the synthesis of the membrane. Stating pressure, temperature, and polymer concentration during synthesis and membrane fabrication can control the permeability and selectivity. The use of membranes can be an advantage for the separation process. The complex material based on polymer exhibits a particular attachment of a thin layer to a non-selective, cost-effective, thicker coating that demonstrates high mechanical strength [42].

There is tremendous progress of membranes for the separation of CO_2 results in a considerable number of reports in the past ten years. There are reports on polymeric membranes, micro-porous membranes, carbon membranes, polymer blend membranes, ZIF membranes, MOF membranes, facilitated membranes, PEG-containing membranes, mixed matrix membranes, and polyimide membranes ionic membranes liquid-based membranes [43]. These methods for the separation of CO_2 are based on biogas upgrading, petrochemical industry application, olefin/paraffin separation, and natural gas sweetening [44-62]. In most of them, the thickness of the self-standing membrane was around 50-150 μm, and the separation performance was compared with 'Robeson upper bound' [63]. However, there are mainly two reasons for the disparity of such values in contrast to industrial values. Firstly, the prediction of pure gas measurements on an industrial scale is inferior, especially for CO_2. Secondly, the permeation properties of composite membranes depend on the thickness of polymeric films, which results in different separation and transport performances [43,64]. There are numerous reports regarding thin-film composite, primary cornerstones on inorganic materials with a selective layer for gas separation. Therefore, it is necessary to furnish a recent advance for the separation of CO_2 using multilayer composite membranes [43]. Polymeric separation demonstrates a very high permeability of CO_2 with lower selectivity values comparing with other separation methods. However, for the CO_2 separation, membranes only allow CO_2 to pass through the membrane, excluding the other components present in flue/fuel gas streams. Fabrication of ultra-thin, defect-free composite achieves a high gas flux. It ensures an economically viable membrane for the separation of gas. The main advantage of using composite membranes over other asymmetric membranes is the material because of its mechanical processability and properties. A minimal amount of the material is needed for deposition in the composite membrane (0.1-2 g/m^2).

Advanced Functional Membranes

Materials Research Foundations **120** (2022) 267-314

Materials Research Forum LLC

https://doi.org/10.21741/9781644901816-9

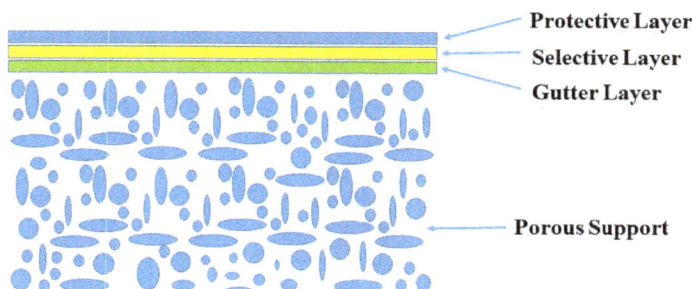

Figure 3 Structure of multilayer composite.

Composite layers contribute different functions to the membranes, and thus requirements of each layer are also other, as shown in Fig. 3. High mechanical strength resulting in low mass transfer resistance and high porosity obtained by porous support of the composite membrane. Additionally, such materials should be cheap as well as quickly processed into a permeable membrane. The selective and absorbent layers contain a gutter layer, which reduces the possible penetration into the membrane's pores of the coating solution and the surface roughness. The main separation properties of the membrane offer by the core part of the membrane is the selective layer.

Along with lifetime, aging, cost, and stability, the first two main criteria for the selectivity of membrane material are high selectivity and gas permeance. If needed, a protective layer can also be employed to protect the membrane during handling and fabrication. The selection of layers such as protective layers, selective layers, gutter layers, and porous support is of significant importance.

2.1 Porous support

The support comprises the phase inversion method immersed in a non-solvent bath involving precipitation of a casting solution. However, gas separation membranes can be primarily prepared by using this method commercially, such as poly-sulfone (PSF) and polyimide (PI). With the control of the synthetic condition of membranes, the supporting material should possess finger-like structures or sponge-type [65]. In the phase inversion method, the polymer consumption is about 50 gm/m^2 for asymmetric porous support where both the support and dense top layer are of the same material. The commercialization of membranes with expensive polymers is challenging if the phase inversion methods work for them. Another technique is used to synthesize the not soluble polymers in commonly used solvents, including heating, annealing, extruding, and stretching. As a non-porous polymer film, heating is the melting stage followed by extruding the polymer. For the improvement of the crystalline structure of the polymer film (e.g., membranes like porous polytetrafluoroethylene (PTFE) and polypropylene (PP)), annealing is used. This stretching step leads to easy pore formation, and the annealed film was extended to form micropores [66]. As porous support, only a handful of commercial polymeric membranes are available. The most common porous supports are PSF, PAN, CA, PEI, PVDF, PPO, PP, PTFE, Teflon, etc. The coated porous support should have at least ten times much higher gas flux than the coated one, which corroborates the resistance of the selective layer. The main criterion of porous support is a smooth and clean surface for uniform and detect-free coating. Based on the intrinsic properties, supporting medium have also contribution to the selectivity of the process. Simultaneously, porous support present on the material surface can diminish the resistance of the mass transfer.

2.2 Gutter layer

The porous and selective layers contain a gutter layer between them and contributes different characteristics for different cases. The primary purpose is to avert the diluted polymer solution from blocking pores and penetrating the porous structure. However, this layer should be smooth to coat and make it defect-free easily. The most routinely used polymer for the fabrication of such layer is the poly(1-(trimethylsilyl)-1-propyne) (PTMSP) and polydimethylsiloxane (PDMS) due to their high porosity. Nevertheless, PTMSP loses its performance, although highly porous, within a concise period [67,68]. During an operation of 14 days, 80 % reduction was observed in the PTMSP composite membrane casted on polyacrylonitrile (PAN) support. A drop of about 5 % was observed using PDMS. The most commonly used polymer for a gutter layer is the PDMS, as it is rubbery and is reasonably stable. To increase the long-term stability cross-linking PDMS polymer is most frequently used. The physical aging issue of PTMSP can also be solved by cross-linking.

2.3 Selective layer

In the separation of gases, many contrasting polymers have been studied in the last decades in the form of dense, thick membranes (thickness~50-150 μm) like polyamides, polyimides, hydrophilic polymers, and block copolymers with copious amino groups. There are many advantages of using a thin-film composite membrane and having commercial applications [69]. A thin film's superior gas transport properties can be lost by physical aging in a short period [70]. Poor contact in a thick membrane at the interface of inorganic filler/polymer causes the interfacial defects to bring enormous challenges in maintaining gas selectivity. This layer offers the function of separation and is the most crucial part of the composite membrane. A well-known route is the mixed matrix membrane to enhance the properties of the polymeric membrane. Mixed Matrix Membranes are fabricated by inorganic filler which will be incorporated into a polymeric matrix. The material used should be easily constructed in to different membrane modules, which founds to be more competitive with the methods that were already reported for separation and efficiency in cost. The main criteria for the preference of such layer of a composite membrane are: (i) high gas permeability, (ii) high selectivity of CO_2, (iii) resistance to aging, (iv) resistance to CO_2 plasticization, (v) good thermal and chemical stability.

Mechanism of gas transport can be divided into two types: one is facilitated transport of gases and the other one is solution-diffusion mechanism. There are no chemical reactions involved in the gas transport mechanism. Both solubility and diffusion of the gas contribute to the permeability of the gas in the solution diffusion method. Such a membrane type is usually engaged in the trade-off between selectivity and gas permeability. [71]. The transport mechanism is another mechanism involved, where a reversible reaction between functional groups in the membrane and CO_2 occurs.

Comparing these two membranes with relatively low driving force indicates that facilitated membranes have higher selectivity and permeability. Observed pressure-dependent CO_2 permeability is another characteristic. At elevated partial pressure of CO_2, there is a chance that the mobile carriers will be saturated and because of this, mobile carriers have the probability to lose their transport properties, including selectivity and permeability of CO_2 over other gases. The most commonly used facilitated transport carriers are the amino groups which includes primary amine, secondary amine, or any sterically hindered amine). Polyvinyl Amine (PVAm) is the most widely and intensively explored polymeric material. Such polymeric material is reactive reversibly and facilitates CO_2 transport where amino moieties are explained as 'fixed site carriers' [72,73]. Such transport includes the reaction of CO_2 with water, promoted by the functional group present, and then convey rapidly in HCO_3^- ions.

In contrast, gases like CH_4, H_2, and N_2, i.e., the non-reactive gases, will exclusively transferred via the membranes through the solution diffusion mechanism. Thus, transport and high selectivity of CO_2 are more significant as compared to the other gases. There are also reports where this transport affects the transport of ions such as potassium (K^+), calcium (Ca^{2+}), carboxylate group (COO^-), and carbonate group (CO_3^{2-}) [74-76]. CO_2 transport in membranes has also been reported to facilitate by mimic enzymes with a metal activation center [77,78].

2.4   Protective layer

In a multilayer composite selective layer can be coated with a protective layer if needed. The primary function of such a layer is to improve the selectivity by plug the minor defects in the membranes and protecting a distinct soft layer (e.g., Pebax, polyactive). Such modification can protect the membrane from being defaced in the fabrication process as well as handling, mainly in the module membrane construction with high membrane packing density [79,80].

3.   Ionic liquid based membrane

In the past few years, these liquids are considered to be one of the most developing candidates for separation and capture of CO_2 owing to their structural features as they consist of anions and cations, and unique functional groups and have some exciting properties such as non-volatility, design ability, selectivity, and higher CO_2 solubility [81]. Exploration of ionic liquids (ILs) in CO_2 separation is based on their inherent structural tunability, good affinity with CO_2, and novelty, which led to unprecedented interest from both industries and academics. ILs can capture the CO_2 efficiently released from saturated solvents due to the absence of volatility lead to a decrease in energy consumption and environmental concerns than the traditional one [82]. The IL-based CO_2 capture, in single/ multi stage, have the ability to reduce the total energy contrast to the well-known methyl diethanolamine (MDEA) process, including thermal energy and electricity, by 42.8 % and 66.04 %, respectively [83]. ILs can be tuned based on their structure, creating a degree of freedom for designing solvents with distinct properties. Thus, designable ILs are considered to be efficient for cost-effective and energy-efficient capture of CO_2. It can replace the nonvolatile solvent and circumventing solvents to go in the ejected gas in order to intercept the environmental pollution originated from the organic materials [84-86]. However, in comparison to the other organic solvents, tuned ILs are unique in nature [87]. The physicochemical properties of ILs can be tailored by many possible functional groups, anions, and cations, their CO_2 separation performances [88-90]. The structural framework for the capture of CO_2 by ILs is depicted in Fig. 4.

3.1 CO_2 selectivity in ionic liquids

Various impurities are found to be present in natural gas, biogas, and flue gas such as CO, H_2S, SO_2, O_2, CH_4, N_2, CO_2 and H_2. Therefore, the separation and purification of such gases are also crucial. It is of up most importance to explore the selectivity in comparison to other gases because the CO_2 solubility data is inadequate to examine the adsorbents separation performance [91].

Advanced Functional Membranes   Materials Research Forum LLC
Materials Research Foundations 120 (2022) 267-314   https://doi.org/10.21741/9781644901816-9

Anderson *et al.,* in their report, used 1-hexyl-3-methyl pyridinium bis(trifluoromethylsulfonyl)imide [Hmpy][Tf$_2$N], which indicate the different solubilities of gases. They have shown that the selectivity of SO$_2$, CO$_2$ is high compared to N$_2$, O$_2$, etc. [92]. The ideal gases selectivity in ILs can be determined using Henry's law under the same temperature for different gases [93]. Thus, selectivity for [Hmpy][Tf$_2$N] of CO$_2$/O$_2$ and CO$_2$/N$_2$ is higher than the CO$_2$/hydrocarbon selectivity because of the lower O$_2$ and N$_2$.

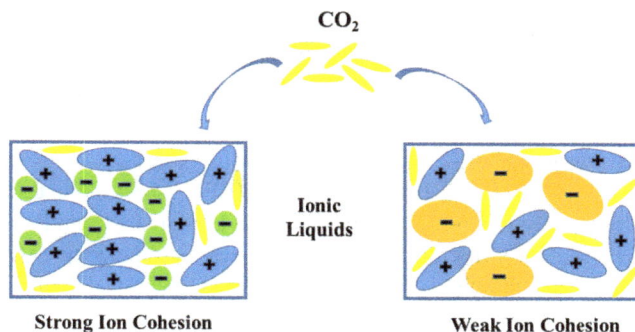

Figure 4 Schematic representation for the mechanism of CO$_2$ absorption in ILs.

3.1.1 CO$_2$/hydrocarbon selectivity

Zheng *et al.,* and Noble *et al.,* have delineated the CH$_4$ and CO$_2$ solubility in imidazolium-based traditional ILs i.e., [Bmim][Tf$_2$N], [Bmim][dca], [Bmim][NO$_3$], and [Bmim][BF$_4$] [94-98]. This shows the higher solubility of CO$_2$ as compared to the CH$_4$ in some of the ILs. Though the selectivity value is lower than 12 in case of CO$_2$/CH$_4$, still inadequate to take part with the traditional physical absorbents sulfonate and rectisol.

Ramdin *et al.,* studied the solubility as well as selectivity of CO$_2$/CH$_4$ in different ILs based on ammonium, pyrrolidinium, piperidinium, and phosphonium. Among all these ILs, selectivity showed for CO$_2$/CH$_4$ are 1-butyl-1-methylpyrrolidinium dicyanamide [Bmpyrr][dca] and 1-allyl-3-methylimidazolium dicyanamide [Amim][dca] as the CH$_4$ solubility is low. Compared with the imidazole-based ILs, selectivity in different ILs is not significantly improved due to the simultaneous increase of CO$_2$ and CH$_4$ solubility [99]. However, with an increase in the molecular weight and free volume of ILs, the selectivity of CO$_2$ is also increases. At the same time, the solubility of CH$_4$ is not dependent on the molecular weight of ILs and is mainly determined by the interaction between the two [100]. Thus, there is a quid-pro-quo between the CO$_2$ and CO$_2$/CH$_4$ for their solubility and selectivity.

3.1.2 Selectivity of CO_2/Diatomic gas mixture

The solubilities of different gases, viz. H_2, CO, N_2, O_2 in ILs were found to be lowest for H_2 and highest for CO_2, which indicates the selectivity towards CO_2.

Jacquemin *et al.,* under different conditions, determined the solubilities of various gases in [Hmim][PF_6] and found some dissimilarity in their solubility with other reported works [101]. Despite such discrepancies, these gases with low solubilities in ILs also have followed the similar trend as cited above. Due to the presence of free volume, which is shown by [PF_6] anion, these above cited gases are better soluble in [Bmim][PF_6] than those in [Bmim][BF_4], directing to higher selectivity of CO_2/CO, CO_2/N_2, and CO_2/O_2 in [Bmim][PF_6] [102].

Finotello and his team mainly focused on the solubility of gases like CO_2, CH_4, H_2, and N_2 as well as the selectivity of CO_2 in ILs such as [Emim][BF_4], 1-hexyl-3-methylimidazolium bis(trifluoromethylsulfonyl)imide ([Hmim]-[Tf_2N]), 1,3-dimethyl imidazolium methyl sulfate ([mmim][$MeSO_4$]) and, [Emim][Tf_2N] at divergent temperatures. According to their studies, the selectivity of gas pairs is directly related to the smaller molar volumes of ILs [96].

Shi *et al.,* in their report, confirmed that the molar volumes of the ILs plays a major role in the solubility of H_2, established with the help of experimental as well as theoretical studies. According to their report, the use of ILs with small molar volumes such as 1-ethyl-3-methylimidazolium acetate ([Emim][Ac]) and intense interaction with CO_2 to attain high CO_2 selectivity and solubility. The result of H_2 solubility in [Emim][Ac] is six times smaller as compared to [Hmim][Tf_2N] which was confirmed by theoretical studies. The main reason behind the low solubility of H_2 in [Emim][Ac] is the smaller volume of IL and the weak interaction between IL and H_2 [103].

4.   Metal-organic framework (MOF)

MOFs, due to their versatileness, are the potential candidates to be implemented in real-time applications for carbon capture. Omar Yaghi was the first who introduced MOF to the scientific community in the 1990s, after that MOF becomes the most important material that has gained momentum being cited in renowned journals and gains countless patents with applications in variable fields [104-107]. Metal-Organic Frameworks are framed by linking of metal ions through some organic linkers forming 1d-, 2d-, 3d-frameworks with appealing features. The basic MOF framework along with its synthetic procedure is displayed in Fig. 5. The most advantageous feature about the MOFs is that they can be engineered to have a desirable shape with ultra-high porosity, different pore sizes, surface area, and they can be tuned at ease for any application [108]. Due to the above-mentioned

Advanced Functional Membranes   Materials Research Forum LLC
Materials Research Foundations **120** (2022) 267-314   https://doi.org/10.21741/9781644901816-9

advantages, MOFs becomes the potential candidate for variable applications viz. gas storage [109,110], separation of gases [111,112], catalysis [113], sensing [114], drug delivery [115], semiconductor [116], photovoltaic device fabrication [117], etc.

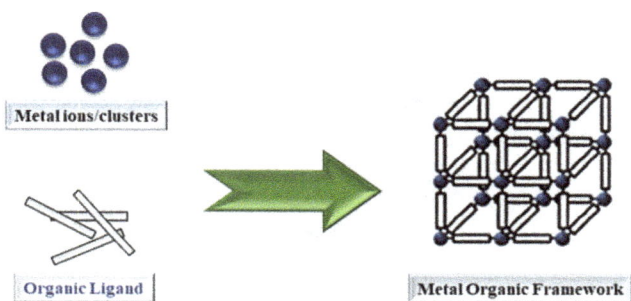

Figure 5 General Synthesis Scheme for Metal-Organic Framework.

Concentrating on process, Carbon Capture and Sequestration (CCS), due to the highly tunable nature of Metal organic frameworks, they become the most appealing candidate, because MOFs can be specifically fabricated to capture CO_2 at variable conditions with high selectivity. MOFs, in comparison to other porous materials, are superior as MOFs structure can be tuned easily to fulfill the desired application, whereas other traditional porous materials such as zeolites have limited structure motifs. Therefore, logical design [118], theoretical model development [119-122], and synthesis schemes with high throughput instrumentation [123-126] are the prime focus of several evaluations to predict systematic properties and characteristics for CCS and other applications as well.

4.1   MOF membranes

Membranes are the potential candidate that can be utilized in real-time applications because of their most advantageous features. The most appealing feature is that the membranes does not have any moving parts and the membranes can be utilized for continuous separation process. Membranes can be employed for solubility, diffusion, and adsorption of analytes for separation. There are several materials used for membrane fabrication which are explored for carbon capture application that includes both porous as well as non-porous structure [127]. The porous/non-porous membranes incorporated with polymeric matrices either discrete or on a support comprising of a combination of dissimilar matter, which are often denoted as hybrid membranes. These hybrid membranes can be synthesized with a couple of polymers mixed or phase-separated, can also include the filler particles.

The shift of powder MOFs to membranes [128] or thin films [129] has magnetized the attention of scientific community for variable utilization which includes Carbon Capture and Sequestration (CCS) [127,130-132], liquid separations [133-136], gas separations [137-140], and sensor related devices [141-143]. The emphasis of the aforementioned applications entails harvesting the characteristics of a metal organic framework to fabricate the membrane/device and examined their execution for real time applications. The tailored MOF membranes for carbon capture are envisaged to have a higher rate of selectivity for CO_2 over N_2 in high flux. Such characteristics are found to be difficult to achieved in capturing in the post-combustion conditions; the MOF membranes should be highly selective to take out at lower concentration of CO_2 in the flue gas. Moreover, when flux flow rate is large, a large surface area is required to perform the separation application. Moreover, specific designs of module and packing of membranes occurred as challenge to scientific community for carbon capture specifics [144].

4.2   Performance of MOF membranes

The implementation of gas separation membranes is governed by three factors: penetrability, permeation, and selectivity [127,145-148]. Penetration or permeability can be defined as the ability of fluids to pass through a porous material and is the change in concentration value of different gases that can be separated in between both the sides of membranes, and it is always reported by Barrer units (10^{-6} cm^3 (STP)/cm^2 s cm Hg). With utilization of chemical modification technique in membrane moieties, the permeability can be changed. Moreover, the porous/non-porous membranes can be compared with the help of permeability values. Concerning the solution-diffusion model, solubility (S) and diffusivity (D) factors are employed to calculate the permeability (P) by the equation,

$$P = S \times D$$

Solubility can be defined as the value which indicates the extent of a gas that solubilizes in the matrix of membrane, whereas the diffusivity of gas is directly associated to the volume that are free for polymer and gas particles of a gas stream. Permeation or permeance can be related to the coatings of membrane on flat surface or on the hollow fiber sheets where the important aspect is the width of the respective sheet. Permeance is also considered to compare membrane performance and is always denoted in gas permeation units (GPU). Other units are also present which are widely used. The conversion relationship between different units are shown in Table 1. The gas permeation unit is related to Barrer units as Barrer/Thickness. The separation factor/selectivity is defined as the correlation between more and the less permeable gas, that can be evaluated by using ideal gas permeabilities of two-variable gases (say A and B),

$$\alpha A/B = PA/PB$$

Advanced Functional Membranes   Materials Research Forum LLC
Materials Research Foundations **120** (2022) 267-314   https://doi.org/10.21741/9781644901816-9

Moreover, for a mixture of gas, the selectivity can be evaluated by calculating the modified equation-

$$\alpha A/B = (X_i/X_j)_{permeate}/(X_iX_j)_{feed}$$

MTR (Membrane Technology and Research) Center's PolarisTM is the most widely utilized membrane for post-combustion carbon capture [144,149]. Polaris's membrane is a thin polymeric membrane with high porosity designed to support high carbon capture. In their reports, MTR mentioned that their membrane captures 10 times higher than other membranes that are available commercially, the membranes show permeance of CO_2 of 1000-2000 GPU at 23 °C and a CO_2/N_2 selectivity of 50-60 [144]. In addition to this, MTR has increased their membrane surface area (up to 1 meter) and achieved a steady carbon capture, conducted at a pilot plant that emits huge amount of CO_2 (~1 ton per day) [149]. Merkel *et al.*, have demonstrated that when selectivity of CO_2/N_2 is more than 30, membrane permeance is the most prominent factor for reduction of capture cost, more membrane permeance, less will be the area of membrane required which simultaneously leads to reduce the capture cost [144].

Table 1 Relation between different Gas Permeation Units

	(GPU) $10^{-6}cm^3$ (STP) $cm^{-2}s^{-1}cmHg^{-1}$	$10^{-7}cm^3$ (STP) $cm^{-2}s^{-1}kPa^{-1}$	$10^{-10}mol$ $m^{-2}s^{-1}Pa^{-1}$	$10^{-3}m^3$(STP) $m^{-2}h^{-1}bar^{-1}$
(GPU) $10^{-6}cm^3$ (STP) $cm^{-2}s^{-1}cmHg^{-1}$	1	7.50	3.35	2.70
$10^{-7}cm^3$ (STP) $cm^{-2}s^{-1}kPa^{-1}$	0.133	1	0.447	0.360
$10^{-10}mol$ $m^{-2}s^{-1}Pa^{-1}$	0.299	2.24	1	0.806
$10^{-3}m^3$(STP) $m^{-2}h^{-1}bar^{-1}$	0.365	2.78	1.24	1

4.3   Design of MOF membranes

With the introduction of customized MOF for CCS into CO_2-permeable polymers, hybrid material can be created that will show high performance which reflects the properties of both. So, with the evolution of this, researchers are quite confident of mixing the organic polymer with inorganic MOF to fabricate a membrane which has the ability to cross the Roberson upper bound for CO_2/N_2 [150]. The diagram below (Fig. 6) will demonstrate the

CO_2 capture by MOF membranes. The blending of porous filler moiety and organic polymers to produce hybrid material is also called mixed matrix membrane (MMM). MMMs are fabricated by utilizing varying procedures such as direct mixing [134,151], and in-situ methods for growing [152,153]. But the blending of polymer with MOF are challenging as the two layers interface are often not compatible with each other. Koros *et al.,* [154] have demonstrated five non-ideal interfaces that are often observed during the fabrication of MMM: (1) stiffen of the organic polymer around the MOF; (2) large voids associated with MOFs within the MOF/polymer dispersion; (3) tiny gaps; (4) clogged MOF pore; (5) combination of all the above due to MOF particulate slows the gas permeability. For a membrane to overpass the limit of Robeson upper bound, the membrane should be fabricated with no or zero defects and should avoid the five interface cases. In this context, computational models are frequently employed to know the insight of the interface and compatibility of MOF and polymer [155,156]. Various instruments are also employed to show the interfacial defects such as Raman Spectroscopy, etc. [131].

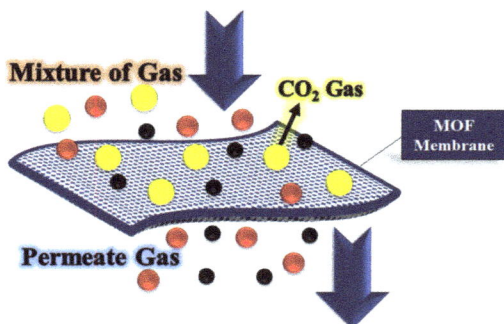

Figure 6 Demonstration of CO_2 capture by MOF Membranes.

Universally there are few reports present till date to understand fully the interaction between the polymer and MOF composite. Moreover, all the MMM systems have different characteristics, due to which the idea to create membranes with no defects may not work universally for all the systems. Due to which a large number of fabrication techniques that involve chemical modification to both the MOF and the polymers are presented by different research groups to get a homogenous membrane where both the moieties mixed perfectly. However, post-synthetic adjustment of the MOFs is quite useful to enhance the interactions between both the moieties [151] which can lead to enhanced performance. There are other modification techniques present for the development of MOF, especially for CCS, which includes MOF particle priming [157], polymer backbone modification [152], and reactive seeding of MOF precursors [139]. To get more detailed background of fabrication technologies of membranes, we hereby direct the reader to the references [145,158,159].

Asymmetric MOF membranes, in compared to the conventional MMMs, are also becoming potential separation membranes. Here, MOFs for thin film are deposited onto polymeric or ceramic supports. We hereby requested the readers to the references [135,137,140,160-165] to gain insight into different processes which are utilized to prepare asymmetric MOF membranes. There are some downfalls while fabricating asymmetric membranes on supports like ceramic (alumina), as this process can be discovered as more costly and mechanically weak. While the fabrication of asymmetric MOF membranes on polymer support is quite easy and is cost-efficient. However, controlling the MOF and polymer interface can be challenging. In this context, a continuous flow synthesis method has been developed which has the potential to reduce the steps of the MOF preparation, which simultaneously reduces the cost also. This method was first implemented on scale-up MOF synthesis materials [166-168], and then it is applied in membrane processing [169,170]. Marti *et al.,* [171] have illustrated the continuous flow process in the ZIF-8 MOF synthesis on the facade of Torlon Hollow fibers. Torlon is referred to as upper-level support, as it is mainly composed of polyimide which can resist the organic solvents and is stable at high pressures. ZIF-8 precursors are allowed to flow through the shell or the bore of the fiber with known concentration. The rate of flow generates fine membrane surface of MOF on the outer side of support. These membranes can show a high selectivity of CO_2/N_2 of 52, and 22 GPU high permeance. Comparing this work to others, it shows more efficient performance and strictly follows the green chemistry protocols. This is an example for many researchers to make such membranes that have application in various fields and can be utilized in CCS technology for higher performance values.

4.4   MOF membranes for CCS

Nowadays, the membrane separation technique emerges as a special separation method that can be used in CCS. It has many advantages as compared to other methods which include low cost, highly energy-efficient, recycling, and ease of operating. The insight mechanism of the separation method is based on molecular sieving towards the gas molecules. As the characteristics of the MOFs can be tuned easily to direct size-selective sieving effects, membranes of MOFs are evolved as the highly promising candidate for carbon capture, gas separation, and separation processes. The efficiency of the membranes of MOF towards carbon capture can be enhanced by inclusion of different functionalities to guide their variable interactions [172] and the post-synthetic functionalization can also be found as important to get improved carbon capture [172].

Lai *et al.*, have synthesized the first MOF membrane [174]. Zhu and Qui have synthesized the HKUST-1 MOF-membrane in a support of oxidized copper nets, which shows a good selectivity of H_2 gas over the other gases such as CH_4, CO_2, N_2, etc. as H_2 molecules due to their small size in nature, can easily permeate through the membrane. The same thing is also confirmed by one more group [175]. Ben and his team have synthesized a stainless-steel net/ PMMA-PMAA-supported HKUST-1 membrane. The same is also utilized in H_2/CO_2 separations from gases [176].

Zeolitic Imidazolate Frameworks (ZIF), which is nothing but a sub-branch of Metal Organic Framework, appeared as a potential moiety in the construction of membrane for the application in gas/hydrocarbon separation, which has a superior thermal as well as chemical stability and has a decent pore size of around 0.3-0.5 nm [177]. To date, different ZIF membranes are synthesized which have the potential for gas separation [178]. Caro and his team were the initial group to develop a ZIF-8 membrane and have showed Knudsen selectivity for H_2/CO_2 separation [179]. The same group has studied the membrane of ZIF-7 for selectivity and has concluded that they have prepared one novel rigid ZIF-7 membrane having nearby ~3 Å pore size [180]. A clear separation of both the gases is observed when H_2/CO_2 are in equal proportion, which are able to pass through the membrane of ZIF-7 [181]. The molecular sieving effect for the aforementioned system gives the separation factor (6.5) and gives a high value of H_2/CO_2 selectivity (6.7), higher than that of 4.7 (Knudsen separation factors). Moreover, they tried to upgrade the gas separation, H_2/CO_2 by a factor of 8.4 where the mixture contains equimolar amount of H_2/CO_2 at a temperature of 200 °C by changing the synthesis procedure. Additionally, they proved the separation of H_2/CO_2 gases through the ZIF-7 membrane can be varied with variable temperatures [182]. The results of the experiment were that the H_2 gas permeance quickly with an increase in temperature, but the permeance of CO_2 remains constant, which leads to the enhancement of separation factor from 5.4 (50 °C) to 13.6 (220 °C). Contrastingly, membrane of ZIF-22 which has the similar to ZIF-7 in case of pore size (~3.0 Å) exhibits more molecular sieving performance on separation of H_2/CO_2 and other gases as well [183]. ZIF-90 was often functionalized with different functionalities to attain selectivity. In comparison to the ZIF-90 membrane, post-synthetic functionalization of the same [86,184] by ethanolamine can significantly enhance separation factor of H_2/CO_2 and other gases as well to ~15.7. Later on, they synthesized APTES-functionalized ZIF-90 molecular sieving membranes that have improved separation of H_2/CO_2 (20) at 225 °C and 1 bar, which shows a high thermal and hydrothermal stability [185]. ZIF-90/Torlon membranes are obtained by Nair *et al.,* where membranes are formed with no defects and have a complete surface coverage, with a separation factor of 1.5 for CO_2/CH_4 and 3.5 for CO_2/N_2 [186].

MMMs, mixed matrix membranes, are nothing but the blending of MOF moieties with the polymer, has been explored and it emerges as an alternative strategy to obtained MOF membranes. Musselman *et al.*, have successfully synthesized MMM where ZIF-8 is the filler phase and polymer used is Matrimid®. When the ZIF-8 loading increased to 40% (w/w), the permeability also continuously increases for H_2, N_2, CO_2, O_2, C_3H_8, CH_4. Moreover, higher loading ratios up to 60% (w/w) of ZIF-8 into polymer Matrimid® enhances the selectivity for many gases. Matrimid® alone show a 2.58 factor for selectivity in a 1:1 H_2/CO_2 mixture, that increases to a factor of 7.01 at 60 % loading of ZIF-8. Matrimid® alone shows selectivity of 42, which was further enhanced to 89 by a 50 % loading ratio of ZIF-8. Mixed matrix membranes framed with PVAc (Polyvinyl-Acetate) and Cu-TPA framework, show higher selectivity in case of CO_2/N_2 as well as CO_2/CH_4 separations [187]. Car *et al.*, have synthesized the HKUST-1/PDMS and HKUST-1/PSf membranes with 10% (w/w) loading of HKUST-1 enhances the selectivity for CO_2/N_2 and CO_2/CH_4.

5.   Water facilitated mixed matrix membrane for CO_2 capture

The concept of mixed matrix membrane (MMM) emerges from the inclination of mixing of polymers and molecular sieves which are implemented for gas separation. The concept of MMM was enlarged in 2011 [188], and in the same time, organic/inorganic fillers are developed simultaneously. Mixed Matrix Membranes are elucidated as the inclusion of disseminate part onto a steady polymer part, to merge and produce new benefits. For Water Facilitated CO_2 capture (WFCC) MMMs, the polymer employed should be an ordinary polymer or high-performance hydrogel or the packing should be in such a way that it can furnish more transport highways for carbon dioxide or can enhance the further profits of the membrane. Here, the minimum criteria are that at least one part should show a good performance in the hydrated state.

Initial category of WFCC MMM was framed by using one inert inorganic filler and a hydrophilic polymer. Here "inert" refers to the filler that does not have any WFCC property. The thermal and mechanical stability of the filler is important so the filler is mainly composed of inorganic moieties [189-192]. The role that the polymer matrix does will be followed by the water (support) in this context. But the consequence of filler is complex. Xing *et al.*, [189] have submerged fumed silica into the carrier containing the PVA-POS network. While increasing the loading ratios of fumed silica from 4 wt.% to 17 wt.%, the permeability of CO_2 increases along with the enhancement of selectivity of CO_2/H_2 from 65 to 87. From the above results, they concluded that the fused silicas retarded the loading of the polymer chains, which is in accordance with reverse selectivity [193,194]. Deng *et al.,* have demonstrated the carbon nanotubes (CNTs) in PVAm-PVA mixture showed two functions, viz., at lower pressure, CNTs slightly enhances the swelling of membrane due to the nano-spacer function and at higher pressure, it increases the swelling of membrane up to a higher strength due to the resistance against compaction effect. In general, CNTs improved the snit-compaction property, increases the mechanical strength, and enhances the separation performance of CO_2.

The second category of WFCC MMM was framed by using one active filler and an ordinary polymer. Here, the active filler can absorb a high quantity of water and can permeate more CO_2 from it. Liu *et al.,* [195] have prepared for the first time the polyzwitterion@CNT (polyzwitterion coated CNTs), which was further coated in Matrimid® 5218. Here the filler shows to enhance the water uptake capacity of the membrane which was further attributed by strong hydrophilicity of polyzwitterion. Whereas Li *et al.,* [196] have incorporated poly(N-isopropyl acrylamide) nanohydrogels in the aforesaid matrix. The nanohydrogels inserted in the Matrimid® matrix act as pool for water which provides water for CO_2 dissolution and also gives pathways for CO_2 transport. The aforementioned Matrimid®/nanohydrogel mixed matrix membranes show an enhance in CO_2/N_2 and CO_2/CH_4 selectivity of 52 and 61, with CO_2 penetrability of 278 Barrer, which is much better than previous MMM that shows a selectivity factor of 36 for CO_2/CH_4 and 103 Barrer CO_2 penetrability. The reason behind the enhanced property is that the nanohydrogels are filled with water and are highly swollen, and main effective hydration layer was found at the CNT surface. The idea of blending nanohydrogels with a commercial polymer seems interesting because of the commercialization prospects as well as the mass transfer model.

The third category of WFCC MMM was framed by using an active filler and a hydrophilic polymer. Here the no mobile carriers are present on the polymer. The mobile carriers are provided by the active fillers to the inner channels. The active fillers also have roles in the structure modifier. Dry MMMs are prepared successfully and the hydrated MMMs should also include inner channels. Few researches have reported the microporous filler-based hydrated MMMs that show better separation of CO_2. Wu *et al.,* have disseminate the functionalized poly-ethylenimine with MCM-41 (PEI-MCM-41) into Pebax 1657. Poly-ethylenimine were present in the channels of pore as well as on the MCM-41 surface, which provides amine carriers for transport of CO_2. Comparing with the MCM-41, modified version of MCM-41, viz., PEI-MCM-41, results in more stiffness in the polymer chain at the polymer-packing interface. By referring mesoporous fillers-based morphological structure of MMMs [197], such interfacial structure enhances the selectivity and permeability. In the same context, Xin *et al.,* by using the facile vacuum-assisted method, have dispersed PEI into a mesoporous MIL-101(Cr), and then incorporate the obtained compound, i.e., PEI@MIL-101(Cr) into SPEEK. For the advantageous feature of SPEEK like high mechanical strength, the loading ratio of 40 can be realized. PEI@MIL-101(Cr)-doped membranes show a 2490 Barrer for the CO_2 permeability, which is the highest as compared to the pristine SPEEK membrane by a 4.6-fold. In comparison to the pristine MIL-101(Cr), permeability, and selectivity of gases such as CO_2/CH_4 and CO_2/N_2 of PEI@MIL-101(Cr)-doped membranes show improvement by 62 %, 128 %, and 102 %. She *et al.,* [198] has also tried by adding sulfonated MIL-101(Cr) to SPEEK. They have found that separation data of CO_2 with comparison to that of the PEI@MIL-101(Cr) shows a little lower. The inclusion of sulfonated MIL-101(Cr) in membrane enhances the solubility of CO_2 which was further accounted for the sorption enhancement effect of adsorbed water. Moreover, new experiments were exploited by using TiO_2, a non-porous filler, and found that despite the surface have amines in it, still it shows the highest CO_2 permeability of 1629 Barrer only [199]. These results directly demonstrate the advantages provided by mesoporous fillers and incorporating such fillers in the hydrophilic matrix. So, by taking this into account, various researches have also been done with microcapsules as filler, due to their hollow structure, it can construct even more CO_2 transport pathways as compared to mesoporous fillers [200].

By considering the above context, PVAm based MMMs are also exploited. Zhao *et al.,* have demonstrated the disruption of the PVAm chain packing by polyaniline nanosheets and can furnish few weaker carriers, which simultaneously show a much higher CO_2/N_2 selectivity of 120 along with higher CO_2 permeance, 1200 GPU. The advantage of this is that, by decreasing the wet coating thickness of polyaniline nanorods, one can easily increase the CO_2/N_2 selectivity and permeance of CO_2 up to 240 and 3080 GPU [201]. Liao *et al.,* have successfully demonstrated an HT (hydrotalcite) channel in PVAm [202]. Here, the flexible carbonate acted as a semi-mobile carrier and shows a CO_2/N_2 selectivity and permeance of CO_2 by 296 and 3187 GPU respectively.

6.   Hollow fiber membranes

Spinning is a process where polymer can be converted to fiber. In hollow fiber (HF) fabrication technique, there is the continuous production of single as well as multiple fiber by means of spinneret followed by returning to the solid state through solidification process [203]. The varying properties of fiber form during this fabrication process depends on several parameters. There are two common methods used in spinning process, solution spinning and melt spinning [204]. The common features used in both of these methods is the spinneret through where a polymer sample solution as well as solvent were extruded. After extruding, the polymer sample retain a hollow cylindrical shape through the solidification process generally called as the phase inversion. Melt spinning process is used for those polymer materials which can be melt easily. The process includes, in an extruder, heating of the polymer flakes or pellets until it melts completely which later pumps the liquid polymer across a spinneret to form liquid spinning dope. The filament in the spinneret is cooled to undergo solidification and with the help of a godet roller, final fiber velocity can be fixed [203,205]. Finally, for the storage, the fiber is wounded on a spool like a sewing thread [206].

On the other hand, solution spinning process [207] can be classified in two categories, wet and dry-wet spinning process. Wet spinning process consist of the methods where the dope solutions were ejected to an external bath that contains a non-solvent liquid which on later exchange themselves, referred to as phase inversion, that leads to the complete dismissal of solvent which is further followed by solidification/precipitation of the fiber. In solidification process, sometimes due to the improper mass transfer through the nonsolvent-dope interface, several voids as well as cross-sectional irregularities were observed which can be reduced by applying the air gap between the spinneret and end of the coagulation bath. This above discussed is called as dry-wet process [206]. Advantages of a Hollow fiber membrane over a flat membrane are that they have self-supporting properties which help them in reducing the complexity in fabrication of the component and the higher productivity, that can be related to its high area of surface with excessive packing density [208,209,210]. Different types of parameters that cause instability during the spinning of HF are draw resonance, cross-section, necking, and irregularity in fabrication of the fibers.

Membrane contactor, as the name suggests, is a mechanical device consisting of a membrane that employs separation or chemical changes between two phases using selective mass transfer. For example, membrane present in gas-liquid contactor is used to separate two phases of gas and liquid respectively, providing sufficient contact with two phases without mixing them directly. The solubility of the components in the liquid phase provides the selectivity of the membrane and for this reason, most gas-liquid contactors are employed with porous membranes with high mass transfer ability. In some cases, the composite or asymmetric membrane with thin non-porous polymer material is also used in the high-pressure separation process [211,212]. To minimize the resistance during mass transfer, a porous hydrophobic membrane is used which allowed the gas phase to enter the pore instead of penetration of liquid as shown in the Fig. 7. In the membrane separation technique, the contact area plays a vital role that can be illustrated as the area of the membrane pore mouths. The operating pressure in systems must be controlled to avoid the intermixing between gas and liquid phase. Especially when the operating pressure is not equal for both phases, gas bubbles dispersion may arise which affects the rate of mass transfer [213]. The mass transfer ability of a porous membrane is also affected when the liquid layer is immobilized in the pore. Wang *et al.,* [214] suggest that a 20 % decrease of

Advanced Functional Membranes
Materials Research Foundations **120** (2022) 267-314

Materials Research Forum LLC
https://doi.org/10.21741/9781644901816-9

mass transfer coefficient is observed when membrane pore is 5 % wetting, while it increases up to 6 % when 2 % wetting of membrane pores.

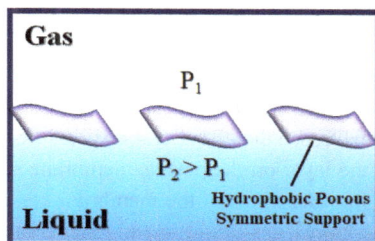

Figure 7 Mass transfer in a membrane contactor: symmetric hydrophobic porous support.

If the pressure of the liquid phase surpasses a critical level, the non-wetted mode can't be provided by the membrane's hydrophobicity [215,216,217]. Therefore, pressure plays a crucial role in the mass transfer process and is dependent on the surface tension and also on the contact angle at stated operating condition. Using Laplace's equation, pressure can be easily quantified-

$$\Delta P = \frac{2\sigma Cos\theta}{r_{p,max}}$$

where, ΔP is the critical transmembrane pressure, $r_{p,max}$ is maximum pore radius, σ is the surface tension of the liquid, and θ is the contact angle.

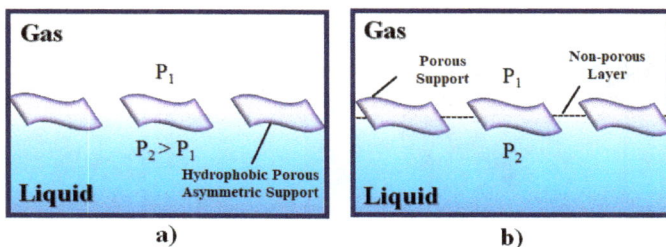

Figure 8 Mass transfer in a membrane contactor: a) Hydrophobic Porous Asymmetric Support.; b) Thin non-porous layer present in asymmetric composite support.

Advanced Functional Membranes   Materials Research Forum LLC
Materials Research Foundations **120** (2022) 267-314   https://doi.org/10.21741/9781644901816-9

Asymmetrical membranes are those where pore size is different for opposite sides. In that case, without dispersion also it is possible to contact two phases from one phase to another even when operating pressure exceed from the larger side pore. As a result, partial wetting takes place from the side of large pore and gas-liquid interface appears inside the pore space. To increase the operating pressure of the contactor, composite membrane with polymer support is generally used which prevent the perforation of the liquid phase into membrane pore space. It can be seen that the operating pressure is further increase by using non-porous layer and laid the layer on the porous membrane surface (Fig. 8), however it must be highly permeable in order to avoid the membrane resistance to the mass transfer process [218-220]. Furthermore, gas-liquid membrane contactor can be utilized for absorption or desorption of the respective moiety in liquid absorbent (Fig. 9). For that hollow fiber membrane, be a promising candidate since it provides large surface area per apparatus volume, thus making the equipment compact.

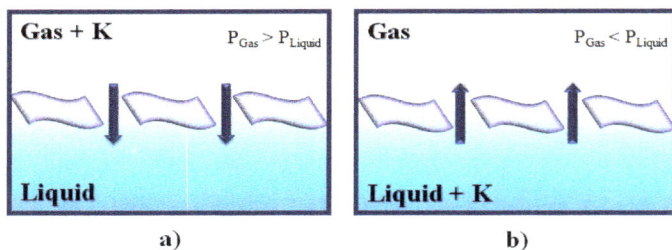

Figure 9 Mass transfer in a membrane contactor: a) Component absorbed from gaseous phase.; b) Component desorbed from the liquid phase.

In addition to all the above literature, the main properties of pre- and post-combustion CO_2 capture by different membranes are summarizes in Table 2.

Table 2 Division and Properties of Membranes for Pre- and Post-combustion CO_2 Capture.

Type of Membrane		Membrane Material	Driving Force	Selectivity	Permeation
Polymer		Glassy & Rubbery	Solution Diffusion	Selectivity towards CO_2 or H_2	Low
		Mixed Matrix			Moderate
Inorganic Porous	Meso (2-50 nm)	Amine Grafted Silicas	Adsorption/Diffusion	Selectivity towards CO_2	Low/Moderate
	Micro (<2 nm)	Amine functionalised Silicas	Adsorption/Diffusion	Selectivity towards CO_2	High
		Stabilized Silicas	Molecular Sieving/Activated Diffusion	Selectivity towards H_2	High
		Zeolites	Adsorption/Diffusion	Selectivity towards CO_2 or H_2	High
		MOFs/ZIFs	Adsorption/Diffusion	Selectivity towards CO_2 or H_2	Moderate/High
Hollow Fiber Membranes		Thin Polymeric Tubes	Phase Inversion	H_2/CO_2	Moderate

Summary

In the past few years, there has been a steady work going on based on polymer and their composite membranes and different conventional ILs based on blending and functionalization. The most widely used polymeric membrane for the separation of CO_2 is the facilitated material-based membranes and glassy rubbery block polymers (e.g., Pebax, Polyactive). The primary mechanism of using such polymeric material is based on solution-diffusion. To make such polymeric membranes more efficient and improve their performances, additives like nanoparticles, PEG, and amines with low molecular weight can be used. Great efforts are needed to industrialize gas separation using polymeric membranes. Some prospective to make these membranes efficient are developing high-performance membrane material, eco-friendly membrane fabrication processes, testing under ambient conditions, long-term stability, and durability of such membranes.

Along with polymers, ILs also played an essential role in the absorption and separation of CO_2. Novelty and structural tunability make ILs a valuable media to utilized in separation/capture of CO_2. The absorption and separation of CO_2 especially depend on permeability and selectivity, and the mechanism can be understood with experimental methods and molecular simulation. However, the ILs based membranes are still new and are yet to be explored to persuade industrial needs. Therefore, more researches are required to explore the field thoroughly. The main criteria for efficient CO_2 capture are understanding the systematic and comprehensive relationship between ILs and CO_2 and their physicochemical and absorption/desorption properties. Instead of rapid progress in this field, there is still a significant research gap. To overcome such difficulties, the main possible thing is to design and prepare a composite membrane of ILs and porous material with high selectivity, permeability, and stability.

MOFs, due to their highly tunable nature, make them a potential candidate for CCS both as sorbents or membranes. MOFs due to their higher porosity and surface area, with tremendous choices of metals and ligands, show potential application in CO_2 capture and separation technology. Various researches reveal that the chemical nature of the MOFs can be changed on the surface, which makes them advantageous compared to other porous entities. Selectivity, Capacity, and regeneration are the most important factors that play role in the CCS process. However, investigations of these materials are ongoing, still many different processes and creative synthesis approaches are appearing to date. For this, the blending of both the chemical design and engineering strategies is still an appealing feature that can be implemented in real-time analysis. Within this chapter, we have discussed the importance of MOF membranes for CCS technology, different parameters, and how the appealing material can be fabricated/designed perfectly to use in the real world to mitigate the CO_2 from the environment. In addition, we have included one new process called continuous flow.

The capture of CO_2 efficiently in the presence of moisture by solid medium is relatively new content, yet it is of up most necessary for real time applications. Various impurities are present in adsorbents of CO_2 capture where water is quite known for the most unexpected impurity. In the same context, water also enables facilitated transport in some CO_2 capture membranes. So, the use of water in CO_2 capture material is worthy, moreover it gives the opportunity to many researchers to explore the WFCC materials. Various strategies are proposed to avoid the negative effects of water, but in the same time, WFCC membranes are widely explored. Among the various materials present till date, rigid framework materials are the most appealing candidate for excellent WFCC membranes/adsorbents. A well-established WFCC membranes have the ability to integrate hierarchical structures with different transport mechanism.

Advanced Functional Membranes
Materials Research Foundations **120** (2022) 267-314

Materials Research Forum LLC
https://doi.org/10.21741/9781644901816-9

Acknowledgement

AK acknowledged Department of Science and Technology, Government of India for financial support under the DST-INSPIRE faculty scheme (Award No. DST/INSPIRE/04/2018/000445). Institute of Advanced Study in Science and Technology (IASST) Guwahati, Assam is acknowledged for all the supports provided as host institutions. AcSIR is acknowledged for Ph.D. guideship awarded to AK.

Reference

[1] R. Y. Cui, N. Hultman, M. R. Edwards, L. He, A. Sen, K. Surana, H. McJeon, G. Iyer, P. Patel, S. Yu, T. Nace, C. Shearer, Quantifying operational lifetimes for coal power plants under the Paris goals, Nat. Commun. 10 (2019), 4759. https://doi.org/10.1038/s41467-019-12618-3

[2] R. B. Mansour, M. A. Habib, O.E. Bamidele, M. Basha, N. A. A. Qasem, A. Peedikakkal, T. Laoui, M. Ali, Carbon capture by physical adsorption: materials, experimental investigations and numerical modeling and simulations–a review. Appl. Energy. 161 (2016), 225-255. https://doi.org/10.1016/j.apenergy.2015.10.011

[3] E. S. Sanz-Pérez, C. R. Murdock, S. A. Didas, C. W. Jones, Recent advances in multilayer composite polymeric membranes for CO_2 separation: A review. Green Energy & Environ. 1 (2016), 102-128. https://doi.org/10.1016/j.gee.2016.08.001

[4] D. D. Iarikov, P. Hacarlioglu, S. T. Oyama, Supported room temperature ionic liquid membranes for CO_2/CH_4 separation, Chem. Eng. Journal, 166 (2011), 401-406. https://doi.org/10.1016/j.cej.2010.10.060

[5] A. Ghoufi, L. Gaberova, J. Rouquerol, D. Vincent, P. L. Llewellyn, G. Maurin, Adsorption of CO_2, CH_4 and their binary mixture in Faujasite NaY: a combination of molecular simulations with gravimetry–manometry and microcalorimetry measurements, Microporous and Mesoporous Mater. 119 (2009), 117-128. https://doi.org/10.1016/j.micromeso.2008.10.014

[6] G. George, N. Bhoria, S. AlHallaq, A. Abdala, V. Mittal, Polymer membranes for acid gas removal from natural gas, Sep. Purif. Technol. 158 (2016), 333-356. https://doi.org/10.1016/j.seppur.2015.12.033

[7] K. Huang, J.-Y. Zhang, X.-B. Hu, Y.-T. Wu, Absorption of H_2S and CO_2 in aqueous solutions of tertiary-amine functionalized protic ionic liquids, Energy Fuels. 31 (2017), 14060-14069. https://doi.org/10.1021/acs.energyfuels.7b03049

[8] K. Huang, D. N. Cai, Y. L. Chen, Y. T. Wu, X. B. Hu, Z. B. Zhang, Dual lewis base functionalization of ionic liquids for highly efficient and selective capture of H_2S, Chem Plus Chem, 79 (2014), 241. https://doi.org/10.1002/cplu.201300365

[9] K. Huang, X. M. Zhang, X. B. Hu, Y. T. Wu, Hydrophobic protic ionic liquids tethered with tertiary amine group for highly efficient and selective absorption of H_2S from CO_2. Alchem Journal, 62 (2016), 4480-4490. https://doi.org/10.1002/aic.15363

[10] L. Riboldi, O. Bolland, Comprehensive analysis on the performance of an IGCC plant with a PSA process integrated for CO_2 capture. Int. J. Greenh. Gas Control. 43 (2015), 57-69. https://doi.org/10.1016/j.ijggc.2015.10.006

[11] J. T. Culp, Flexible solid sorbents for CO_2 capture and separation. Novel Materials for Carbon Dioxide Mitigation Technology, 2015, 149. https://doi.org/10.1016/B978-0-444-63259-3.00005-7

[12] H. Schultz, Climate change and viticulture: a European perspective on climatology, carbon dioxide and UV-B effects. Australian Journal of Grape and Wine Research, 6 (2000), 2-12. https://doi.org/10.1111/j.1755-0238.2000.tb00156.x

[13] L. Riboldi, O. Bolland, Evaluating Pressure Swing Adsorption as a CO_2 separation technique in coal-fired power plants. Int. J. Greenh. Gas Control. 39 (2015), 1-16. https://doi.org/10.1016/j.ijggc.2015.02.001

[14] L. Riboldi, O. Bolland, Pressure swing adsorption for coproduction of power and ultrapure H_2 in an IGCC plant with CO_2 capture. Int. J. Hydrog. Energy, 41(2016), 10646-10660. https://doi.org/10.1016/j.ijhydene.2016.04.089

[15] F. R. Abdeen, M. Mel, M. S. Jami, S. I. Ihsan, A. F. Ismail, A review of chemical absorption of carbon dioxide for biogas upgrading. Chin. J. Chem. Eng. 24 (2016), 693-702. https://doi.org/10.1016/j.cjche.2016.05.006

[16] D. P. Hanak, C. Biliyok, E. J. Anthony, V. Manovic, Modelling and comparison of calcium looping and chemical solvent scrubbing retrofits for CO_2 capture from coal-fired power plant. Int. J. Greenh. Gas Control. 42 (2015), 226-236. https://doi.org/10.1016/j.ijggc.2015.08.003

[17] D. P. Hanak, C. Biliyok, V. Manovic, Rate-based model development, validation and analysis of chilled ammonia process as an alternative CO_2 capture technology for coal-fired power plants. Int. J. Greenh. Gas Control, 34 (2015), 52-62. https://doi.org/10.1016/j.ijggc.2014.12.013

[18]   K. A. Mumford, Y. Wu, K. H. Smith, G. W. Stevens, Review of solvent based carbon-dioxide capture technologies. Front. Chem. Sci. Eng. 9 (2015), 125-141. https://doi.org/10.1007/s11705-015-1514-6

[19]   S. Mirzaei, A. Shamiri, M. K. Aroua, A review of different solvents, mass transfer, and hydrodynamics for post combustion CO_2 capture. Rev. Chem. Eng. 31 (2015), 521-561. https://doi.org/10.1515/revce-2014-0045

[20]   T. Mulukutla, J. Chau, D. Singh, G. Obuskovic, K.K. Sirkar, Novel membrane contactor for CO_2 removal from flue gas by temperature swing absorption. J. Membr. Sci. 493 (2015), 321-328. https://doi.org/10.1016/j.memsci.2015.06.039

[21]   S. Li, T.J. Pyrzynski, N.B. Klinghoffer, T. Tamale, Y. Zhong, J.L. Aderhold, B. Bikson, Scale-up of PEEK hollow fiber membrane contactor for post-combustion CO2 capture. J. Membr. Sci. 527 (2017), 92-101. https://doi.org/10.1016/j.memsci.2017.01.014

[22]   I. Sreedhar, R. Vaidhiswaran, B.M. Kamani, A. Venugopal, Process and engineering trends in membrane-based carbon capture. Renew. Sust. Energ. Rev. 68 (2017), 659-684. https://doi.org/10.1016/j.rser.2016.10.025

[23]   M. Li, X. Jiang, G. He, Application of membrane separation technology in postcombustion carbon dioxide capture process. Front. Chem. Sci. Eng. 8 (2014), 233-239. https://doi.org/10.1007/s11705-014-1408-z

[24]   P. Luis, B. Van der Bruggen, The role of membranes in post-combustion CO_2 capture. Greenh. Gases. 3 (2013), 318-337. https://doi.org/10.1002/ghg.1365

[25]   L. Li, N. Zhao, W. Wei, Y. Sun, A review of research progress on CO_2 capture, storage, and utilization. Fuel, 108 (2013), 112-130. https://doi.org/10.1016/j.fuel.2011.08.022

[26]   M. E. Boot-Handford, J. C. Abanades, E. J. Anthony, M. J. Blunt, S. Brandani, N. Mac Dowell, P.S. Fennell, Carbon capture and storage update. Energy Environ. Sci. 7 (2014), 130-189. https://doi.org/10.1039/C3EE42350F

[27]   G. Hu, N. J. Nicholas, K. H. Smith, K. A. Mumford, S. E. Kentish, G. W. Stevens, Carbon dioxide absorption into promoted potassium carbonate solutions: A review. Int. J. Greenh. Gas Control. 53 (2016), 28-40. https://doi.org/10.1016/j.ijggc.2016.07.020

[28]   B. Zhao, W. Tao, M. Zhong, Y. Su, G. Cui, Process, performance and modeling of CO2 capture by chemical absorption using high gravity: A review. Renew. Sust. Energ. Rev. 65 (2016), 44-56. https://doi.org/10.1016/j.rser.2016.06.059

[29]   Z. Liang, K. Fu, R. Idem, P. Tontiwachwuthikul, Review on current advances, future challenges and consideration issues for post-combustion CO_2 capture using amine-based absorbents. Chin. J. chem. Eng. 24 (2016), 278-288. https://doi.org/10.1016/j.cjche.2015.06.013

[30]   Z. Dai, L. Ansaloni, L. Deng, Recent advances in multilayer composite polymeric membranes for CO_2 separation: A review. Green Energy Environ. 1 (2016), 102-128. https://doi.org/10.1016/j.gee.2016.08.001

[31]   Y. S. Bae, K. L. Mulfort, H. Frost, P. Ryan, S. Punnathanam, L. J. Broadbelt, R. Q. Snurr, Separation of CO_2 from CH_4 using mixed-ligand metal-organic frameworks. Langmuir. 24 (2008), 8592-8598. https://doi.org/10.1021/la800555x

[32]   X. Q. Zhang, W. C. Li, A. H. Lu, Designed porous carbon materials for efficient CO_2 adsorption and separation. New Carbon Mater. 30 (2015), 481-501. https://doi.org/10.1016/S1872-5805(15)60203-7

[33]   M. Wiśniewski, S. Koter, A. P. Terzyk, J. Włoch, P. Kowalczyk, CO_2-Reinforced nanoporous carbon potential energy field during CO_2/CH_4 mixture adsorption. A comprehensive volumetric, in-situ IR, and thermodynamic insight. Carbon. 122 (2017), 185-193. https://doi.org/10.1016/j.carbon.2017.06.057

[34]   N. Jusoh, Y. F. Yeong, K. K. Lau, A. M. Shariff, Fabrication of silanated zeolite T/6FDA-durene composite membranes for CO_2/CH_4 separation. J. Clean. Prod. 166 (2017), 1043-1058. https://doi.org/10.1016/j.jclepro.2017.08.080

[35]   S. R. Venna, M. A. Carreon, Metal organic framework membranes for carbon dioxide separation. Chem. Eng. Sci. 124 (2015), 3-19. https://doi.org/10.1016/j.ces.2014.10.007

[36]   S. Kayal, A. Chakraborty, Activated carbon (type Maxsorb-III) and MIL-101 (Cr) metal organic framework based composite adsorbent for higher CH_4 storage and CO_2 capture. Chem. Eng. J. 334 (2018), 780-788. https://doi.org/10.1016/j.cej.2017.10.080

[37]   G. Bello, R. García, R. Arriagada, A. Sepulveda-Escribano, F. Rodrıguez-Reinoso, Carbon molecular sieves from Eucalyptus globulus charcoal. Microporous and mesoporous materials. 56 (2002), 139-145. https://doi.org/10.1016/S1387-1811(02)00465-1

[38]   R. Arriagada, G. Bello, R. García, F. Rodríguez-Reinoso, A. Sepúlveda-Escribano, Carbon molecular sieves from hardwood carbon pellets. The influence of carbonization temperature in gas separation properties. Microporous and mesoporous materials. 81 (2005), 161-167. https://doi.org/10.1016/j.micromeso.2005.02.005

[39]   M. Lutyński, P. Waszczuk, P. Słomski, J. Szczepański, CO_2 sorption of Pomeranian gas bearing shales–the effect of clay minerals. Energy Procedia. 125 (2017), 457-466. https://doi.org/10.1016/j.egypro.2017.08.153

[40]   F. Gholipour, M. Mofarahi, Adsorption equilibrium of methane and carbon dioxide on zeolite 13X: Experimental and thermodynamic modeling. The Journal of Supercritical Fluids. 111 (2016), 47-54. https://doi.org/10.1016/j.supflu.2016.01.008

[41]   X. Xu, J. Wang, A. Zhou, S. Dong, K. Shi, B. Li, D. O'Hare, High-efficiency CO_2 separation using hybrid LDH-polymer membranes. Nature comm. 12 (2021), 1-10. https://doi.org/10.1038/s41467-021-23121-z

[42]   A. Mukhtar, S. Saqib, N. B. Mellon, M. Babar, S. Rafiq, S. Ullah, M. Chawla, CO_2 capturing, thermo-kinetic principles, synthesis and amine functionalization of covalent organic polymers for CO_2 separation from natural gas: A review. J. Nat. Gas Sci. Eng. 77 (2020), 103203. https://doi.org/10.1016/j.jngse.2020.103203

[43]   Z. Dai, L. Ansaloni, L. Deng, Recent advances in multilayer composite polymeric membranes for CO_2 separation: A review. Green Energy Environ. 1 (2016), 102-128. https://doi.org/10.1016/j.gee.2016.08.001

[44]   Y. W. Jeon, D. H. Lee, Gas membranes for CO_2/CH_4 (biogas) separation: a review. Env. Eng. Sci. 32 (2015), 71-85. https://doi.org/10.1089/ees.2014.0413

[45]   X. Y. Chen, H. Vinh-Thang, A. A. Ramirez, D. Rodrigue, S. Kaliaguine, Membrane gas separation technologies for biogas upgrading. Rsc Adv. 5 (2015), 24399-24448. https://doi.org/10.1039/C5RA00666J

[46]   E. R. Minardi, S. Chakraborty, V. Calabro, S. Curcio, E. Drioli, Membrane applications for biogas production and purification processes: an overview on a smart alternative for process intensification. Rsc Adv. 5 (2015), 14156-14186. https://doi.org/10.1039/C4RA11819G

[47]   M. Scholz, T. Melin, M. Wessling, Transforming biogas into biomethane using membrane technology. Renew. Sust. Energ. Rev. 17 (2013), 199-212. https://doi.org/10.1016/j.rser.2012.08.009

[48]   S. Basu, A.L. Khan, A. Cano-Odena, C. Liu, I.F. Vankelecom, Membrane-based technologies for biogas separations. Chem. Soc. Rev. 39 (2010), 750-768. https://doi.org/10.1039/B817050A

[49]   J. K. Adewole, A. L. Ahmad, S. Ismail, C. P. Leo, Current challenges in membrane separation of CO_2 from natural gas: A review. Int. J. Greenh. Gas Control. 17 (2013), 46-65. https://doi.org/10.1016/j.ijggc.2013.04.012

[50]   Y. Zhang, J. Sunarso, S. Liu, R. Wang, Current status and development of membranes for CO_2/CH_4 separation: A review. Int. J. of Greenh. Gas Control, 12 (2013), 84-107. https://doi.org/10.1016/j.ijggc.2012.10.009

[51]   C. A. Scholes, G. W. Stevens, S. E. Kentish, Membrane gas separation applications in natural gas processing. Fuel. 9 6(2012), 15-28. https://doi.org/10.1016/j.fuel.2011.12.074

[52]   Z. Y. Yeo, T. L. Chew, P. W. Zhu, A. R. Mohamed, S. P. Chai, Conventional processes and membrane technology for carbon dioxide removal from natural gas: A review. J. Nat. Gas Chem. 21 (2012), 282-298. https://doi.org/10.1016/S1003-9953(11)60366-6

[53]   P. Luis, T. Van Gerven, B. Van der Bruggen, Recent developments in membrane-based technologies for CO_2 capture. Prog. Energy Combust. Sci. 38 (2012), 419-448. https://doi.org/10.1016/j.pecs.2012.01.004

[54]   C. A. Scholes, K. H. Smith, S. E. Kentish, G. W. Stevens, CO_2 capture from pre-combustion processes-strategies for membrane gas separation. Int. J. Greenh. Gas Control. 4 (2010), 739-755. https://doi.org/10.1016/j.ijggc.2010.04.001

[55]   K. Ramasubramanian, W. W. Ho, Recent developments on membranes for post-combustion carbon capture. Curr. Opi. Chem. Eng. 1 (2011), 47-54. https://doi.org/10.1016/j.coche.2011.08.002

[56]   X. Zhang, B. Singh, X. He, T. Gundersen, L. Deng, S. Zhang, Post-combustion carbon capture technologies: energetic analysis and life cycle assessment. Int. J. Greenh. Gas Control. 27 (2014), 289-298. https://doi.org/10.1016/j.ijggc.2014.06.016

[57]   F. Gallucci, E. Fernandez, P. Corengia, M. van Sint Annaland, Recent advances on membranes and membrane reactors for hydrogen production. Chem. Eng. Sci. 92 (2013), 40-66. https://doi.org/10.1016/j.ces.2013.01.008

[58]   L. Shao, B. T. Low, T. S. Chung, A. R. Greenberg, Polymeric membranes for the hydrogen economy: contemporary approaches and prospects for the future. J. Membr. Sci. 327 (2009), 18-31. https://doi.org/10.1016/j.memsci.2008.11.019

[59]   N. W. Ockwig, T. M. Nenoff, Membranes for hydrogen separation. Chem. Rev. 107 (2007), 4078-4110. https://doi.org/10.1021/cr0501792

[60]   P. Bakonyi, N. Nemestóthy, K. Bélafi-Bakó, Biohydrogen purification by membranes: an overview on the operational conditions affecting the performance of non-porous, polymeric and ionic liquid-based gas separation membranes. Int. J.

Hydrogen Energy. 38 (2013), 9673-9687.
https://doi.org/10.1016/j.ijhydene.2013.05.158

[61] R. Faiz, K. Li, Olefin/paraffin separation using membrane based facilitated transport/chemical absorption techniques. Chem. Eng. Sci. 73 (2012), 261-284. https://doi.org/10.1016/j.ces.2012.01.037

[62] M. T. Ravanchi, T. Kaghazchi, A. Kargari, Application of membrane separation processes in petrochemical industry: a review. Desalination. 235 (2009), 199-244. https://doi.org/10.1016/j.desal.2007.10.042

[63] L. M. Robeson, The upper bound revisited. J. Membr. Sci. 320 (2008), 390-400. https://doi.org/10.1016/j.memsci.2008.04.030

[64] R. W. Baker, B. T. Low, Gas separation membrane materials: a perspective. Macromolecules. 47 (2014), 6999-7013. https://doi.org/10.1021/ma501488s

[65] S. H. Yoo, J. H. Kim, J. Y. Jho, J. Won, Y. S. Kang, Influence of the addition of PVP on the morphology of asymmetric polyimide phase inversion membranes: effect of PVP molecular weight. J. Membr. Sci. 236(2004), 203-207. https://doi.org/10.1016/j.memsci.2004.02.017

[66] H. Lee, M. Yanilmaz, O. Toprakci, K. Fu, X. Zhang, A review of recent developments in membrane separators for rechargeable lithium-ion batteries. Energy Env. Sci. 7 (2014), 3857-3886. https://doi.org/10.1039/C4EE01432D

[67] T. Masuda, Y. Iguchi, B. Z. Tang, T. Higashimura, Diffusion and solution of gases in substituted polyacetylene membranes. Polymer, 29 (1988), 2041-2049. https://doi.org/10.1016/0032-3861(88)90178-4

[68] L. M. Robeson, W. F. Burgoyne, M. Langsam, A. C. Savoca, C. F. Tien, High performance polymers for membrane separation. Polymer. 35(1994), 4970-4978. https://doi.org/10.1016/0032-3861(94)90651-3

[69] R. W. Baker, B. T. Low, Gas separation membrane materials: a perspective. Macromolecules. 47(2014), 6999-7013. https://doi.org/10.1021/ma501488s

[70] S. Harms, K. Rätzke, F. Faupel, N. Chaukura, P. M. Budd, W. Egger, L. Ravelli, Aging and free volume in a polymer of intrinsic microporosity (PIM-1). The Journal of Adhesion. 88 (2012), 608-619. https://doi.org/10.1080/00218464.2012.682902

[71] J. G. Wijmans, R. W. Baker, The solution-diffusion model: a unified approach to membrane permeation. Materials science of membranes for gas and vapor separation. 1 (2006), 159-189. https://doi.org/10.1002/047002903X.ch5

[72]   R. D. Noble, C. A. Koval, Review of facilitated transport membranes John Wiley
        and Sons: Chichester, England. 2006, 411-435.
        https://doi.org/10.1002/047002903X.ch17

[73]   L. Deng, T. J. Kim, M. B. Hägg, Facilitated transport of CO_2 in novel PVAm/PVA
        blend membrane. J. Membr. Sci. 340(2009), 154-163.
        https://doi.org/10.1016/j.memsci.2009.05.019

[74]   R. D. Noble, C. A. Koval, Review of facilitated transport membranes. John Wiley
        and Sons: Chichester, England, 2006, 411-435.
        https://doi.org/10.1002/047002903X.ch17

[75]   M. Wang, Z. Wang, S. Li, C. Zhang, J. Wang, S. Wang, A high performance
        antioxidative and acid resistant membrane prepared by interfacial polymerization for
        CO_2 separation from flue gas. Energy Environ. Sci. 6 (2013), 539-551.
        https://doi.org/10.1039/C2EE23080A

[76]   Y. Li, Q. Xin, H. Wu, R. Guo, Z. Tian, Y. Liu, Z. Jiang, Efficient CO_2 capture by
        humidified polymer electrolyte membranes with tunable water state. Energy Environ.
        Sci. 7 (2014), 1489-1499. https://doi.org/10.1039/c3ee43163k

[77]   M. Saeed, L. Deng, CO_2 facilitated transport membrane promoted by mimic
        enzyme. J. Membr. Sci. 494 (2015), 196-204.
        https://doi.org/10.1016/j.memsci.2015.07.028

[78]   K. Yao, Z. Wang, J. Wang, S. Wang, Biomimetic material-poly (N-
        vinylimidazole)-zinc complex for CO_2 separation. Chem. Comm. 48 (2012), 1766-
        1768. https://doi.org/10.1039/c2cc16835a

[79]   W. R. Browall, Washington, DC: US Patent and Trademark Office, 1976, US
        Patent No. 3,980,456.

[80]   J. M., Henis, M. K. Tripodi, Washington, DC: US Patent and Trademark Office.
        1980, US Patent No. 4,230,463.

[81]   S. Zeng, X. Zhang, L. Bai, X. Zhang, H. Wang, J. Wang, S. Zhang, Ionic-liquid-
        based CO_2 capture systems: structure, interaction and process. Chem. Rev. 117
        (2017), 9625-9673. https://doi.org/10.1021/acs.chemrev.7b00072

[82]   M. E. Boot-Handford, J. C. Abanades, E. J. Anthony, M. J. Blunt, S. Brandani, N.
        Mac Dowell, P.S. Fennell, Carbon capture and storage update. Energy Env. Sci. 7
        (2014), 130-189. https://doi.org/10.1039/C3EE42350F

[83]   X. Liu, Y. Huang, Y. Zhao, R. Gani, X. Zhang, S. Zhang, Ionic liquid design and process simulation for decarbonization of shale gas. Ind. Eng. Chem. Res. 55 (2016), 5931-5944. https://doi.org/10.1021/acs.iecr.6b00029

[84]   G. Cevasco, C. Chiappe, Are ionic liquids a proper solution to current environmental challenges? Green Chem. 16 (2014), 2375-2385. https://doi.org/10.1039/c3gc42096e

[85]   Z. Z. Yang, Y. N. Zhao, L. N. He, CO_2 chemistry: task-specific ionic liquids for CO_2 capture/activation and subsequent conversion. RSC adv. 1 (2011), 545-567. https://doi.org/10.1039/c1ra00307k

[86]   Giernoth, R. Task-specific ionic liquids. Angewandte Chemie International Edition. 49 (2010), 2834-2839. https://doi.org/10.1002/anie.200905981

[87]   E. I. Privalova, P. Mäki-Arvela, D. Y. Murzin, J. P. Mikkhola, Capturing CO_2: conventional versus ionic-liquid based technologies. Russ. Chem. Rev. 81 (2012), 81(5), 435. https://doi.org/10.1070/RC2012v081n05ABEH004288

[88]   J. F. Brennecke, B. E. Gurkan, Ionic liquids for CO_2 capture and emission reduction. J. Phys. Chem. Lett. 1 (2010), 3459-3464. https://doi.org/10.1021/jz1014828

[89]   L. Zhang, J. Chen, J.X. Lv, S.F. Wang, Y. Cui, Progress and development of capture for CO_2 by ionic liquids-a review. Asian J. Chem. 25 (2013), 2355. https://doi.org/10.14233/ajchem.2013.13552

[90]   C. Wang, X. Luo, X. Zhu, G. Cui, D. E. Jiang, Deng, D. S. Dai, The strategies for improving carbon dioxide chemisorption by functionalized ionic liquids. Rsc Adv. 3 (2013), 15518-15527. https://doi.org/10.1039/c3ra42366b

[91]   S. Zeng, X. Zhang, L. Bai, X. Zhang, H. Wang, J. Wang, S. Zhang, Ionic-liquid-based CO_2 capture systems: structure, interaction and process. Chem. Rev. 117 (2017), 9625-9673. https://doi.org/10.1021/acs.chemrev.7b00072

[92]   J. L. Anderson, J. K. Dixon, J. F. Brennecke, Solubility of CO_2, CH_4, C_2H_6, C_2H_4, O_2, and N_2 in 1-Hexyl-3-methylpyridinium Bis (trifluoromethylsulfonyl) imide: Comparison to other ionic liquids. Acc. Chem. Res. 40 (2007), 1208-1216. https://doi.org/10.1021/ar7001649

[93]   K. Huang, X. M. Zhang, Y. Xu, Y. T. Wu, X. B. Hu, Y. Xu, Protic ionic liquids for the selective absorption of H_2S from CO_2: thermodynamic analysis. AIChE Journal. 60 (2014), 4232-4240. https://doi.org/10.1002/aic.14634

[94]   J. E. Bara, C. J. Gabriel, S. Lessmann, T. K. Carlisle, A. Finotello, D. L. Gin, R. D. Noble, Enhanced CO_2 separation selectivity in oligo (ethylene glycol) functionalized room-temperature ionic liquids. Ind. Eng. Chem. Res. 46 (2007), 5380-5386. https://doi.org/10.1021/ie070437g

[95]   T. K. Carlisle, J. E. Bara, C. J. Gabriel, R. D. Noble, D. L. Gin, Interpretation of CO_2 solubility and selectivity in nitrile-functionalized room-temperature ionic liquids using a group contribution approach. Ind. Eng. Chem. Res. 47 (2008), 7005-7012. https://doi.org/10.1021/ie8001217

[96]   A. Finotello, J. E. Bara, D. Camper, R. D. Noble, Room-temperature ionic liquids: temperature dependence of gas solubility selectivity. Ind. Eng. Chem. Res. 47 (2008), 3453-3459. https://doi.org/10.1021/ie0704142

[97]   A. Finotello, J. E. Bara, S. Narayan, D. Camper, R. D. Noble, Ideal gas solubilities and solubility selectivities in a binary mixture of room-temperature ionic liquids. J. Phys. Chem. B. 112 (2008), 2335-2339. https://doi.org/10.1021/jp0755721

[98]   X. Zhang, S. Zhang, D. Bao, Y. Huang, X. Zhang, Absorption degree analysis on biogas separation with ionic liquid systems. Bioresour. Technol. 175 (2015), 135-141. https://doi.org/10.1016/j.biortech.2014.10.048

[99]   M. Ramdin, A. Amplianitis, S. Bazhenov, A. Volkov, V. Volkov, T. J. Vlugt, T. W. de Loos, Solubility of CO_2 and CH_4 in ionic liquids: ideal CO_2/CH_4 selectivity. Ind. Eng. Chem. Res. 53 (2014), 15427-15435. https://doi.org/10.1021/ie4042017

[100] M. Ramdin, T.W. de Loos, T. J. Vlugt, State-of-the-art of CO_2 capture with ionic liquids. Ind. Eng. Chem. Res. 51 (2012), 8149-8177. https://doi.org/10.1021/ie3003705

[101] J. Jacquemin, P. Husson, V. Majer, M.F.C. Gomes, Low-pressure solubilities and thermodynamics of solvation of eight gases in 1-butyl-3-methylimidazolium hexafluorophosphate. Fluid Phase Equilibria. 240 (2006), 87-95. https://doi.org/10.1016/j.fluid.2005.12.003

[102] J. Jacquemin, M. F. C. Gomes, P. Husson, V. Majer, Solubility of carbon dioxide, ethane, methane, oxygen, nitrogen, hydrogen, argon, and carbon monoxide in 1-butyl-3-methylimidazolium tetrafluoroborate between temperatures 283 K and 343 K and at pressures close to atmospheric. J. Chem. Thermodyn. 38 (2006), 490-502. https://doi.org/10.1016/j.jct.2005.07.002

[103] W. Shi, D.C. Sorescu, D.R. Luebke, M.J. Keller, S. Wickramanayake, Molecular simulations and experimental studies of solubility and diffusivity for pure and mixed

Advanced Functional Membranes

Materials Research Foundations **120** (2022) 267-314

Materials Research Forum LLC

https://doi.org/10.21741/9781644901816-9

gases of H_2, CO_2, and Ar absorbed in the ionic liquid 1-n-Hexyl-3-methylimidazolium Bis(Trifluoromethylsulfonyl) amide ([hmim][Tf$_2$N]). J. Phy. Chem. B. 114 (2010), 6531-6541. https://doi.org/10.1021/jp101897b

[104] L. Hillain, M. Eddaoudi, M. O'Keefe, O.M. Yaghi, Design and synthesis of an exceptionally stable and highly porous metal-organic framework, Nature, 402 (1999), 276-279. https://doi.org/10.1038/46248

[105] O. M. Yaghi, H. Li, Hydrothermal Synthesis of a Metal-organic framework containing large rectangular channels, J. Am. Chem. Soc. 117 (1995), 10401-10402. https://doi.org/10.1021/ja00146a033

[106] S. Qiu, M. Xue, G. Zhu, Metal–organic framework membranes: from synthesis to separation application, Chem. Soc. Rev. 43 (2014), 6116-6140. https://doi.org/10.1039/C4CS00159A

[107] M. Mohanned, N. Devjyoti, I. Hussameldin, H. Amr, Review of Recent Developments in CO_2 Capture Using Solid Materials: Metal Organic Frameworks (MOFs), Greenhouse Gases Bernardo Llamas, IntechOpen, 2016, 115-154.

[108] S. M. Cohen, Postsynthetic methods for the functionalization of metal–organic frameworks, Chem. Rev. 112 (2011), 970-1000. https://doi.org/10.1021/cr200179u

[109] J. Sculley, D. Yuan, H.-C. Zhou, The current status of hydrogen storage in metal–organic frameworks—updated, Energy Environ Sci. 4 (2011), 2721-2735. https://doi.org/10.1039/c1ee01240a

[110] J. A. Mason, M. Veenstra, J. R. Long, Evaluating metal–organic frameworks for natural gas storage, Chem. Sci. 5 (2014), 32-51. https://doi.org/10.1039/C3SC52633J

[111] K. Sumida, D. L. Rogow, J. A. Mason, T. M. McDonald, E. D. Bloch, Z. R. Herm, T. H. Bae, J. R. Long, Carbon dioxide capture in metal–organic frameworks, Chem. Rev. 112 (2012), 724-781. https://doi.org/10.1021/cr2003272

[112] M. S. Denny, J. C. Moreton, L. Benz, S. M. Cohen, Metal–organic frameworks for membrane-based separations, Nat. Rev. Mater. 1 (2016), 16078. https://doi.org/10.1038/natrevmats.2016.78

[113] M. Ranocchiari, J. A. van Bokhoven, Catalysis by metal–organic frameworks: fundamentals and opportunities, Phys. Chem. Chem. Phys., 13 (2011), 6388-6396. https://doi.org/10.1039/c0cp02394a

[114] B. Liu, Metal–organic framework-based devices: separation and sensors, J. Mater. Chem., 22 (2012), 10094. https://doi.org/10.1039/c2jm15827b

[115] A. C. McKinlay, R. E. Morris, P. Horcajada, G. Férey, R. Gref, P. Couvreur, C. Serre, BioMOFs: Metal–organic frameworks for biological and medical applications, Angew. Chem. Int. Ed. 49 (2010), 6260-6266. https://doi.org/10.1002/anie.201000048

[116] C. G. Silva, A. Corma, H. Garcia, Metal–organic frameworks as semiconductors, J. Mater. Chem. 20 (2010), 3141-3156. https://doi.org/10.1039/b924937k

[117] P. Kumar, V. Bansal, A. Deep, K.-H. Kim, Synthesis and energy applications of metal organic frameworks, J. Porous Mater. 22 (2015), 413-424. https://doi.org/10.1007/s10934-015-9910-3

[118] M. Zhang, M. Bosch, I.T. Gentle, H.-C. Zhou, Rational design of metal–organic frameworks with anticipated porosities and functionalities, CrystEngComm. 16 (2014), 4069-4089. https://doi.org/10.1039/C4CE00321G

[119] J. P. Sculley, W. M. Verdegaal, W. Lu, M. Wriedt, H. C. Zhou, High-throughput analytical model to evaluate materials for temperature swing adsorption Processes, Adv. Mater. 25 (2013), 3957-3961. https://doi.org/10.1002/adma.201204695

[120] T. Watanabe, D. S. Sholl, Accelerating applications of metal–organic frameworks for gas adsorption and separation by computational screening of materials, Langmuir. 28 (2012), 14114-14128. https://doi.org/10.1021/la301915s

[121] Y. J. Colón, R. Q. Snurr, High-throughput computational screening of metal–organic frameworks, Chem. Soc. Rev. 43 (2014), 5735-5749. https://doi.org/10.1039/C4CS00070F

[122] D. A. Gomez-Gualdron, Y. J. Colon, X. Zhang, T. C. Wang, Y.-S. Chen, J. T. Hupp, T. Yildirim, O.K. Farha, J. Zhang, R. Q. Snurr, Understanding volumetric and gravimetric hydrogen adsorption trade-off in metal–organic frameworks, Energy Environ. Sci. 9 (2016), 33419-33428. https://doi.org/10.1039/C6EE02104B

[123] D. Feng, K. Wang, Z. Wei, Y.-P. Chen, C.M. Simon, R.K. Arvapally, R.L. Martin, M. Bosch, T.-F. Liu, S. Fordham, D. Yuan, M.A. Omary, M. Haranczyk, B. Smit, H.-C. Zhou, Kinetically tuned dimensional augmentation as a versatile synthetic route towards robust metal–organic frameworks, Nat. Commun. 5 (2014), 5723. https://doi.org/10.1038/ncomms6723

[124] D. Andirova, C.F. Cogswell, Y. Lei, S. Choi, Effect of the structural constituents of metal organic frameworks on carbon dioxide capture, Microporous Mesoporous Mater. 219 (2016), 276-305. https://doi.org/10.1016/j.micromeso.2015.07.029

[125] S. Han, Y. Huang, T. Watanabe, Y. Dai, K.S. Walton, S. Nair, D.S. Sholl, J. C. Meredith, High-throughput screening of metal–organic frameworks for CO_2 separation, ACS Comb. Sci., 14 (2012), 263-267. https://doi.org/10.1021/co3000192

[126] J. A. Mason, T. M. McDonald, T.-H. Bae, J. E. Bachman, K. Sumida, J. J. Dutton, S. S. Kaye, J. R. Long, Application of a high-throughput analyzer in evaluating solid adsorbents for post-combustion carbon capture via multicomponent adsorption of CO_2, N_2, and H_2O, J. Am. Chem. Soc., 137 (2015), 4787-4803. https://doi.org/10.1021/jacs.5b00838

[127] I. Sreedhar, R. Vaidhiswaran, B. M. Kamani, A. Venugopal, Process and engineering trends in membrane-based carbon capture, Renewable Sustainable Energy Rev. 68 (2017), 659-684. https://doi.org/10.1016/j.rser.2016.10.025

[128] M. S. Denny, J. C. Moreton, L. Benz, S. M. Cohen, Metal–organic frameworks for membrane-based separations, Nat. Rev. Mater. 1 (2016), 16078. https://doi.org/10.1038/natrevmats.2016.78

[129] D. Zacher, O. Shekhah, C. Woll, R. A. Fischer, Thin films of metal–organic frameworks, Chem. Soc. Rev. 38 (2009), 1418-1429. https://doi.org/10.1039/b805038b

[130] N. C. Su, D. T. Sun, C. M. Beavers, D. K. Britt, W. L. Queen, J. J. Urban, Enhanced permeation arising from dual transport pathways in hybrid polymer–MOF membranes, Energy Environ. Sci. 9 (2016), 922-931. https://doi.org/10.1039/C5EE02660A

[131] A. Sabetghadam, B. Seoane, D. Keskin, N. Duim, T. Rodenas, S. Shahid, S. Sorribas, C.L. Guillouzer, G. Clet, C. Tellez, M. Daturi, J. Coronas, F. Kapteijn, J. Gascon, Metal-organic framework crystals in mixed-matrix membranes: impact of the filler morphology on the gas separation performance, Adv. Funct. Mater. 26 (2016), 3154-3163. https://doi.org/10.1002/adfm.201505352

[132] T. Rodenas, I. Luz, G. Prieto, B. Seoane, H. Miro, A. Corma, F. Kapteijn, F.X. Llabrés i Xamena, J. Gascon, Metal–organic framework nanosheets in polymer composite materials for gas separation, Nat. Mater. 14 (2015), 48-55. https://doi.org/10.1038/nmat4113

[133] A. Marti, D. Tran, K. Jr. Balkus, Materials and processes for CO_2 capture, conversion, and sequestration, J. Porous Mater. 1 (2015).

[134] M. S. Denny, S. M. Cohen, In situ modification of metal–organic frameworks in mixed-matrix membranes, Angew. Chem. Int. Ed. 54 (2015), 9029-9032. https://doi.org/10.1002/anie.201504077

[135] K. Huang, Q. Li, G. Liu, J. Shen, K. Guan, W. Jin, A ZIF-71 Hollow fiber membrane fabricated by contra-diffusion, ACS Appl. Mater. Interfaces. 7 (2015), 16157-16160. https://doi.org/10.1021/acsami.5b04991

[136] X. Liu, N. K. Demir, Z. Wu, K. Li, Highly water-stable zirconium metal–organic framework UiO-66 membranes supported on alumina hollow fibers for desalination, J. Am. Chem. Soc. 137 (2015), 6999-7002. https://doi.org/10.1021/jacs.5b02276

[137] M. N. Shah, M. A. Gonzalez, M. C. McCarthy, H.-K. Jeong, An unconventional rapid synthesis of high-performance metal–organic framework membranes, Langmuir, 29 (2013), 7896-7902. https://doi.org/10.1021/la4014637

[138] M. Tu, S. Wannapaiboon, R. A. Fischer, Programmed functionalization of SURMOFs via liquid phase heteroepitaxial growth and post-synthetic modification, Dalton Trans. 42 (2013), 16029-16035. https://doi.org/10.1039/c3dt51457a

[139] S. C. Hess, R. N. Grass, W. J. Stark, MOF channels within porous polymer film: flexible, self-supporting ZIF-8 Poly(ether sulfone) composite membrane, Chem. Mater. 28 (2016), 7638-7644. https://doi.org/10.1021/acs.chemmater.6b02499

[140] J. Hou, P.D. Sutrisna, Y. Zhang, V. Chen, Formation of ultrathin, continuous metal–organic framework membranes on flexible polymer substrates, Angew. Chem. Int. Ed. 55 (2016), 3947-3951. https://doi.org/10.1002/anie.201511340

[141] J. Dechnik, F. Mühlbach, D. Dietrich, T. Wehner, M. Gutmann, T. Lühmann, L. Meinel, C. Janiak, K. Müller-Buschbaum, Luminescent metal–organic framework mixed-matrix membranes from lanthanide metal–organic frameworks in polysulfone and matrimid, Eur. J. Inorg. Chem. 27 (2016), 4408-4415. https://doi.org/10.1002/ejic.201600235

[142] W. J. Li, S. Y. Gao, T. F. Liu, L. W. Han, Z. J. Lin, R. Cao, In Situ Growth of Metal–Organic Framework Thin Films with Gas Sensing and Molecule Storage Properties, Langmuir. 29 (2013), 8657-8664. https://doi.org/10.1021/la402012d

[143] C. R. Wade, M. Li, M. Dincă, Facile deposition of multicolored electrochromic metal–organic framework thin films, Angew. Chem. Int. Ed. 52 (2013), 13377-13381. https://doi.org/10.1002/anie.201306162

[144] T. C. Merkel, H. Lin, X. Wei, R. Baker, Power plant post-combustion carbon dioxide capture: An opportunity for membranes, J. Membr. Sci., 359 (2010), 126-139. https://doi.org/10.1016/j.memsci.2009.10.041

[145] B. Zornoza, C. Tellez, J. Coronas, J. Gascon, F. Kapteijn, Metal organic framework based mixed matrix membranes: An increasingly important field of research with a large application potential, Microporous Mesoporous Mater. 166 (2013), 67-78. https://doi.org/10.1016/j.micromeso.2012.03.012

[146] S. Kim, Y. M. Lee, High performance polymer membranes for CO_2 separation, Curr. Opin. Chem. Eng. 2 (2013), 238-244. https://doi.org/10.1016/j.coche.2013.03.006

[147] D. F. Sanders, Z. P. Smith, R. Guo, L. M. Robeson, J. E. McGrath, D. R. Paul, B. D. Freeman, Energy-efficient polymeric gas separation membranes for a sustainable future: A review, Polymer. 54 (2013), 4729-4761. https://doi.org/10.1016/j.polymer.2013.05.075

[148] B. Seoane, J. Coronas, I. Gascon, M. E. Benavides, O. Karvan, J. Caro, F. Kapteijn, J. Gascon, Metal–organic framework based mixed matrix membranes: a solution for highly efficient CO_2 capture? Chem. Soc. Rev. 44 (2015), 2421-2454. https://doi.org/10.1039/C4CS00437J

[149] L. S. White, X. Wei, S. Pande, T. Wu, T. C. Merkel, Extended flue gas trials with a membrane-based pilot plant at a one-ton-per-day carbon capture rate, J. Membr. Sci. 496 (2015), 48-57. https://doi.org/10.1016/j.memsci.2015.08.003

[150] R. Adams, C. Carson, J. Ward, R. Tannenbaum, W. Koros, Metal organic framework mixed matrix membranes for gas separations, Microporous Mesoporous Mater. 131 (2010), 13-20. https://doi.org/10.1016/j.micromeso.2009.11.035

[151] S. R. Venna, M. Lartey, T. Li, A. Spore, S. Kumar, H. B. Nulwala, D. R. Luebke, N. L. Rosi, E. Albenze, Fabrication of MMMs with improved gas separation properties using externally-functionalized MOF particles, J. Mater. Chem. A. 3 (2015), 5014-5022. https://doi.org/10.1039/C4TA05225K

[152] S. Shahid, K. Nijmeijer, S. Nehache, I. Vankelecom, A. Deratani, D. Quemener, MOF-mixed matrix membranes: Precise dispersion of MOF particles with better compatibility via a particle fusion approach for enhanced gas separation properties, J. Membr. Sci. 492 (2015), 21-31. https://doi.org/10.1016/j.memsci.2015.05.015

[153] J. Campbell, G. Szekely, R.P. Davies, D.C. Braddock, A.G. Livingston, Fabrication of hybrid polymer/metal organic framework membranes: mixed matrix

membranes versus in situ growth, J. Mater. Chem. A. 2 (2014), 9260-9271.
https://doi.org/10.1039/C4TA00628C

[154] T. T. Moore, W. J. Koros, Non-ideal effects in organic–inorganic materials for gas separation membranes, J. Mol. Struct. 739 (2005), 87-98.
https://doi.org/10.1016/j.molstruc.2004.05.043

[155] R. Pal, Permeation models for mixed matrix membranes, J. Colloid Interface Sci. 317 (2008), 191-198. https://doi.org/10.1016/j.jcis.2007.09.032

[156] R. Semino, N.A. Ramsahye, A. Ghoufi, G. Maurin, Microscopic model of the metal–organic framework/polymer interface: A first step toward understanding the compatibility in mixed matrix membranes, ACS Appl. Mater. Interfaces. 8 (2016), 809-819. https://doi.org/10.1021/acsami.5b10150

[157] R. Lin, L. Ge, H. Diao, V. Rudolph, Z. Zhu, Ionic liquids as the MOFs/polymer interfacial binder for efficient membrane separation, ACS Appl. Mater. Interfaces. 8 (2016), 32041-32049. https://doi.org/10.1021/acsami.6b11074

[158] M. W. Anjum, F. Vermoortele, A. L. Khan, B. Bueken, D. E. De Vos, I. F. J. Vankelecom, Modulated UiO-66-based mixed-matrix membranes for CO_2 separation, ACS Appl. Mater. Interfaces, 7 (2015), 25193-25201.
https://doi.org/10.1021/acsami.5b08964

[159] E. Perez, C. Karunaweera, I. Musselman, K. Balkus, J. Ferraris, Origins and evolution of inorganic-based and MOF-based mixed-matrix membranes for gas separations, Processes, 4 (2016), 32. https://doi.org/10.3390/pr4030032

[160] F. Cacho-Bailo, G. Caro, M. Etxeberria-Benavides, O. Karvan, C. Tellez, J. Coronas, MOF–polymer enhanced compatibility: post-annealed zeolite imidazolate framework membranes inside polyimide hollow fibers, RSC Advances, 6 (2016), 5881-5889. https://doi.org/10.1039/C5RA26076K

[161] P. Su, W. Li, C. Zhang, Q. Meng, C. Shen, G. Zhang, Metal based gels as versatile precursors to synthesize stiff and integrated MOF/polymer composite membranes, J. Mater. Chem. A. 3 (2015), 20345-20351. https://doi.org/10.1039/C5TA04400F

[162] F. Cacho-Bailo, G. Caro, M. Etxeberria-Benavides, O. Karvan, C. Tellez, J. Coronas, High selectivity ZIF-93 hollow fiber membranes for gas separation, Chem. Commun. 51 (2015), 11283-11285. https://doi.org/10.1039/C5CC03937A

[163] W. Li, Z. Yang, G. Zhang, Z. Fan, Q. Meng, C. Shen, C. Gao, Stiff metal–organic framework–polyacrylonitrile hollow fiber composite membranes with high gas

permeability, J. Mater. Chem. A. 2 (2014), 2110-2118.
https://doi.org/10.1039/C3TA13781C

[164] T. Ben, C. Lu, C. Pei, S. Xu, S. Qiu, Polymer-supported and free-standing metal–organic framework membrane, Chem. Eur. J. 18 (2012), 10250-10253.
https://doi.org/10.1002/chem.201201574

[165] Y. Hu, X. Dong, J. Nan, W. Jin, X. Ren, N. Xu, Y. M. Lee, Metal–organic framework membranes fabricated via reactive seeding, Chem. Commun. 47 (2011), 737-739. https://doi.org/10.1039/C0CC03927F

[166] P. A. Bayliss, I. A. Ibarra, E. Perez, S.Yang, C. C. Tang, M. Poliakoff, M. Schroder, Synthesis of metal–organic frameworks by continuous flow Green Chem. 16 (2014), 3796-3802. https://doi.org/10.1039/C4GC00313F

[167] M. P. Batten, M. Rubio-Martinez, T. Hadley, K.-C. Carey, K.-S. Lim, A. Polyzos, M. R. Hill, Continuous flow production of metal-organic frameworks, Curr. Opin. Chem. Eng. 8 (2015), 55-59. https://doi.org/10.1016/j.coche.2015.02.001

[168] M. Rubio-Martinez, T.D. Hadley, M.P. Batten, K. Constanti-Carey, T. Barton, Marley, D.; Mönch, A.; Lim, K.-S.; Hill, M. R. Scalability of continuous flow production of metal-organic frameworks, ChemSusChem. 9 (2016), 938.
https://doi.org/10.1002/cssc.201501684

[169] A. J. Brown, N.A. Brunelli, K. Eum, F. Rashidi, J.R. Johnson, W.J. Koros, C.W. Jones, S. Nair, Interfacial microfluidic processing of metal-organic framework hollow fiber membranes, Science. 345 (2014), 72-75.
https://doi.org/10.1126/science.1251181

[170] L. Kong, X. Zhang, Y. Liu, S. Li, H. Liu, J. Qiu, K. L. Yeung, In situ fabrication of high-permeance ZIF-8 tubular membranes in a continuous flow system, Mater. Chem. Phys. 148 (2014), 10-16. https://doi.org/10.1016/j.matchemphys.2014.07.036

[171] A. M. Marti, W. Wickramanayake, G. Dahe, A. Sekizkardes, T.L. Bank, D.P. Hopkinson, S.R. Venna, Continuous flow processing of ZIF-8 membranes on polymeric porous hollow fiber supports for CO_2 capture, ACS Appl. Mater. Interfaces. 9 (2017), 5678-5682. https://doi.org/10.1021/acsami.6b16297

[172] F. Zhang, X. Zou, X. Gao, S. Fan, F. Sun, H. Ren, G. Zhu, Hydrogen selective NH_2-MIL-53(Al) MOF membranes with high permeability, Adv. Funct. Mater. 22 (2012), 3583-3590. https://doi.org/10.1002/adfm.201200084

[173] A. Huang, J. Caro, Covalent post-functionalization of zeolitic imidazolate framework ZIF-90 membrane for enhanced hydrogen selectivity, Angew. Chem. Int. Ed. 50 (2011), 4979-4982. https://doi.org/10.1002/anie.201007861

[174] A. Huang, N. Wang, C. Kong, J. Caro, Organosilica-functionalized zeolitic imidazolate framework ZIF-90 membrane with high gas-separation performance, Angew. Chem. Int. Ed. 51 (2012), 10551-10555. https://doi.org/10.1002/anie.201204621

[175] V. V. Guerrero, Y. Yoo, M. C. McCarthy, H. K. Jeong, HKUST-1 membranes on porous supports using secondary growth, J. Mater. Chem. 20 (2010), 3938-3943. https://doi.org/10.1039/b924536g

[176] W. Morris, C. J. Doonan, H. Furukawa, R. Banerjee, O. M. Yaghi, Crystals as molecules: Postsynthesis covalent functionalization of zeolitic imidazolate frameworks, J. Am. Chem. Soc. 130 (2008), 12626-12627. https://doi.org/10.1021/ja805222x

[177] H. Bux, F. Liang, Y. Li, J. Cravillon, M. Wiebcke, J. Caro, Zeolitic imidazolate framework membrane with molecular sieving properties by microwave-assisted solvothermal synthesis, J. Am. Chem. Soc. 131 (2009), 16000-16001. https://doi.org/10.1021/ja907359t

[178] Z. Bao, L. Yu, Q. Ren, X. Lu, S. Deng, Adsorption of CO_2 and CH_4 on a magnesium-based metal organic framework, J. Colloid Interface Sci. 353 (2011), 549-556. https://doi.org/10.1016/j.jcis.2010.09.065

[179] Y.-S. Li, F.-Y. Liang, H. Bux, A. Feldhoff, W.-S. Yang, J. Caro, Molecular sieve membrane: supported metal–organic framework with high hydrogen selectivity, Angew. Chem., Int. Ed. 49 (2010), 548-551. https://doi.org/10.1002/anie.200905645

[180] Y. S. Li, H. Bux, A. Feldhoff, G. L. Li, W. S. Yang, J. Caro, Controllable synthesis of metal-organic frameworks: From MOF nanorods to oriented MOF membranes, Adv. Mater. 22 (2010), 3322-3326. https://doi.org/10.1002/adma.201000857

[181] Y. S. Li, F. Y. Liang, H. G. Bux, W. S. Yang, J. Caro, Zeolitic imidazolate framework ZIF-7 based molecular sieve membrane for hydrogen separation, J. Membr. Sci. 354 (2010), 48-54. https://doi.org/10.1016/j.memsci.2010.02.074

[182] A. Huang, H. Bux, F. Steinbach, J. Caro, New progress of microporous metal-organic frameworks in CO_2 capture and separation, Angew. Chem., Int. Ed. 49 (2010), 4958-4861. https://doi.org/10.1002/anie.201001919

[183] A. Huang, J. Caro, Covalent post-functionalization of zeolitic imidazolate framework ZIF-90 membrane for enhanced hydrogen selectivity, Angew. Chem. Int. Ed. 50 (2011), 4979-4982. https://doi.org/10.1002/anie.201007861

[184] A. Huang, N. Wang, C. Kong, J. Caro, Organosilica-functionalized zeolitic imidazolate framework ZIF-90 membrane with high gas-separation performance, Angew. Chem. Int. Ed. 51 (2012), 10551-10555. https://doi.org/10.1002/anie.201204621

[185] A. J. Brown, J. R. Johnson, M. E. Lydon, W. J. Koros, C. W. Jones, S. Nair, Continuous polycrystalline zeolitic imidazolate framework-90 membranes on polymeric hollow fibers, Angew. Chem. Int. Ed. 51 (2012), 10615-10618. https://doi.org/10.1002/anie.201206640

[186] M. J. C. Ordonez, K. J. Balkus Jr., J. P. Ferraris, I. H. Musselman, Molecular sieving realized with ZIF-8/Matrimid® mixed-matrix membranes, J. Membr. Sci. 361 (2010), 28-37. https://doi.org/10.1016/j.memsci.2010.06.017

[187] A. Car, C. Stropnik, K.-V. Peinemann, Hybrid membrane materials with different metal–organic frameworks (MOFs) for gas separation, Desalination. 200 (2006), 424-426. https://doi.org/10.1016/j.desal.2006.03.390

[188] R. D. Noble, Perspectives on mixed matrix membranes, J. Membr. Sci. 378 (2011), 393-397. https://doi.org/10.1016/j.memsci.2011.05.031

[189] R. Xing, W. S. W. Ho, Crosslinked polyvinylalcohol–polysiloxane/fumed silica mixed matrix membranes containing amines for CO_2/H_2 separation, J. Membr. Sci. 367 (2011), 91-102. https://doi.org/10.1016/j.memsci.2010.10.039

[190] Y. Zhao, B. T. Jung, L. Ansaloni, W. S. W. Ho, Multiwalled carbon nanotube mixed matrix membranes containing amines for high pressure CO_2/H_2 separation, J. Membr. Sci. 459 (2014), 233-243. https://doi.org/10.1016/j.memsci.2014.02.022

[191] L. Ansaloni, Y. Zhao, B. T. Jung, K. Ramasubramanian, M. G. Baschetti, W. S. W. Ho, Facilitated transport membranes containing amino-functionalized multi-walled carbon nanotubes for high-pressure CO_2 separations, J. Membr. Sci. 490 (2015), 18-28. https://doi.org/10.1016/j.memsci.2015.03.097

[192] L. Deng, M.-B. Hägg, Carbon nanotube reinforced PVAm/PVA blend FSC nanocomposite membrane for CO_2/CH_4 separation, Int. J. Greenh. Gas Con. 26 (2014), 127-134. https://doi.org/10.1016/j.ijggc.2014.04.018

[193] T. C. Merkel, B. D. Freeman, R. J. Spontak, Z. He, I. Pinnau, P. Meakin, A. J. Hill, Ultrapermeable, reverse-selective nanocomposite membranes, Science. 296 (2002), 519-522. https://doi.org/10.1126/science.1069580

[194] C. H. Lau, P. Li, F. Li, T.-S. Chung, D. R. Paul, Reverse-selective polymeric membranes for gas separations, Prog. Polym. Sci. 38 (2013), 740-766. https://doi.org/10.1016/j.progpolymsci.2012.09.006

[195] Y. Liu, D. Peng, G. He, S. Wang, Y. Li, H. Wu, Z. Jiang, Enhanced CO_2 permeability of membranes by incorporating polyzwitterion@CNT composite particles into polyimide matrix, ACS Appl. Mater. Interfaces, 6 (2014), 13051-13060. https://doi.org/10.1021/am502936x

[196] X. Li, M. Wang, S. Wang, Y. Li, Z. Jiang, R. Guo, H. Wu, X. Cao, J. Yang, B. Wang, Constructing CO2 transport passageways in Matrimid® membranes using nanohydrogels for efficient carbon capture, J. Membr. Sci., 474 (2015), 156-166. https://doi.org/10.1016/j.memsci.2014.10.003

[197] S. A. Hashemifard, A. F. Ismail and T. Matsuura, Mixed matrix membrane incorporated with large pore size halloysite nanotubes (HNTs) as filler for gas separation: morphological diagram, Chem. Eng. J. 172 (2011), 581-590. https://doi.org/10.1016/j.cej.2011.06.031

[198] Q. Xin, T. Liu, Z. Li, S. Wang, Y. Li, Z. Li, J. Ouyang, Z. Jiang, H. Wu, Mixed matrix membranes composed of sulfonated poly (ether ether ketone) and a sulfonated metal–organic framework for gas separation, J. Membr. Sci. 488 (2015), 67-78. https://doi.org/10.1016/j.memsci.2015.03.060

[199] Q. Xin, H. Wu, Z. Jiang, Y. Li, S. Wang, Q. Li, X. Li, X. Lu, X. Cao, J. Yang, SPEEK/amine-functionalized TiO_2 submicrospheres mixed matrix membranes for CO2 separation, J. Membr. Sci. 467 (2014), 23-35. https://doi.org/10.1016/j.memsci.2014.04.048

[200] Q. Xin, H. Liu, Y. Zhang, H. Ye, S. Wang, L. Lin, X. Ding, B. Cheng, Y. Zhang, H. Wu, Z. Jiang, Widening CO_2-facilitated transport passageways in SPEEK matrix using polymer brushes functionalized double-shelled organic sub-microcapsules for efficient gas separation, J. Membr. Sci. 525 (2017), 330-341. https://doi.org/10.1016/j.memsci.2016.12.007

[201] S. Zhao, Z. Wang, Z. Qiao, X. Wei, C. Zhang, J. Wang, S. Wang, Gas separation membrane with CO_2-facilitated transport highway constructed from amino carrier containing nanorods and macromolecules, J. Mater. Chem. A. 1 (2013), 246-249. https://doi.org/10.1039/C2TA00247G

[202] J. Liao, Z. Wang, C. Gao, M. Wang, K. Yan, X. Xie, S. Zhao, J. Wang, S. Wang, A high performance PVAm–HT membrane containing high-speed facilitated transport channels for CO_2 separation, J. Mater. Chem. A. 3 (2015), 16746-16761. https://doi.org/10.1039/C5TA03238E

[203] I. L. Alsvik, M.-B. Hägg, Pressure retarded osmosis and forward osmosis membranes: Materials and methods, Polymers. 5 (2013), 303-327. https://doi.org/10.3390/polym5010303

[204] R. Asmatulu, W. S. Khan, Chapter 1 - Introduction to electrospun nanofibers, Synthesis and Applications of Electrospun Nanofibers, Micro and Nano Technologies, 2019, Pages 1-15. https://doi.org/10.1016/B978-0-12-813914-1.00001-8

[205] H. Qu, M. Skorobogatiy, Conductive polymer yarns for electronic textile, Electronic Textiles Smart Fabrics and Wearable Technology, 2015, Pages 21-53. https://doi.org/10.1016/B978-0-08-100201-8.00003-5

[206] A. L. Ahmad, T. A. Otitoju, B. S. Ooi, Hollow fiber (HF) membrane fabrication: A review on the effects of solution spinning conditions on morphology and performance, J Ind Eng Chem. 70 (2019), 35-50. https://doi.org/10.1016/j.jiec.2018.10.005

[207] V. B. Gupta, Solution-spinning processes, Manufactured Fibre Technology, 1997, 124-138. https://doi.org/10.1007/978-94-011-5854-1_6

[208] T. A. Otitoju, M. Ahmadipour, S. Li, N. F. Shoparwe, L. X. Jie, A. L. Owolabi, Influence of nanoparticle type on the performance of nanocomposite membranes for wastewater treatment, J. Water Process. Eng. 36 (2020), 101356. https://doi.org/10.1016/j.jwpe.2020.101356

[209] O. A. Jimoh, P. U. Okoye, T. A. Otitoju, K. S. Ariffin, Aragonite precipitated calcium carbonate from magnesium rich carbonate rock for polyethersulfone hollow fibre membrane application, J. Clean. Prod. 195 (2018), 79-92. https://doi.org/10.1016/j.jclepro.2018.05.192

[210] T. A. Otitoju, A. L. Ahmad, B. S. Ooi, Polyvinylidene fluoride (PVDF) membrane for oil rejection from oily wastewater: A performance review, J. Water Process. Eng. 14 (2016), 41-59. https://doi.org/10.1016/j.jwpe.2016.10.011

[211] Z. Qi, Microporous hollow fibers for gas absorption, J. Membr. Sci. 23 (1985) 333-345. https://doi.org/10.1016/S0376-7388(00)83150-6

[212] A. Gabelmana, S.-T. Hwang, Hollow fiber membrane contactors, J. Membr. Sci. 159 (1999) 61-106. https://doi.org/10.1016/S0376-7388(99)00040-X

[213] G. A. Dibrov, V. V. Volkov, V.P. Vasilevsk, A. A. Shutova, S. D. Bazhenov, V. S. Khotimsky, A. van de Runstraat, E. L. V. Goetheer, A.V. Volkov, Robust high-permeance PTMSP composite membranes for CO_2 membrane gas desorption at elevated temperatures and pressures, J. Membr. Sci. 470 (2014) 439-450. https://doi.org/10.1016/j.memsci.2014.07.056

[214] R. Wang, H. Y. Zhang, P. H. M. Feron, D. T. Liang, Influence of membrane wetting on CO_2 capture in microporous hollow fiber membrane contactors, Sep. Purif. Technol. 46 (2005) 33-40. https://doi.org/10.1016/j.seppur.2005.04.007

[215] H. Kreulen, H. Kreulen, G. Versteeg, C. A. Smolders, C. A. Smolders, W. P. M. van Swaaij, Determination of mass transfer rates in wetted and non-wetted microporous membranes, Chem. Eng. Sci. 48 (1993), 2093-2102. https://doi.org/10.1016/0009-2509(93)80084-4

[216] A. Malek, K. Li, W. K. Teo, Modeling of microporous hollow fiber membrane modules operated under partially wetted conditions, Ind. Eng. Chem. Res. 36 (1997), 784-793. https://doi.org/10.1021/ie960529y

[217] A. Mansourizadeh, A. F. Ismail, Hollow fiber gas-liquid membrane contactors for acid gas capture: a review, J. Hazard. Mater. 171 (2009), 38-53. https://doi.org/10.1016/j.jhazmat.2009.06.026

[218] K. Li, W. K. Teo, An ultrathin skinned hollow fibre module for gas absorption at elevated pressures, Chem. Eng. Res. Des. 74 (1996), 856-862. https://doi.org/10.1205/026387696523157

[219] P. T. Nguyena, E. Lasseuguette, Y. M. Gonzalez, J. C. Remigy, D. Roizarda, E. Favrea, A dense membrane contactor for intensified CO_2 gas/liquid absorption in post-combustion capture. J. Membr. Sci. 377 (2011), 261-272. https://doi.org/10.1016/j.memsci.2011.05.003

[220] C. A. Scholes, S. E. Kentish, G. W. Stevens, D. de Montigny, Comparison of thin film composite and microporous membrane contactors for CO_2 absorption into monoethanolamine. Int. J. Greenh. Gas Control. 42 (2015), 66-74. https://doi.org/10.1016/j.ijggc.2015.07.032

Advanced Functional Membranes   Materials Research Forum LLC
Materials Research Foundations **120** (2022) 315-333   https://doi.org/10.21741/9781644901816-10

Chapter 10

Advanced Functional Membranes for Sensor Technologies

A.R. Hernandez-Martinez[1]*

Centro de Física Aplicada y Tecnología Avanzada (CFATA), Universidad Nacional Autónoma de México (UNAM), Blvd. Juriquilla 3000, Querétaro, México

angel.ramon.hernandez@gmail.com*

Abstract

Sensors are modern data acquisition devices intended for detecting a specific change in the surrounding area and responding with an output signal. Among them, biological sensors (or biosensors) are of great interest due to their capacity for detecting molecules that indicate changes in people's health. This chapter provides an overview of the polymeric membranes that have novel or advanced functions in applications for the development of lab-on-chip or sensor technologies. Important approaches in this matter include the improvement of membranes as supporting medium of recognition element of the sensor, advances in membrane composition for protecting the integrity of target molecules, or the option to filter undesired components. The chapter is divided into three sections: membrane fabrication, molecular probes and platforms for reading sensor devices.

Keywords

Active Sensor, Target Molecule, Transducer, Wearable Technology

Contents

1.   Introduction

Any device with the ability to detect and respond to a physical stimulus from the environment can be considered a sensor. The specific stimulus could be light, heat, movement, humidity, pressure, or other parameters that define the physicochemical state of the environment (system). The response is generally a signal that is transmitted on a user-readable screen.

In this sense, a device that responds to the presence of a particular chemical selectively and transforms the chemical quantity input into an output signal is known as a chemical sensor. That was defined by IUPAC in 1991 as a device that transforms chemical information [total sample concentration or specific component (analyte) concentration] into an analytically useful signal [1]. Chemical sensors consist basically of two components connected in series: a chemical recognition system (or receptor) and a physico-chemical transducer. The receptor is capable of selectively recognizing a certain analyte through a chemical interaction with it. The transducer is responsible for physically transforming the chemical signal, which is the result of the interaction between the receptor and the analyte, into another signal that is more easily quantifiable, measurable and manipulable. On the other hand, if the chemical sensor involves some kind of biological material as a receptor, then it is called a biosensor [2–4].

In the technological development of chemical sensors (including biosensors), membranes have played an important role, despite not being included in the IUPAC definition among the components of chemical sensors. A membrane should be understood as an interface acting as a selective barrier between two adjacent phases [5]. Its main function is to regulate the exchange of substances between the two adjacent phases, through the transport of matter within itself. The main advantage of the membrane technology is the selectivity of transport through the membrane, which depends on the chemical and physical structure. The regulation of the exchange of substances through the membranes does not require additives and can be carried out isothermally at low temperatures, which requires low energy consumption compared to other thermal separation processes. Hence, membranes are intended to be used as filters for concentrating and isolating analytes, but also, they have been used as platforms to immobilize, encapsulate or protect the receptors and bioreceptors, and they proved to be useful.

This chapter reviews recent literature on membranes specifically for their use in the development of sensor technologies. The chapter is divided in three sections, membrane fabrication, molecular probes and platforms for reading sensor devices.

2.   About sensors

Sensors are common devices in modern data acquisition systems. Human beings use them everywhere on a daily basis, even if not noticing their presence, because many of them are smaller than a hair. They are devices or subsystems whose task is to detect a specific change in the environment and respond with a specific output signal. I.e., a sensor transforms a physical or chemical phenomenon into a measurable, easily interpreted, human-readable analogue or digital signal or transmitted for reading by another component or device, for further processing. A couple of examples of sensors that we use in everyday life, without having them in the mind as sensors are: *(i)* the microphone, which is a device that converts sound energy into an electrical signal, which can be amplified, broadcast or recorded; and *(ii)* a mercury thermometer, that is an old-fashioned sensor for measuring temperature. In a smartphone, there is an example of a system integrated by several sensors, it is a device that is not a sensor itself, but is loaded with several sensors such as; a fingerprint reader, an accelerometer, a magnetometer, or a proximity or ambient light sensor.

Human technology has progressed dramatically and many critical processes could not be possible without sensor data. Sensor technology is everywhere in our homes, hospitals and shopping centres. Sensors manufactured for applications in hotels, homes and smart rooms enhance artificial environments experience and imply a reduction of services costs. This reduction is achieved through intelligent energy management using automatic light and temperature controls by using thermostats and occupancy sensors [6–11].

Today, there are many types of human-made sensors because we have put our skills and imagination into the design of an overwhelming variety of sensors; therefore, a classification of them is useful, as is shown in Fig. 1.

In various research reports sensors are classified in various ways, including a review of the different categories of sensors[12–14]. Sensors were first classified as active and passive; this group is included in the category "Based on collecting data technique" in Fig. 1. Active sensors need external excitation energy for data acquisition, while passive sensors have no need for external energy to provide an output response [15,16].

Another category used is "Based on the Application", which considers the technological field or the activity of its applicability, e.g.; automotive, domestic application, telecommunication, space, agriculture, manufacturing, for daily life, or quality of life and health. However, the most used category is "Based on the detection principle used", where

Advanced Functional Membranes   Materials Research Forum LLC
Materials Research Foundations **120** (2022) 315-333   https://doi.org/10.21741/9781644901816-10

sensors are grouped into physical; chemical; thermal; and biological (biosensors, in abbreviated form).

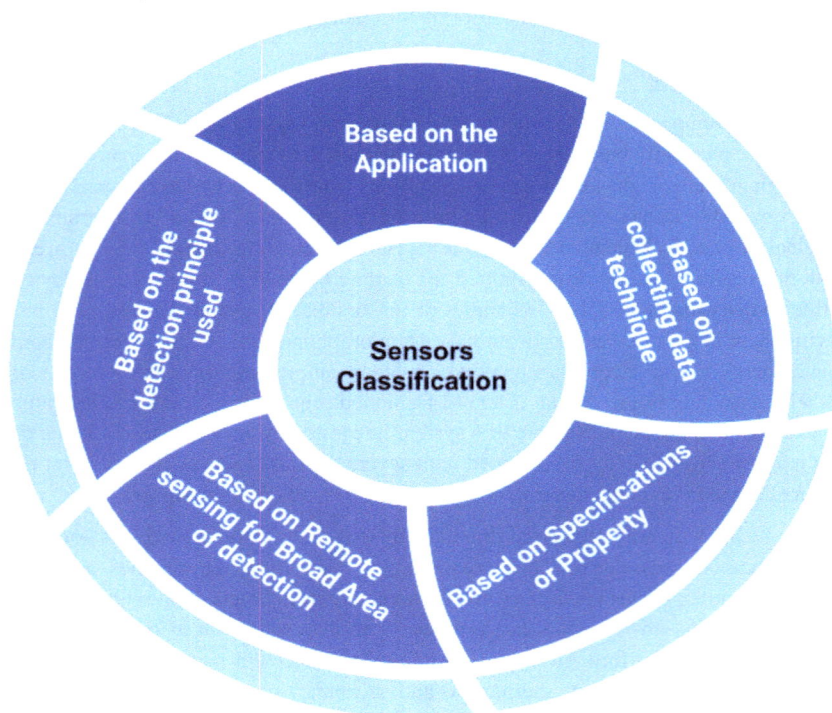

Figure 1 Sensors Classification

Physical sensors are characterized for measuring a physical property such as force, pressure, acceleration, flow rate, mass, volume, or density. I.e., microelectromechanical systems (MEMS) are physical sensors used in the biomedical field. Whereas chemical sensors are defined as devices that chemically interact with a target molecule or analyte. According to the IUPAC, they also have two parts; the recognition element, in charge of identifying the analyte; and a transducer that transforms chemical information into an analytically useful signal (analyte concentration or/and total composition analysis)[17].

Biosensors are also defined by the IUPAC in a similar way to chemical sensors, with the main difference in the nature of the recognition element or probe; which must be formed by a biomolecule such as an antibody, antigen, DNA fragment, enzymatic, or polysaccharide. Therefore, biosensors specialize in the detection of biomolecules of interest in the diagnosis of diseases [1,3,4,17–19]. Chemical sensors and biosensors, both are used to obtain information about the activity or concentration of the relevant chemical species in the gas or liquid phase, with importance for the control of environmental contamination, food analysis, drug analysis, control organophosphate compound assays, or for clinical diagnostic purposes.

For the physical sensors in some cases the membranes are an important part of these devices, as in the case of the sensors based on Quartz Crystal Microbalance (QCM). But it has been chemical sensors and biosensors that have made the best use of the sales that membrane technology offers to sensor development. So we will focus on QCM-type sensors, biosensors and chemical sensors.

2.1   The importance of membranes as sensor components

As mentioned in the introduction, chemical sensors have basic elements such as a receptor and a transducer. The receptor interacts with the analyte (target molecule) to transform chemical information into chemical energy and transfer it to the transducer. The transducer transforms the chemical energy into an energy of another physicochemical nature. This process is shown in Fig. 2, considering a chemical sensor interacting with a sample with n-components. Each component has a different chemical identity and only one of them is the molecule of interest for sensing, which corresponds to the analyte or target molecule. The analyte must approach the probes or "recognition elements" which are chemical molecules capable of recognizing the target. These probes can be synthesized in the laboratory or extracted from a natural source. The interaction between the analyte and the probe can be chemical or physical. If the interaction is chemical, it can be destructive as a chemical reaction, or non-destructive as a non-covalent interaction between the analyte and the recognition element. This process will generate specific chemical information from the interaction between analyte functional groups electrons and probe electrons. In this sense, biosensors are chemical sensors (they are a subgroup) whose chemical probes are biomolecules such as nucleic acid fragments, enzymes, proteins, or polysaccharides; according to the IUPAC definition [4,5,17,18].

The chemical energy related to the interaction between the probe and the analyte must be transformed into the energy of easy handling to generate a useful analytical signal. The transducer will convert chemical energy into light energy, fluorescent energy, acoustic energy, calorific energy, electrochemical energy, etc. Therefore, a user-friendly signal will be produced at the transducer, as illustrated in Fig. 2. This signal must be processed later by a properly designed electronic system. Hence, Fig. 2 is the general representation of chemical sensors and biosensors depending on the nature of the probe, but in which the role of the membranes is emphasized. In Fig. 2, a membrane has the function of being the supporting medium of the recognition element, which must be immobilized and ready to work. In the case of biosensors, the biomolecules generally used are sensitive to the oxygen in the atmosphere and to other chemical compounds that may be present in the sample. Therefore, probes based on biomolecules must be protected from possible chemical attacks or unwanted interactions, and here the membranes have another great role; to encapsulate the probes to protect them.

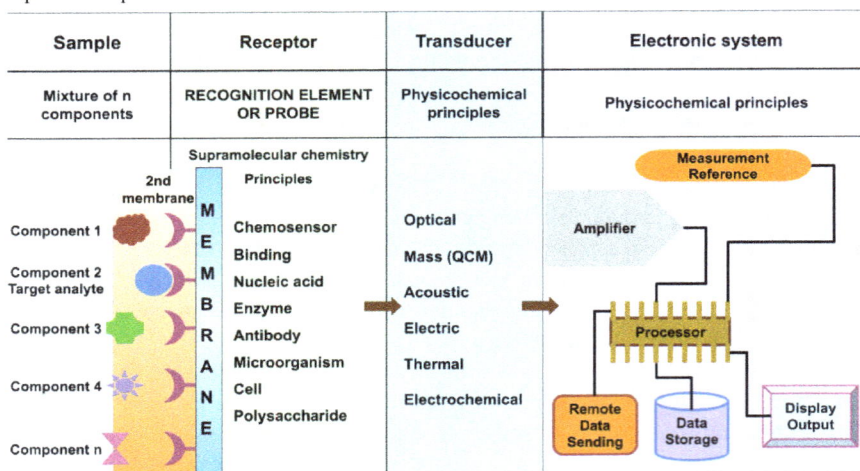

Figure 2 Elements and components of a typical chemistry sensor or biosensor

A third role of the membranes also has been represented in Fig. 2, by placing the different components of the sample at different distances from the probes. This is intended to represent the importance of membranes to filter or separate uninteresting components from the target or analyte.

Advanced Functional Membranes   Materials Research Forum LLC
Materials Research Foundations **120** (2022) 315-333   https://doi.org/10.21741/9781644901816-10

3.   Membranes for sensor technologies

The IUPAC defines membranes as structures that have lateral dimensions significantly greater than their thickness, through which mass transfer can occur under a variety of driving forces. They are continuous layers composed of a semi-permeable material, with controlled permeability. Therefore, the membranes can separate the components of the test or sample solution [20,21].

Biomedical applications of membranes are extensive, such as the semi-permeable membranes that only allow target molecules to pass through them. As well as platforms or support structures for other layers, to immobilization, and/or to encapsulation of sensing probes as illustrated in Fig. 3. They are also used as a support structure for proteins, DNA, and RNA within techniques such as western, southern, and northern blots, or indirect epifluorescence (DEFT). For this reason, device developers integrate membranes into sensors (particularly biosensors). Detection of pathogens or diseases (through biomarkers) has been achieved using some of these sensors. A classic example is the glucometer or glucose sensor, which can have more than two different types of membranes.

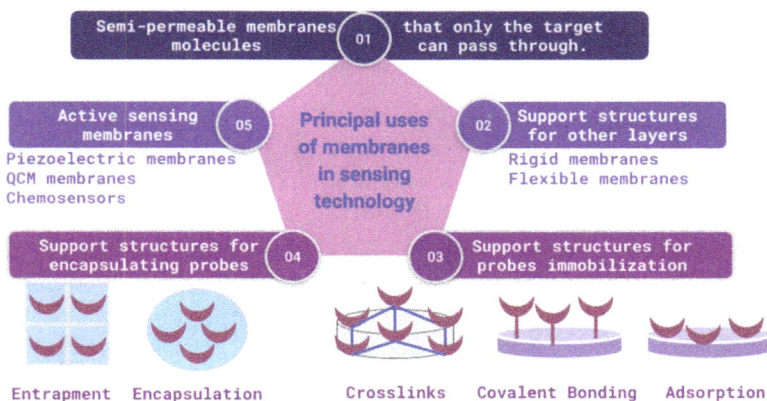

Figure 3 Applications and uses of membranes in sensing technology

Given the variety of applications and the different important roles that membranes must fulfil, their design requires a high control of various parameters, which can be grouped into two major focus areas. On one hand, the physical parameters such as surface area and roughness, porosity (pore size and distribution), and membrane thickness. On the other hand, chemical parameters or characteristics such as chemical composition, functional groups (type and their distribution), chemical affinity, swelling, and chemisorption properties.

The physical parameters of the membrane are decisive for filtration and separation applications; while the chemical parameters are essential for structural support. For the chemical sensor design, both sets of parameters are important.

Chemical properties can be studied taking advantage of the pragmatic division that has been used for years, for the study of chemical compounds; i.e., membranes can be classified as organic, inorganic, and hybrid (composite). Some examples of the materials used in the manufacture of those membranes are mentioned in Table 1.

The main function of inorganic membranes has been as a support structure, by increasing the sensor surface area, to carry out a capillary action, or as mediators of fluid transport by capillary action. As mentioned in Table 1, glass membranes also are good examples of inorganic membranes that have been reported [38–40] for biosensor applications. In this table, more examples of inorganic membranes are given; nevertheless, the reports of organic membranes in sensing applications are much more numerous.

Organic membranes are used primarily as a support structure to immobilize sensing probes and as a filter medium. However, they are also used to microencapsulate and protect sensing probes or biorecognition elements such as enzymes. But they have also been used as a functional part of the recognition element, such as fluorescent polymeric membranes that are both support and chemosensor [41–43].

Another main difference between organic membranes and inorganic ones is the great variety of refilled materials for organic membranes. Cellulose derivatives, as seen in Table 1, are one of the first materials reported for the manufacture of organic membranes used in the development of sensors, again, the glucometer is a good example of this. In addition to the materials already mentioned (Table 1), polyethersulfone, polydimethylsiloxane, polycarbonate, polyvinyl chloride, polyamine, polyvinylidene fluoride, nylon, 2-hydroxyethyl methacrylate, polyacrylamide, polylactic acid, among others, are also used with that purpose.

Table 1 Examples of materials for membranes manufacturing

Type of membrane	Material	References
Inorganic	Aluminum or aluminum oxide	[22–25]
	Gold	[26–28]
	Silver	[29–31]
	Titanium oxide	[32–34]
	Silicon nitride	[35–37]
	Glass membranes	[38–40]
Organic	Cellulose nitrate, nitrocellulose, cellulose acetate	[2–4]
	Polyurethane	[1][2–4]
Hybrid (composite)	Titanium Carbide ($Ti_3C_2T_x$) modified with Ag nanoparticles	[31]

Sometimes it is desirable to have both organic and inorganic membrane properties. Hence, it is not unusual to find hybrid membranes for sensing applications, which are composed of organic and inorganic components that effectively merge. This type of membrane can be of two types; *(i)* Organic membranes covered by a thin layer of inorganic material, usually, a conductor such as gold [31,40-45]; and *(ii)* Membranes synthesized by sol-gel methods to obtain structures of inorganic nanoparticles with polymer chains[13,26,30,46]. Finally, composite membranes are multiple layers of membranes that are stacked together side-to-side or vertically to comprise a whole sensor[32–38,47].

4. Wearable membranes sensor

The prospects for sensor technology point towards its integration with wearable technology and flexible electronics; with the aim of developing portable sensors and biosensors, as less invasive as possible and with continuous data collection, to monitor user activities. Challenges in this area include standardization, device connectivity, interaction between multiple devices, compatibility between the gadgets and the analysis software. On the other hand, the growth of technology has allowed portable devices to become an everyday and versatile tool. Portable devices currently have several integrated sensors with which they acquire the data necessary to carry out the tasks for which they were created. Statistical data is even collected to study our habits and preferences with the intention of improving the user experience[48–50].

Wearable or portable devices make use of wireless strategies for non-invasive, real-time monitoring in many applications. One of these applications that have gained special relevance and interest is the personalized monitoring of parameters related to health and fitness. These portable devices can be designed as common wearables; i.e., smartwatches, bracelets, hearing aids, electronic or optical tattoos, sports glasses, screens, subcutaneous sensors, electronic footwear and textiles. In this context, there is a keen interest in membrane development that improves these technologies. To achieve the efficient applicability of membrane technology (continuous, efficient and without losing its general properties), the membranes must meet specific requirements in terms of body adaptation and biocompatibility. Traditionally, the common approach to developing sensing technology was miniaturized electronics mounted on rigid platforms [13,40,49,51–53].

However, they have a low body adaptation capacity due to their low resistance to deformation and, therefore, flexible materials and membranes come into play in this new trend of developing wearables. To achieve the desired mechanical stability, the electrodes and, in general, the electronics associated with the sensors and biosensors must be flexible, therefore, they require the introduction of characteristics such as the ability to stretch to withstand mechanical deformations and the associated mechanical stress during use. Furthermore, these membranes must be non-toxic and biocompatible to minimize the risk of epithelialization or allergic reactions. In this context, the use of enzymes [54–57], hydrogels [6,41,58–60], and conductive polymers such as polyaniline (PANI) [61–64], polypyrrole (PPy) [65–67], polythiophene (PT) [68–70], or polyethylene dioxide thiophene (PEDOT) [43,71–74] has been reported for the development of bioelectrodes due to their biocompatibility, conductivity, and the presence of tunable functionalities. However, its application in the development of biosensors is limited by different reasons and challenges to overcome, such as its relative fragility; its decrease in properties against variations in the pH of the medium, possible degradation in toxic or reactive compounds; and in some cases the use of synthesis processes and complex manufacturing procedures that are difficult to implement outside the laboratory scale [70–76].

Conclusions

The chapter describes the importance of membranes for sensing technological applications, including specialized selectivity, encapsulation features, and structural supporting properties. Materials for manufacturing membranes are used depending on the purpose of the specific sensor or the selected target molecule. Polymeric materials had been combined with inorganic materials for obtaining a greater efficiency in the sensor design. Physical parameters (surface area and roughness, porosity, and membrane thickness) and chemical parameters (chemical composition and affinity, swelling, and chemisorption properties) determine membrane performance in sensing techniques. The prospects for sensor technology point towards its integration with wearable technology and flexible electronics, aiming to develop less invasive biosensors with continuous data collection to monitor user activities and sending those data to a health professional on a real-time basis. The development of membrane technology is crucial to extend the technological frontier for these sensing/biosensing devices and platforms. It is the polymeric membranes that will help take the step to the front of the wearable and portable biosensors, which promise a great advance in the monitoring of biometric parameters to improve the quality of life of the modern human.

References

[1]   J. Janata, Chemical sensors, Anal. Chem. 64 (1992) 196–219. https://doi.org/10.1021/ac00036a012

[2]   A. Ruiu, M. Vonlanthen, E.G. Morales-Espinoza, S.M. Rojas-Montoya, I. González-Méndez, E. Rivera, Pyrene chemosensors for nanomolar detection of toxic and cancerogenic amines, J. Mol. Struct. 1196 (2019) 1–7. https://doi.org/10.1016/j.molstruc.2019.06.061

[3]   A. Turner, I. Karube, G.S. Wilson, Biosensors : Fundamentals and Applications, Oxford University Press, 1987. http://urn.kb.se/resolve?urn=urn:nbn:se:liu:diva-92007 (accessed August 17, 2021).

[4]   D.R. Thévenot, K. Toth, R.A. Durst, G.S. Wilson, Electrochemical biosensors: recommended definitions and classification. International Union of Pure and Applied Chemistry: Physical Chemistry Division, Commission I.7 (Biophysical Chemistry); Analytical Chemistry Division, Commission V.5 (Electroanalytical Chemistry).1, Biosens. Bioelectron. 16 (2001) 121–131. https://doi.org/10.1016/S0956-5663(01)00115-4

[5]   S. J. Chalk., IUPAC. Compendium of Chemical Terminology, 2nd ed. (the "Gold Book"). Compiled by A. D. McNaught and A. Wilkinson. Blackwell Scientific Publications, Oxford (1997). Online version (2019-), (2019). https://doi.org/10.1351/goldbook.M03823

[6]   H. Zhou, M. Wang, X. Jin, H. Liu, J. Lai, H. Du, W. Chen, A. Ma, Capacitive Pressure Sensors Containing Reliefs on Solution-Processable Hydrogel Electrodes,

ACS Appl. Mater. Interfaces. 13 (2021) 1441–1451.
https://doi.org/10.1021/acsami.0c18355

[7]　H. Zhou, Z. Wang, W. Zhao, X. Tong, X. Jin, X. Zhang, Y. Yu, H. Liu, Y. Ma, S. Li, W. Chen, Robust and sensitive pressure/strain sensors from solution processable composite hydrogels enhanced by hollow-structured conducting polymers, Chem. Eng. J. 403 (2021) 126307. https://doi.org/10.1016/j.cej.2020.126307

[8]　M. Batool, A. Jalal, K. Kim, Sensors Technologies for Human Activity Analysis Based on SVM Optimized by PSO Algorithm, in: 2019 Int. Conf. Appl. Eng. Math. ICAEM, 2019: pp. 145–150. https://doi.org/10.1109/ICAEM.2019.8853770

[9]　Z.L. Wang, Triboelectric nanogenerators as new energy technology and self-powered sensors – Principles, problems and perspectives, Faraday Discuss. 176 (2015) 447–458. https://doi.org/10.1039/C4FD00159A

[10]　A. Nait Aicha, G. Englebienne, K.S. van Schooten, M. Pijnappels, B. Kröse, Deep Learning to Predict Falls in Older Adults Based on Daily-Life Trunk Accelerometry, Sensors. 18 (2018) E1654. https://doi.org/10.3390/s18051654

[11]　B. Perriot, J. Argod, J.-L. Pepin, N. Noury, A network of collaborative sensors for the monitoring of COPD patients in their daily life, in: 2013 IEEE 15th Int. Conf. E-Health Netw. Appl. Serv. Heal. 2013, 2013: pp. 299–302. https://doi.org/10.1109/HealthCom.2013.6720689

[12]　R.M. White, A Sensor Classification Scheme, IEEE Trans. Ultrason. Ferroelectr. Freq. Control. 34 (1987) 124–126. https://doi.org/10.1109/T-UFFC.1987.26922

[13]　V. Naresh, N. Lee, A Review on Biosensors and Recent Development of Nanostructured Materials-Enabled Biosensors, Sensors. 21 (2021) 1109. https://doi.org/10.3390/s21041109

[14]　M. Javaid, A. Haleem, S. Rab, R. Pratap Singh, R. Suman, Sensors for daily life: A review, Sens. Int. 2 (2021) 100121. https://doi.org/10.1016/j.sintl.2021.100121

[15]　S. Lindsay, Introduction to Nanoscience, OUP Oxford, 2010.

[16]　A.R. Hernandez-Martinez, Chapter 10 - Remote sensing for environmental analysis: Basic concepts and setup, in: Inamuddin, R. Boddula, A.M. Asiri (Eds.), Green Sustain. Process Chem. Environ. Eng. Sci., Elsevier, 2021: pp. 209–224. https://doi.org/10.1016/B978-0-12-821883-9.00012-6

[17]　A. Hulanicki, S. Glab, F. Ingman, Chemical sensors: definitions and classification, Pure Appl. Chem. 63 (1991) 1247–1250. https://doi.org/10.1351/pac199163091247

[18] D.R. Thevenot, K. Tóth, R.A. Durst, G.S. Wilson, Electrochemical Biosensors: Recommended Definitions and Classification, Pure Appl. Chem. 71 (1999) 2333–2348. https://doi.org/10.1351/pac199971122333

[19] J. Labuda, R.P. Bowater, M. Fojta, G. Gauglitz, Z. Glatz, I. Hapala, J. Havliš, F. Kilar, A. Kilar, L. Malinovská, H.M.M. Sirén, P. Skládal, F. Torta, M. Valachovič, M. Wimmerová, Z. Zdráhal, D.B. Hibbert, Terminology of bioanalytical methods (IUPAC Recommendations 2018), Pure Appl. Chem. 90 (2018) 1121–1198. https://doi.org/10.1515/pac-2016-1120

[20] W.J. Koros, Y.H. Ma, T. Shimidzu, Terminology for membranes and membrane processes (IUPAC Recommendations 1996), Pure Appl. Chem. 68 (1996) 1479–1489. https://doi.org/10.1351/pac199668071479

[21] T.I.U. of P. and A. Chemistry (IUPAC), IUPAC - membrane (M03823), (n.d.). https://doi.org/10.1351/goldbook.M03823

[22] W.W. Ye, J.Y. Shi, C.Y. Chan, Y. Zhang, M. Yang, A nanoporous membrane based impedance sensing platform for DNA sensing with gold nanoparticle amplification, Sens. Actuators B Chem. 193 (2014) 877–882. https://doi.org/10.1016/j.snb.20153.09

[23] L. Wang, Q. Liu, Z. Hu, Y. Zhang, C. Wu, M. Yang, P. Wang, A novel electrochemical biosensor based on dynamic polymerase-extending hybridization for E. coli O157:H7 DNA detection, Talanta. 78 (2009) 647–652. https://doi.org/10.1016/j.talanta.2008.12.001

[24] V. Rai, J. Deng, C.-S. Toh, Electrochemical nanoporous alumina membrane-based label-free DNA biosensor for the detection of Legionella sp, Talanta. 98 (2012) 112–117. https://doi.org/10.1016/j.talanta.2012.06.055

[25] V. Rai, H.C. Hapuarachchi, L.C. Ng, S.H. Soh, Y.S. Leo, C.-S. Toh, Ultrasensitive cDNA Detection of Dengue Virus RNA Using Electrochemical Nanoporous Membrane-Based Biosensor, PLOS ONE. 7 (2012) e42346. https://doi.org/10.1371/journal.pone.0042346

[26] G. Rossi, L. Monticelli, Gold nanoparticles in model biological membranes: A computational perspective, Biochim. Biophys. Acta BBA - Biomembr. 1858 (2016) 2380–2389. https://doi.org/10.1016/j.bbamem.2016.04.001

[27] L. Zhang, Y. Wang, M. Chen, Y. Luo, K. Deng, D. Chen, W. Fu, A new system for the amplification of biological signals: RecA and complimentary single strand

DNA probes on a leaky surface acoustic wave biosensor, Biosens. Bioelectron. 60 (2014) 259–264. https://doi.org/10.1016/j.bios.2014.04.037

[28]   S. Sang, H. Witte, A novel PDMS micro membrane biosensor based on the analysis of surface stress, Biosens. Bioelectron. 25 (2010) 2420–2424. https://doi.org/10.1016/j.bios.2010.03.035

[29]   R.V. den Hurk, S. Evoy, A Review of Membrane-Based Biosensors for Pathogen Detection, Sensors. 15 (2015) 14045–14078. https://doi.org/10.3390/s150614045

[30]   I.A. Dimulescu (Nica), A.C. Nechifor, C. Bărdacă (Urducea), O. Oprea, D. Pașcu, E.E. Totu, P.C. Albu, G. Nechifor, S.G. Bungău, Accessible Silver-Iron Oxide Nanoparticles as a Nanomaterial for Supported Liquid Membranes, Nanomaterials. 11 (2021) 1204. https://doi.org/10.3390/nano11051204

[31]   R.P. Pandey, K. Rasool, V.E. Madhavan, B. Aïssa, Y. Gogotsi, K.A. Mahmoud, Ultrahigh-flux and fouling-resistant membranes based on layered silver/MXene (Ti3C2Tx) nanosheets, J. Mater. Chem. A. 6 (2018) 3522–3533. https://doi.org/10.1039/C7TA10888E

[32]   F. He, S. Liu, Detection of P. aeruginosa using nano-structured electrode-separated piezoelectric DNA biosensor, Talanta. 62 (2004) 271–277. https://doi.org/10.1016/j.talanta.2003.07.007

[33]   S. Shanmugam, K. Ketpang, Md.A. Aziz, K. Oh, K. Lee, B. Son, N. Chanunpanich, Composite polymer electrolyte membrane decorated with porous titanium oxide nanotubes for fuel cell operating under low relative humidity, Electrochimica Acta. 384 (2021) 138407. https://doi.org/10.1016/j.electacta.2021.138407

[34]   O.A. Sadik, N. Du, I. Yazgan, V. Okello, Chapter 6 - Nanostructured Membranes for Water Purification, in: A. Street, R. Sustich, J. Duncan, N. Savage (Eds.), Nanotechnol. Appl. Clean Water Second Ed., William Andrew Publishing, Oxford, 2014: pp. 95–108. https://doi.org/10.1016/B978-1-4557-3116-9.00006-8

[35]   K. Yazda, K. Bleau, Y. Zhang, X. Capaldi, T. St-Denis, P. Grutter, W.W. Reisner, High Osmotic Power Generation via Nanopore Arrays in Hybrid Hexagonal Boron Nitride/Silicon Nitride Membranes, Nano Lett. 21 (2021) 4152–4159. https://doi.org/10.1021/acs.nanolett.0c04704

[36]   J.L. Braun, S.W. King, E.R. Hoglund, M.A. Gharacheh, E.A. Scott, A. Giri, J.A. Tomko, J.T. Gaskins, A. Al-kukhun, G. Bhattarai, M.M. Paquette, G. Chollon, B. Willey, G.A. Antonelli, D.W. Gidley, J. Hwang, J.M. Howe, P.E. Hopkins, Hydrogen

effects on the thermal conductivity of delocalized vibrational modes in amorphous silicon nitride…, Phys. Rev. Mater. 5 (2021) 035604. https://doi.org/10.1103/PhysRevMaterials.5.035604

[37]   S. Jiang, T. Isik, C.Y. Akkaya, S. Kumari, V. Ortalan, Evaluation of Gallium Ion\Xe Plasma Beam for Patterning of Suspended Silicon Nitride Membranes, Microsc. Microanal. 27 (2021) 438–439. https://doi.org/10.1017/S1431927621002075

[38]   O. Sayginer, E. Iacob, S. Varas, A. Szczurek, M. Ferrari, A. Lukowiak, G.C. Righini, O.S. Bursi, A. Chiasera, Design, fabrication and assessment of an optomechanical sensor for pressure and vibration detection using flexible glass multilayers, Opt. Mater. 115 (2021) 111023. https://doi.org/10.1016/j.optmat.2021.111023

[39]   S.E. Henkelis, S.J. Percival, L.J. Small, D.X. Rademacher, T.M. Nenoff, Continuous MOF Membrane-Based Sensors via Functionalization of Interdigitated Electrodes, Membranes. 11 (2021) 176. https://doi.org/10.3390/membranes11030176

[40]   F. Vivaldi, P. Salvo, N. Poma, A. Bonini, D. Biagini, L. Del Noce, B. Melai, F. Lisi, F.D. Francesco, Recent Advances in Optical, Electrochemical and Field Effect pH Sensors, Chemosensors. 9 (2021) 33. https://doi.org/10.3390/chemosensors9020033

[41]   J.L. Pablos, S. Vallejos, S. Ibeas, A. Muñoz, F. Serna, F.C. García, J.M. García, Acrylic Polymers with Pendant Phenylboronic Acid Moieties as "Turn-Off" and "Turn-On" Fluorescence Solid Sensors for Detection of Dopamine, Glucose, and Fructose in Water, ACS Macro Lett. 4 (2015) 979–983. https://doi.org/10.1021/acsmacrolett.5b00465

[42]   M. Gómez-García, J.M. Benito, A.P. Butera, C.O. Mellet, J.M.G. Fernández, J.L.J. Blanco, Probing Carbohydrate-Lectin Recognition in Heterogeneous Environments with Monodisperse Cyclodextrin-Based Glycoclusters, J. Org. Chem. 77 (2012) 1273–1288. https://doi.org/10.1021/jo201797b

[43]   A.R. Hernandez-Martinez, Chapter 11 - Materials science and lab-on-a-chip for environmental and industrial analysis, in: Inamuddin, R. Boddula, A.M. Asiri (Eds.), Green Sustain. Process Chem. Environ. Eng. Sci., Elsevier, 2021: pp. 225–236. https://doi.org/10.1016/B978-0-12-821883-9.00011-4

[44]   M. Khan, T. Li, A. Hayat, A. Zada, T. Ali, I. Uddin, A. Hayat, M. Khan, A. Ullah, A. Hussain, T. Zhao, A concise review on the elastomeric behavior of electroactive polymer materials, Int. J. Energy Res. 45 (2021) 14306–14337. https://doi.org/10.1002/er.6747

[45]   J.-S. Jang, L.R. Winter, C. Kim, J.D. Fortner, M. Elimelech, Selective and sensitive environmental gas sensors enabled by membrane overlayers, Trends Chem. 3 (2021) 547–560. https://doi.org/10.1016/j.trechm.2021.04.005.

[46]   C. Gao, J. Liao, J. Lu, J. Ma, E. Kianfar, The effect of nanoparticles on gas permeability with polyimide membranes and network hybrid membranes: a review, Rev. Inorg. Chem. 41 (2021) 1–20. https://doi.org/10.1515/revic-2020-0007

[47]   A. Michalke, H.-J. Galla, C. Steinem, Channel activity of a phytotoxin of Clavibacter michiganense ssp. nebraskense in tethered membranes, Eur. Biophys. J. 30 (2001) 421–429. https://doi.org/10.1007/s002490100154

[48]   M. Domb, Wearable Devices and their Implementation in Various Domains, IntechOpen, 2019. https://doi.org/10.5772/intechopen.86066

[49]   J.J. Ferreira, C.I. Fernandes, H.G. Rammal, P.M. Veiga, Wearable technology and consumer interaction: A systematic review and research agenda, Comput. Hum. Behav. 118 (2021) 106710. https://doi.org/10.1016/j.chb.2021.106710

[50]   A. Ometov, V. Shubina, L. Klus, J. Skibińska, S. Saafi, P. Pascacio, L. Flueratoru, D.Q. Gaibor, N. Chukhno, O. Chukhno, A. Ali, A. Channa, E. Svertoka, W.B. Qaim, R. Casanova-Marqués, S. Holcer, J. Torres-Sospedra, S. Casteleyn, G. Ruggeri, G. Araniti, R. Burget, J. Hosek, E.S. Lohan, A Survey on Wearable Technology: History, State-of-the-Art and Current Challenges, Comput. Netw. 193 (2021) 108074. https://doi.org/10.1016/j.comnet.2021.108074

[51]   L. Nayak, S. Mohanty, S. Kumar Nayak, A. Ramadoss, A review on inkjet printing of nanoparticle inks for flexible electronics, J. Mater. Chem. C. 7 (2019) 8771–8795. https://doi.org/10.1039/C9TC01630A

[52]   M. Buaki-Sogó, L. García-Carmona, M. Gil-Agustí, M. García-Pellicer, A. Quijano-López, Flexible and Conductive Bioelectrodes Based on Chitosan-Carbon Black Membranes: Towards the Development of Wearable Bioelectrodes, Nanomaterials. 11 (2021) 2052. https://doi.org/10.3390/nano11082052

[53]   Z. Qi, M. Zhou, Y. Li, Z. Xia, W. Huo, X. Huang, Reconfigurable Flexible Electronics Driven by Origami Magnetic Membranes, Adv. Mater. Technol. 6 (2021) 2001124. https://doi.org/10.1002/admt.202001124

[54]   M. Guo, J. Chi, C. Zhang, M. Wang, H. Liang, J. Hou, S. Ai, X. Li, A simple and sensitive sensor for lactose based on cascade reactions in Au nanoclusters and enzymes co-encapsulated metal-organic frameworks, Food Chem. 339 (2021) 127863. https://doi.org/10.1016/j.foodchem.2020.127863

Advanced Functional Membranes
Materials Research Foundations **120** (2022) 315-333

Materials Research Forum LLC
https://doi.org/10.21741/9781644901816-10

[55]   S. Park, H. Kim, S.-H. Paek, J.W. Hong, Y.-K. Kim, Enzyme-linked immuno-strip biosensor to detect Escherichia coli O157:H7, Ultramicroscopy. 108 (2008) 1348–1351. https://doi.org/10.1016/j.ultramic.2008.04.063

[56]   M.F.M. Shakhih, A.S. Rosslan, A.M. Noor, S. Ramanathan, A.M. Lazim, A.A. Wahab, Review-Enzymatic and Non-Enzymatic Electrochemical Sensor for Lactate Detection in Human Biofluids, J. Electrochem. Soc. 168 (2021) 067502. https://doi.org/10.1149/1945-7111/ac0360

[57]   T. Ozer, C.S. Henry, Review—Recent Advances in Sensor Arrays for the Simultaneous Electrochemical Detection of Multiple Analytes, J. Electrochem. Soc. 168 (2021) 057507. https://doi.org/10.1149/1945-7111/abfc9f

[58]   Z. Jia, M. Müller, T. Le Gall, M. Riool, M. Müller, S.A.J. Zaat, T. Montier, H. Schönherr, Multiplexed detection and differentiation of bacterial enzymes and bacteria by color-encoded sensor hydrogels, Bioact. Mater. 6 (2021) 4286–4300. https://doi.org/10.1016/j.bioactmat.2021.04.022

[59]   Q. Wang, J. Guo, X. Lu, X. Ma, S. Cao, X. Pan, Y. Ni, Wearable lignin-based hydrogel electronics: A mini-review, Int. J. Biol. Macromol. 181 (2021) 45–50. https://doi.org/10.1016/j.ijbiomac.2021.03.079

[60]   Y. Xiong, X. Zhang, X. Ma, W. Wang, F. Yan, X. Zhao, X. Chu, W. Xu, C. Sun, A review of the properties and applications of bioadhesive hydrogels, Polym. Chem. 12 (2021) 3721–3739. https://doi.org/10.1039/D1PY00282A

[61]   I. Falina, N. Loza, S. Loza, E. Titskaya, N. Romanyuk, Permselectivity of Cation Exchange Membranes Modified by Polyaniline, Membranes. 11 (2021) 227. https://doi.org/10.3390/membranes11030227

[62]   M. Sairam, S.K. Nataraj, T.M. Aminabhavi, S. Roy, C.D. Madhusoodana, Polyaniline Membranes for Separation and Purification of Gases, Liquids, and Electrolyte Solutions, Sep. Purif. Rev. 35 (2006) 249–283. https://doi.org/10.1080/15422110600859727

[63]   M. Beygisangchin, S. Abdul Rashid, S. Shafie, A.R. Sadrolhosseini, H.N. Lim, Preparations, Properties, and Applications of Polyaniline and Polyaniline Thin Films—A Review, Polymers. 13 (2021) 2003. https://doi.org/10.3390/polym13122003

[64]   Q.N. Al-Haidary, A.M. Al-Mokaram, F.M. Hussein, A.H. Ismail, Development of polyaniline for sensor applications: A review, J. Phys. Conf. Ser. 1853 (2021) 012062. https://doi.org/10.1088/1742-6596/1853/1/012062

[65]   S.A. Gupta, J.S. Singh, A Study of Conducting Electrochemical Sensors Based on Molecularly Imprinted Polymer on Carbon Nanostructure Using Polypyrrole Film: A Review, J. Sci. Res. 65 (2021) 110–115. https://doi.org/10.37398/JSR.2021.650222

[66]   M. Das, S. Roy, Polypyrrole and associated hybrid nanocomposites as chemiresistive gas sensors: A comprehensive review, Mater. Sci. Semicond. Process. 121 (2021) 105332. https://doi.org/10.1016/j.mssp.2020.105332

[67]   S. Sriprasertsuk, S.C. Mathias, J.R. Varcoe, C. Crean, Polypyrrole-coated carbon fibre electrodes for paracetamol and clozapine drug sensing, J. Electroanal. Chem. 897 (2021) 115608. https://doi.org/10.1016/j.jelechem.2021.115608

[68]   G. Prunet, F. Pawula, G. Fleury, E. Cloutet, A.J. Robinson, G. Hadziioannou, A. Pakdel, A review on conductive polymers and their hybrids for flexible and wearable thermoelectric applications, Mater. Today Phys. 18 (2021) 100402. https://doi.org/10.1016/j.mtphys.2021.100402

[69]   T. Minami, W. Tang, K. Asano, Chemical sensing based on water-gated polythiophene thin-film transistors, Polym. J. (2021) 1–9. https://doi.org/10.1038/s41428-021-00537-4

[70]   A. Husain, S. Ahmad, S.P. Ansari, M.O. Ansari, M.M.A. khan, DC electrical conductivity retention and acetone/acetaldehyde sensing on polythiophene/molybdenum disulphide composites, Polym. Polym. Compos. (2021) 09673911211002781. https://doi.org/10.1177/09673911211002781

[71]   G.E. Fenoy, O. Azzaroni, W. Knoll, W.A. Marmisollé, Functionalization Strategies of PEDOT and PEDOT:PSS Films for Organic Bioelectronics Applications, Chemosensors. 9 (2021) 212. https://doi.org/10.3390/chemosensors9080212

[72]   Gbolahan Joseph Adekoya, Rotimi Emmanuel Sadiku, Suprakas Sinha Ray, Nanocomposites of PEDOT:PSS with Graphene and its Derivatives for Flexible Electronic Applications: A Review, Macromol. Mater. Eng. 306 (2021) 716–24. https://doi.org/10.1002/mame.202000716

[73]   N. Gao, J. Yu, Q. Tian, J. Shi, M. Zhang, S. Chen, L. Zang, Application of PEDOT:PSS and Its Composites in Electrochemical and Electronic Chemosensors, Chemosensors. 9 (2021) 79. https://doi.org/10.3390/chemosensors9040079

[74]   L. Vigna, A. Verna, S.L. Marasso, M. Sangermano, P. D'Angelo, F.C. Pirri, M. Cocuzza, The effects of secondary doping on ink-jet printed PEDOT:PSS gas sensors for VOCs and NO2 detection, Sens. Actuators B Chem. 345 (2021) 130381. https://doi.org/10.1016/j.snb.2021.130381

[75]   A.L. Ramos-Jacques, J.A. Lujan-Montelongo, C. Silva-Cuevas, M. Cortez-Valadez, M. Estevez, A.R. Hernandez-Martínez, Lead (II) removal by poly(N,N-dimethylacrylamide-co-2-hydroxyethyl methacrylate), Eur. Polym. J. 101 (2018) 262–272. https://doi.org/10.1016/j.eurpolymj.2018.02.032

[76]   A Review on Materials and Technologies for Organic Large-Area Electronics - Buga - 2021 - Advanced Materials Technologies - Wiley Online Library, (n.d.), (accessed August 24, 2021). https://onlinelibrary.wiley.com/doi/full/10.1002/admt.202001016?casa_token=Tr-T90osS4YAAAAA%3Ahkq68JH4Xl-1-8_aDoyVLNan-cPtGZJMSpDygbfbi2jJxfOhBvbaKuuXkEXRPDfymG5GCUCoFJa_NGgC

[76]   Buga, Cláudia S., and Júlio C. Viana. "A Review on Materials and Technologies for Organic Large-Area Electronics." Advanced Materials Technologies (2021): 2001016.

Keyword Index

About the Editors

Dr. Inamuddin is working as Assistant Professor at the Department of Applied Chemistry, Aligarh Muslim University, Aligarh, India. He obtained Master of Science degree in Organic Chemistry from Chaudhary Charan Singh (CCS) University, Meerut, India, in 2002. He received his Master of Philosophy and Doctor of Philosophy degrees in Applied Chemistry from Aligarh Muslim University (AMU), India, in 2004 and 2007, respectively. He has extensive research experience in multidisciplinary fields of Analytical Chemistry, Materials Chemistry, and Electrochemistry and, more specifically, Renewable Energy and Environment. He has worked on different research projects as project fellow and senior research fellow funded by University Grants Commission (UGC), Government of India, and Council of Scientific and Industrial Research (CSIR), Government of India. He has received Fast Track Young Scientist Award from the Department of Science and Technology, India, to work in the area of bending actuators and artificial muscles. He has completed four major research projects sanctioned by University Grant Commission, Department of Science and Technology, Council of Scientific and Industrial Research, and Council of Science and Technology, India. He has published 186 research articles in international journals of repute and nineteen book chapters in knowledge-based book editions published by renowned international publishers. He has published 144 edited books with Springer (U.K.), Elsevier, Nova Science Publishers, Inc. (U.S.A.), CRC Press Taylor & Francis Asia Pacific, Trans Tech Publications Ltd. (Switzerland), IntechOpen Limited (U.K.), Wiley-Scrivener, (U.S.A.) and Materials Research Forum LLC (U.S.A). He is a member of various journals' editorial boards. He is also serving as Associate Editor for journals (Environmental Chemistry Letter, Applied Water Science and Euro-Mediterranean Journal for Environmental Integration, Springer-Nature), Frontiers Section Editor (Current Analytical Chemistry, Bentham Science Publishers), Editorial Board Member (Scientific Reports-Nature), Editor (Eurasian Journal of Analytical Chemistry), and Review Editor (Frontiers in Chemistry, Frontiers, U.K.) He is also guest-editing various special thematic special issues to the journals of Elsevier, Bentham Science Publishers, and John Wiley & Sons, Inc. He has attended as well as chaired sessions in various international and national conferences. He has worked as a Postdoctoral Fellow, leading a research team at the Creative Research Initiative Center for Bio-Artificial Muscle, Hanyang University, South Korea, in the field of renewable energy, especially biofuel cells. He has also worked as a Postdoctoral Fellow at the Center of Research Excellence in Renewable Energy, King Fahd University of Petroleum and Minerals, Saudi Arabia, in the field of polymer electrolyte membrane fuel cells and computational fluid dynamics of polymer electrolyte membrane fuel cells. He is a life member of the Journal of the Indian

Chemical Society. His research interest includes ion exchange materials, a sensor for heavy metal ions, biofuel cells, supercapacitors and bending actuators.

Dr. Tariq Altalhi joined Department of Chemistry at Taif University, Saudi Arabia as Assistant Professor in 2014. He received his doctorate degree from University of Adelaide, Australia in the year 2014 with Dean's Commendation for Doctoral Thesis Excellence. He was promoted to the position of the head of Chemistry Department at Taif university in 2017 and Vice Dean of Science college in 2019 till now. His group is involved in fundamental multidisciplinary research in nanomaterial synthesis and engineering, characterization, and their application in molecular separation, desalination, membrane systems, drug delivery, and biosensing. In 2015, one of his works was nominated for Green Tech awards from Germany, Europ's largest environmental and business prize, amongst top 10 entries.

His interest lies in developing advanced chemistry-based solutions for solid and liquid municipal (both organic and inorganic) waste management. In this direction, he focuses on transformation of solid organic waste to valuable nanomaterials & economic nanostructure. His research work focuses on conversion of plastic bags to carbon nanotubes, fly ash to efficient adsorbent material, etc. Another stream of interests looks at natural extracts and their application in generation of value- added products such as nanomaterials, incense, etc. Through his work as an independent researcher, he has gathered strong management and mentoring skills to run a group of multidisciplinary researchers of various fields including chemistry, materials science, biology, and pharmaceutical science. His publications show that he has developed a wide network of national and international researchers who are leaders in their respective fields. In addition, he has established key contacts with major industries in Kingdom of Saudi Arabia.

Dr. Mohd Imran Ahamed received his Ph.D degree on the topic "Synthesis and characterization of inorganic-organic composite heavy metals selective cation-exchangers and their analytical applications", from Aligarh Muslim University, Aligarh, India in 2019. He has published several research and review articles in the journals of international recognition. Springer (U.K.), Elsevier, CRC Press Taylor & Francis Asia Pacific and Materials Research Forum LLC (U.S.A). He has completed his B.Sc. (Hons) Chemistry from Aligarh Muslim University, Aligarh, India, and M.Sc. (Organic Chemistry) from Dr. Bhimrao Ambedkar University, Agra, India. He has co-edited more than 20 books with Springer (U.K.), Elsevier, CRC Press Taylor & Francis Asia Pacific, Materials Research Forum LLC (U.S.A) and Wiley-Scrivener, (U.S.A.). His research work includes ion-exchange chromatography, wastewater treatment, and analysis, bending actuator and electrospinning.

Dr. Mohammad Luqman has 12+ years of post-PhD experience in Teaching, Research, and Administration. Currently, he is serving as an Assistant Professor of Chemical Engineering in Taibah University, Saudi Arabia. Before joining here, he served as an Assistant Professor in College of Applied Science at A'Sharqiyah University, Oman, and in College of Engineering at King Saud University, Saudi Arabia. He served as a Research Engineer in SAMSUNG Cheil Industries, South Korea. Moreover, he served as a post-doctoral fellow at Artificial Muscle Research Center, Konkuk University, South Korea, in the field of Ionic Polymer Metal Composites for the development of Artificial Muscles, Robotic Actuators and Dynamic Sensors. He earned his PhD degree in the field of Ionomers (Ion-containing Polymers), from Chosun University, South Korea. He successfully served as an Editor to three books, published by world renowned publishers. He published numerous high-quality papers, and book chapters. He is serving as an Editor and editorial/review board members to many International SCI and Non-SCI journals. He has attracted a few important research grants from industry and academia. His research interests include but not limited to Development of Ionomer/Polyelectrolyte/non-ionic Polymer Nanocomposites/Blends for Smart and Industrial/Engineering Applications.

www.ingramcontent.com/pod-product-compliance
Lightning Source LLC
Chambersburg PA
CBHW071322210326
41597CB00015B/1314